Law and Policy for the Quantum Age

It is often said that quantum technologies are poised to change the world as we know it, but cutting through the hype, what will quantum technologies actually mean for countries and their citizens? In Law and Policy for the Quantum Age, Chris Jay Hoofnagle and Simson L. Garfinkel explain the genesis of quantum information science (QIS) and the resulting quantum technologies that are most exciting: quantum sensing, computing, and communication. This groundbreaking, timely text explains how quantum technologies work, how countries will likely employ QIS for future national defense and what the legal landscapes will be for these nations, and how companies might (or might not) profit from the technology. Hoofnagle and Garfinkel argue that the consequences of CIS are so profound that we must begin planning for them today.

Chris Jay Hoofnagle is professor of law in residence at the University of California, Berkeley and affiliated faculty with the Simons Institute for the Theory of Computing. He is an elected member of the American Law Institute, and author of Federal Trade Commission Privacy Law and Policy (Cambridge University Press 2016). Hoofnagle is of counsel to Gunderson Dettmer Stough Villeneuve Franklin & Hachigian, LLP, and serves on boards for Constella Intelligence and Palantir Technologies.

Simson L. Garfinkel is a pioneer in digital forensics, with a career in technology spanning starting a local internet service provider in 1995 to academia and government service. Garfinkel holds a Ph.D. in computer science from the Massachusetts Institute of Technology, was a tenured professor of computer science at the Naval Postgraduate School, and now is a Senior Data Scientist at the U.S. Department of Homeland Security, a part-time faculty member at the George Washington University in Washington, DC, and a member of the Association for Computing Machinery's US Technology Policy Committee (ACM USTPC). He has authored and edited 16 books, over 100 scholarly articles, and is a fellow of both the ACM and the Institute of Electrical and Electronics Engineers (IEEE). This book is written in Simson Garfinkel's personal capacity and does not reflect the views or policy of the U.S. Government, the U.S. Department of Homeland Security, the U.S. Department of Commerce, or the U.S. Census Bureau.

Law and Policy for the Quantum Age

Chris Jay Hoofnagle
University of California, Berkeley

Simson L. Garfinkel

CAMBRIDGE
UNIVERSITY PRESS

CAMBRIDGE
UNIVERSITY PRESS

University Printing House, Cambridge CB2 8BS, United Kingdom

One Liberty Plaza, 20th Floor, New York, NY 10006, USA

477 Williamstown Road, Port Melbourne, VIC 3207, Australia

314–321, 3rd Floor, Plot 3, Splendor Forum, Jasola District Centre,
New Delhi – 110025, India

103 Penang Road, #05–06/07, Visioncrest Commercial, Singapore 238467

Cambridge University Press is part of the University of Cambridge.

It furthers the University's mission by disseminating knowledge in the pursuit of
education, learning, and research at the highest international levels of excellence.

www.cambridge.org
Information on this title: www.cambridge.org/9781108835343
DOI: 10.1017/9781108883719

First published 2022

A catalogue record for this publication is available from the British Library.

ISBN 978-1-108-83534-3 Hardback
ISBN 978-1-108-79317-9 Paperback

Contents

Part 10 Shaping the Quantum Future · 303

8 Quantum Technologies and Possible Futures · 305

List of Figures

List of Tables

Preface

THIS book is the result of a chance meeting between the authors in the summer of 2019 on a 12-hour international flight. This was not a case of quantum superposition, but it certainly demonstrates the power of chance.

The *Oxford English Dictionary* defines *quantum* as "A discrete quantity of electromagnetic energy proportional in magnitude to the frequency of the radiation it represents."[1] In this book, we use the term *quantum technologies* to mean tools that use those discrete quantities of energy to provide some utility. Classical technologies are indeed made of those discrete quanta of energy, but when we use a hammer, or fly in an airplane, or even use a computer, we do not concern ourselves with quanta-level energy or effects. Quantum technologies focus on the smallest quanta of energy and their effects, and this focus is what makes quantum technologies so surprising: mastering the physics of the small, has surprisingly large implications. We classify quantum technologies into quantum sensing and metrology, computing, and communications.

In the chapters on computing we distinguish the words *calculation* and *computation*. We use the word calculation to describe rote mathematical processes that are data independent – that is, that can be performed without concern to the numbers being acted upon. We use the word computation to describe all other processing of information, be it mathematical or otherwise. Calculation, such as doubling a number, or determining the number of days in a year by fetching the value from an almanac, can be performed with a simple device. Computation requires a more complex device that can read, execute, and modify its own program. In the academic literature the terms *finite state machine* and *pushdown automata* are used to describe

---

[1] "quantum, n. and adj.", definition A.5.a, OED Online, Oxford University Press, December 2020.

devices that perform what we call calculation, and *Turing machine* to describe what we call computation.

In this book we use the `courier typewriter font` to present computer code and pseudocode, as well as specific base-10 numbers used in computer algorithms. We use the stylized numbers `0` and `1` when we are referring to binary digits. Thus, `13` = `1101`. Occasionally we may indicate the base using a subscript following the number, or use the Python programming language convention for hexadecimal numbers, such that `1101` $=1101_2=0D_{16}=0x0D$.

We endeavor to list companies, countries, people, and other proper nouns in alphabetical order unless there is a specific reason to list them otherwise. When order is meant to convey importance, we make this clear. So if we write that China, Russia, and the United States are all world powers, we are sorting the countries alphabetically. If we say that the world's most populous countries as of January 1, 2021 are China, India, the United States, Indonesia, and Pakistan, you can assume that China's population is the largest, Pakistan ranks fifth, and you should expect us to cite our source.[2] When numeric order is relevant, we will number using hash-marks, such as when Step #1 is followed by Step #2.

We have a few chemical formulas in this book, and when we present a molecule, we will include the hydrogen atoms and attempt to present the formula in a manner that conveys its structure. That is, ethanol is CH_3CH_2OH and not C_2H_5OH.

Currencies, unless otherwise stated, are in US dollars. When comparing spending across time, we convert to inflation-adjusted US 2020 dollars using the US Labor Department's Bureau of Labor Statistics Consumer Price Index (CPI) and the calculator at www.usinflationcalculator.com.

---

[2]US Census Bureau, "US and World Population Clock" (2021).

Acknowledgments

Law and Policy for the Quantum Age has been a fascinating and challenging book to research and write. We went long on the history of technology, as we believe that experience with the introduction of previous game-changing technologies offers important context for making decisions about such technologies today. We believe that good technology policy can only be made with a rough-and-accurate understanding of the underlying technology. We are determined to correct much of the misinformation that is present in the popular literature of quantum information science today.

Quantum technologies was a new topic for both authors. Author Hoofnagle decided to research the field after conversations with Lily Lin (Berkeley MIMS 2019), whose narrative made clear that quantum sensing was much more interesting than cryptanalysis. Then, the good folks at Delta Airlines seated author Garfinkel together with Hoofnagle on a long flight back from Tel Aviv in the summer of 2019. Together we discussed the national security implications of quantum technologies and formed plans to write this book.

As part of researching this book, the authors downloaded and reviewed over 1500 scientific articles, popular articles, and books pertaining to the topics discussed herein. We purchased sheets of polarizing material, 3D glasses, and large optical grade calcite crystals so that we could experience first-hand the mysteries of superposition at the macro scale. We haunted online forums, emailed with a Nobel Laureate, and tried the very best we could to make up for the fact that neither of us had studied quantum physics in college.

This book would not have been possible without the thoughtful engagement from many experts in quantum information science, who gave generously of both their time and counsel. We owe many thanks to those who helped us with difficult material, and acknowledge that any mistakes remaining are our own.

We would like to explicitly thank those who answered questions about technology and history while working on this project, including Scott Aaronson, Ross Anderson, Syed Assad, Holger Axelsen, Philip Ball, Charles Bennett, Scott Bradner, Steve Crocker, Nike Dattani, Peter Denning, Edward Fredkin, Joyce Fredkin, Michael Frank, Shohini Ghose, David Grier, Aram Harrow, Marco Lanzagorta, Seth Lloyd, Chao-Yang Lu, Norman Margolus, Henry Minsky, Margaret Minsky, Christopher Monroe, Jian-Wei Pan, Daniel Polanic, Peter W. Shor, Adam Shostack, Bill Silver, Tommaso Toffoli, Rainer

Weiss, and Stephen Wolfram. Tom Slee reviewed our section on chemistry and found several errors, which we attempted to correct; those that remain are ours, not his.

Mark Horowitz at Stanford worked with us to resolve several lingering questions regarding the National Academies report on the potential of quantum computing. Overall, we found the report useful as an initial, highly technical introduction to this complicated topic; we strongly recommend the report to anyone looking for a description of quantum computing that is more technical than the one we present here.

At IBM, we would like to thank Chris Nay for working with us over the course of more than a year in both answering questions and arranging interviews with Charles Bennett, John Smolin, and Bob Sutor.

We would like to thank Diane Carr at ColdQuanta for answering our questions and providing us with overview information; Misha Rindisbacher at Launch Squad for her help in answering questions about the D-Wave computer; and Jason Freidenfelds at Google for answering our questions about Google's efforts.

We especially benefited from Bill Silver sharing with us his notes and recollections of Ed Fredkin's course on Digital Physics, and from Charles Bennett sharing with us his photograph of the 1981 Conference on Physics and Computation.

We also benefited from commentary and support from Geoff Cohen, Andrew Grosso, Burt Kaliski Jr., Darrell Long, Hartmut Häffner and Stuart Schechter..

The text benefited from workshops held by the Haifa Center for Law and Technology (with commentary by Amnon Reichman, Tal David, Shay Gueron, Michal Gal, and Orr Dunkelman); the Sandia National Laboratory (with thanks to Andrew Reddie); and the Ohio State University (with thanks to Dennis Hirschman).

In its final form, various chapters of this book benefited from in-depth reviews by Ross Anderson, Michael Grant, Ted Huffmire, and Bill Silver.

Author Hoofnagle's family, doctors Jay and Mark Hoofnagle and Cheryl Winchell, read drafts and provided commentary; Hoofnagle's father-in-law, Jon Wilbrecht, provided commentary and suggested the ultimate organization of the work. Yasemin Acar helped us with German translations.

We are grateful to our editor, Matthew Gallaway, and colleagues Cameron Daddis, Jadyn Fauconier-Herry, and Chloe Quinn of Cambridge University Press. Our copy editor, Ken Moxham, our proofreader Ian Pickett, and the TeX Support team at SPi Content Solutions, particularly Suresh Kumar, caught and fixed many problems with the text and formatting. Those that remain are ours, not theirs.

This work started as an article that received comment from Lily Lin, Evan Wolff, and Peter Swire; the participants in the Future of Cybersecurity Working Group, particularly Andrew Reddie and Kristy Cappelli; the participants of the Berkeley faculty scholarship workshop, including Jennifer Urban, Sonia Katyal, Christopher Kutz, Andrea Roth, Amnon Reichman, Mark Gergen, Nicole Ozer, Steven Sugarman, Eugene Bardach, Peter Schuck, and Christopher Slobogin; the participants of the Yale Information Society Project paper series, including Laurin Weissinger, Sam Hayek, Nikolas Guggenberger, Tiffany Li, and Mason Marks; the participants of the Privacy Law Scholars Conference including Andrea Matwyshyn, Sue Glueck, Andrew Odlyzko, Maria Brincker, Ot Van Daalen, Katherine Strandburg, Alex Deane, Ari Waldman, Aaron Massey, Pam Dixon, Jeff Brueggeman, Mary Madden, and Nick Merrill.

Special thanks are due to the contributors of *tex.stackexchange.com* who answered our LaTeX questions, as well as to our respondents on *quantumcomputing.stackexchange.com*, *chemistry.stackexchange.com*, *crypto.stackexchange.com*, and *hsm.stackexchange.com* (history of science and mathematics).

The appearance of US Department of Defense (DOD) visual information in this work does not imply or constitute DOD endorsement.

Publication with a *CC-BY-NC* license was made possible in part by support from the Berkeley Research Impact Initiative (BRII) sponsored by the University of California, Berkeley Library and by the Berkeley Center for Law and Technology. We are grateful to Timothy Vollmer, Rachael Samberg, Margaret Phillips, and James X. Dempsey for their support in making an open access version of this work available.

0	1	0	0	0	1	1	0	0	1	1	0	0	0	0	1
0	1	1	0	1	0	0	1	0	1	1	1	0	1	0	0
0	0	1	0	0	0	0	0	0	1	0	0	1	1	0	0
0	1	1	1	0	1	0	1	0	1	1	1	1	0	0	0

Introduction

W̱E are at the cusp of a technological revolution, one where technologists master the special physics of the smallest particles; a revolution that promises to provide capabilities that are, somewhat paradoxically, extraordinarily large.

Quantum mechanics explains the interaction of mass and energy at the smallest scales – why a molecule of water gets hot in a microwave oven, or how a uranium atom splits in a nuclear reactor. The rules of quantum mechanics are often counterintuitive and seem incompatible with our everyday experiences. Over the past century, deeper understanding of quantum mechanics has given scientists better control of the quantum world and quantum effects. This control provides technologists with new ways to acquire, process, and transmit information as part of a new scientific field known as *quantum information science* (QIS).

QIS combines quantum mechanics and information theory. QIS is not new – its roots go back to the 1960s. In recent years, however, technologists have made advances in quantum information acquisition, processing, and transmission, discussed in this book as *quantum sensing, quantum computing,* and *quantum communications*. Advances in these three classes of technology have moved discussions of QIS from the world of academic journals to corporate boardrooms and government offices. As the capabilities of quantum technologies have become clearer, both governments and companies have increased investment.

As quantum technologies arrive, we need both a clearer understanding of their implications for stakeholders and an open discussion of policies dealing with the impact of quantum technologies. Quantum technologies have *strategic* implications for nation states, they present challenges for decisionmakers such as investors, and they have many practical implications for individuals' lives. This book ex-

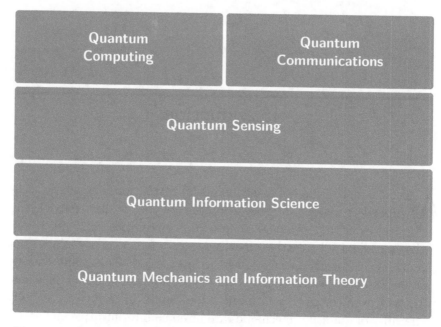

From quantum mechanics and information theory to quantum technologies. Quantum sensing is a precursor technology to computing and communications.

plains the political relevance of quantum technologies and begins a policy discussion for their management.

The strategic implications of quantum technologies have ignited a technology race among stakeholders:

- **China and Europe** see QIS as an opportunity to leapfrog US technological superiority. In particular, nations see deployment of quantum technologies an an opportunity to counter the asymmetric advantages the US has gained from inventing the Internet. Seeking superiority carries with it themes of sovereign technology politics, and as a result, the risk of less scientific openness.

  Research groups in China and Europe have achieved fundamental, state-of-the-science gains in some quantum technology fields, renewing calls for large government investment in quantum technologies by the US and other countries. Reports of quantum-enhanced sonar and radar capabilities by Chinese scientists have rattled some US policymakers. Meanwhile, Germany, the United Kingdom, and the European Union (EU) as a collective are also making major investments in quantum technologies, often with an emphasis on quantum networking and quantum key distribution. These are strategic emphases, because quantum communications could potentially narrow the aperture of foreign intelligence agencies.

- **Corporations** see the potential for billions in profits from the development and use of quantum computing, but the path to success is not clear and is fraught with risk. The most direct path to profit is to use quantum simulators to reduce research and development costs and to enable new discoveries, particularly in chemistry, pharmaceuticals, and materials science. Quantum computing may also enable breakthroughs in operations research and the optimization of business decisions, although existing classical alternatives are superior and may remain so for some time.

For companies and investors, key issues include: whether quantum computing is a *winner-take-all* technology, that is, does a company have to be the first to develop a quantum computer, or can profit be realized by innovators in second and third place? Companies are also concerned whether paths to profit will be constrained by government technology superiority goals. Governments' competition over technology has already imposed export controls and demands for secrecy. Those controls and secrecy might make it more difficult to recruit the best workers. Companies are also concerned that their hard work will be copied or stolen by other nations or by competitors.

The good news for companies is that the barriers to entry in quantum technologies are falling, thanks to the development and commercial availability of devices that produce and measure quantum effects, such as single-photon emitters and detectors. Hundreds of companies have some significant emphasis in quantum technologies, some have even brought quantum technologies to the marketplace that you can buy online today.

Quantum technologies present opportunity and investment risk. Investors need to understand that the complexity and promise of quantum innovations make specious claims of profit and success difficult to evaluate. Given that investors were swindled by miracle narratives in less complex fields, we should be ready for the charismatic business leader to emerge promising billions based on wondrous yet unsound quantum technology concepts.

- **The US government** views quantum technologies as *dual-use* (both peaceful and military) and as important to the nation's strategic posture. Those invested in maintaining US technological superiority are worried about advances in quantum technologies made outside the nation.

  The US government has promised billions in funding for QIS and is in the process of awarding research projects through the research agencies of the armed forces and through the Department of Energy's National Laboratories. This funding, which represents a strong *industrial policy* approach, will stimulate both basic and applied research in all manner of quantum technologies. Quantum technology development policy is thus like the history of computing, the Apollo Space Program, and the Global Positioning Satellite network – projects as uncertain in benefit as they were costly to the taxpayer. But in each of these projects, unforeseen technologies were developed that eventually devolved to the private business community and to the average consumer.

Quantum technologies are heating fever dreams for nations' technological superiority goals. However, achieving superiority may be much harder in quantum technologies than in nuclear and aerospace programs. Quantum technologies are not in the exclusive control of any individual nation. Not only that, government strategies seeking technological superiority must anticipate the innovative power of academia and resource-rich private companies, as both have basic and applied research programs in quantum technologies.

Quantum technologies are expensive to develop, and require expertise that is in short supply. Much of that expertise is concentrated within organizations that have a commitment to open research and the free flow of ideas. Many of the teams working on quantum technologies are multinational, and virtually all of them have incentives to commercialize quantum technologies, complicating the task of developing tools that would be restricted to use by militaries. Indeed, some quantum technology innovators are shunning public funding to avoid the strings attached to government patronage.

Tomorrow's likely developments in QIS will have consequences for how we will measure and sense the world, for how we will communicate, and for how computing will work for us. These consequences are so profound that we should begin planning for them today.

This book summarizes the state of QIS today in the form of quantum sensing, computing, and communications with the purpose of elucidating policy contours.

Outline of the Book

Part 01, "Quantum Technologies," begins with the highest-level concepts one needs to grasp in order to understand QIS and quantum technologies. Chapter 1 briefly covers what we consider to be the three ideas central to the field: uncertainty, entanglement, and superposition.

Readers wanting deeper treatment of quantum effects in Chapter 1 could turn to the appendixes of this book. We wrote the appendixes to provide policymakers, investors, and others who have to make critical decisions, with the scientific context relevant to today's policy issues. Appendix A provides an explanation of the quantum world: its size, how it is measured, and the meaning of the quantum scale. Appendix B continues the exploration of quantum theory with an exploration of the quantum state and how one measures at the quantum level. This material is presented with a historical lens, summarizing the debates and questions that animated decades of empirical and theoretical research in quantum mechanics.

Part 01 proceeds with the state of the science in quantum technologies. Quantum technologies sometimes provide improvements on classical methods, and in other cases create new capabilities. Quantum sensing is the most promising quantum technology, and thus we begin our journey in Chapter 2 focusing on it. Quantum metrology and quantum remote sensing are the first large-scale deployments of quantum technologies. *Metrology* is the scientific study of measurement (not to be confused with meteorology, the study of weather), while *quantum remote sensing* (or simply *quantum sensing*) refers specifically to the measurement of things in the distance. This chapter explains how the exquisite sensitivity of quantum states make it possible to perform precise measurements on things that are nearby or in the distance (underground, in the sky, or even in Earth's orbit).

Nuclear weapons provided the first significant – and horrific – demonstration of quantum technology. Today, the most visible use of that technology comes in the form of nuclear power plants. During the same period that nuclear weapons were developed, quantum sensing contributed to the diagnosis and treatment of untold numbers of people. The physics of nuclear magnetic resonance (NMR) spectroscopy was worked out in the late 1940s;[3] commercial NMR spectrometers were offered for sale just a few years later, and in 1977 the first two-dimensional image of a person's chest was produced.

NMR spectroscopy and magnetic resonance imaging (MRI) were game-changers for chemistry and medicine, and examining the history of these technologies from our twenty-first-century vantage point gives us a template for understanding the impact that quantum sensing technologies might have in the future. Quantum sensing possesses a number of affordances that make its strategic value apparent: first, quantum sensing can be stealthy, that is, it is possible to deploy quantum sensors in ways that an adversary may not detect them, making quantum sensors very different than long-distance radar arrays. Second, quantum sensors resist existing electronic warfare countermeasures, thus making it possible to determine one's position, engage in navigation, or make highly accurate measurements of time in the presence of jamming. Third, quantum sensors create several new capabilities, such as the ability to determine one's location underwater or underground (that is, when lacking a clear view of the sky to catch a GPS signal). Fourth, quantum sensing make it possible to detect objects that are obscured by barriers such as walls or those that are buried. This capability makes quantum sensing a potentially destabilizing technology for submarine and aircraft stealth. Finally, quantum sensing includes a curious application called ghost imaging, a technique so sensitive that it enables detection of things not in the direct line-of-sight of a sensor.

Quantum sensing is a precursor technology to quantum computing and communications. That is, in order to have a quantum computer or a workable quantum network, one must first develop control and readout systems focused on sensing individual particles. Some believe that a large-scale quantum computer will never be built. But when it comes to quantum sensors, there have been decades of successful development, continuing refinement, and even commercial availability.

---

[3]Edward Mills Purcell at Harvard University and Felix Block at Stanford University shared the 1952 Nobel Prize in Physics for its discovery.

For all these reasons, quantum sensing, in our view, is the "killer app" of quantum technologies for at least the next decade. Particularly in the medical field, quantum sensing will benefit humankind in palpable, direct ways. The application of quantum sensing to intelligence, military, and law enforcement uses is more disruptive and harder to address with countermeasures, and thus warrants significant policy attention.

The following four chapters unpack quantum computing – the quantum technology that is most discussed in the media and also most challenging to realize.

To understand quantum computers, it helps to have a foundation in the history of classical computing. This history elucidates many parallels and lessons for quantum computing. Chapter 3 summarizes humankind's development of calculation technologies and the rise of the earliest computers. Like many other technologies, computing required the creation of wildly expensive prototypes and was followed by periods of refinement in both theory and engineering. Over time, these refinements resulted in cost-cutting, and democratization of the technology to large businesses, and ultimately, the consumer. We will show the success of American and British computing prowess as a result of state patronage, and contrast it with a

Quantum Sensing
Uses quantum effects to acquire data.
Capabilities
Measurement of magnetic fields, electric fields, gravity, temperature, pressure, rotation, acceleration, and time.
Near-term applications
Could change every strategically important industry: aerospace, intelligence, military, law enforcement, extractive industries, medical, and others.
Outlook
Highly optimistic because of multitudinous commercial applications, government investment because of strategic applications, relative simplicity, and increasing commercial availability of components.

cutting-edge technology that Germany possessed before World War II that withered for lack of government support.

Quantum computing is a family of approaches for building computers that switch information with quantum interactions, rather than with the electronic interactions that power today's computers. Chapter 4 presents an in-depth history of quantum computing, including the genesis of the field's foundational concepts. Many provocative ideas and engineering projects have a shared genesis with quantum computing including theories of time, theories of emergent complexity, and even whether our own existence is a kind of computer simulation. These ideas were incubated among researchers awash with government support; that support gave them the time and academic freedom to connect the concepts of physics and computing.

Encouraged by thinkers in this environment, Richard Feynman crystallized a vision for quantum computing: that only a computer based on quantum interactions could simulate the complex and probabilistic nature of reality. The *Feynman vision* unifies physics and computing in an effort to understand physical processes. If realized the payoff would be life-changing for humans. Examples abound and are discussed later in this book, but for now consider just one example that could change the prospects for all of humanity: if humans could better understand the basis of a physical process like photosynthesis (one that naturally takes advantage of quantum effects to capture energy efficiently in ways humans have not been able to replicate), we might find ways to harness energy from the Sun far beyond the capacity of existing solar cells. The same insights might allow us to store that energy for when we need it, and then use that energy to grow more food and ultimately feed more people. The Feynman vision is our lodestar for quantum technologies, as it is the most compelling one to support more life and at a higher standard of living.

Not long after Feynman's insight, a different vision for quantum computing arose when scientists discovered quantum algorithms likely to undo encryption systems. These discoveries ignited new interest and investment in quantum computing. They also altered the field's narrative from Feynman's science and exploration vision to something darker: a world where quantum computers are developed to help the world's intelligence agencies discover secrets. Predictions based on this vision hold that quantum computing will bring about a fundamental change to data privacy, a crisis where secrets can no longer be kept. This *dark vision* for quantum computing is often accompanied by privacy doomsday scenarios that are not in touch with technological and practical realities.

We think the Feynman vision is more likely to take hold, and base our argument in the likely applications flowing from quantum computing in Chapter 5. As a starting matter, the Feynman vision presents more opportunities for profit. Just as importantly, the Feynman vision can be used to scale larger quantum computers. That is, by simulating fundamental processes in chemistry and materials science, an innovator might discover insights making it possible to build a larger quantum computer.

Large quantum computers do not currently exist and the path to build one is unclear. The encryption-ending vision for quantum computing requires large devices, but also is subject to practical limits that make simple narratives of a privacy

Quantum Computing
Uses nondeterministic nature of quantum interactions to process data.
Capabilities
Simulation in biology, chemistry, materials sciences; will perform some computations dramatically faster than classical computers.
Near-term applications
Simulation of natural processes, optimization, improvements in search.
Outlook
Most challenging and complex quantum technology; requires fundamental science advance to scale devices to have universal, fault-tolerant computing. In the near term, quantum simulators will be the most significant kind of quantum computer.

doomsday unlikely. In fact, we believe that privacy crisis scenarios, ones defined by shifts in the fundamental assumptions about the power to collect and use data, are likely to come from quantum *sensing*. Quantum sensing is the bigger threat because the technologies are maturing, easier to deploy, and in some cases, countermeasures are out of reach.

We also dispel popular notions about the capabilities and powers of quantum computers. For instance, quantum computers will not "consider all possible solutions to problems" and magically make all computing tasks blindly faster. As we currently understand them, quantum speedups will be limited to a small number of important problems; classical computers will remain in use for all others. Indeed, as they are currently imagined, quantum computers are better thought of as specialized processors bolted onto the side of conventional computers, there to perform specific functions.

Today, some researchers are merely attempting to demonstrate that quantum computers can compute things that conventional computers cannot – what is termed, controversially and somewhat misleadingly, as *quantum supremacy*. Chapter 6 canvasses the state of the science in today's quantum computing landscape.

Quantum computing is still at an early stage: researchers are building the first working prototypes, and others are arguing about whether these machines will ever be more than research curiosities. The fundamental challenge is one of scale: the transistor allowed classical computers to scale for decades. A similar, but so far elusive, breakthrough is necessary to manage the more difficult challenge of scaling a machine that masters quantum states. This chapter discusses the different kinds of quantum computers that have been built to date, their accomplishments, and speculates on what tomorrow's quantum computers might bring.

Quantum communications could be thought of as a merger of quantum sensing and computing. The purpose of this union is to send messages across distances with fundamentally stronger security. Chapter 7 explains the applications and implications of quantum communications. We distinguish between two technologies often combined under the term "quantum communications": quantum key distribution and quantum networking. Quantum key distribution (QKD) involves distributing keys that are information-theoretic secure, thus enabling classical communication over the Internet that is resilient even against an attack with a quantum computer.

The second technology is quantum networking or "quantum internet." Quantum networking involves reengineering network layers to communicate using entangled photons. If achieved, quantum networking will have benefits for confidentiality and integrity; for instance, users would no longer have to rely on network trust as communications become end-to-end. The quantum internet would also eliminate metadata surveillance, a key advance for communications secrecy.

A quantum network will enable the interconnection of different quantum computers. Interconnection means that one path to building a large quantum computer might be to interconnect several smaller ones over a quantum network.

The outlook for quantum communications is a mixed bag. On one hand, classical alternatives for securing codes against quantum computers – so-called post-quantum cryptography – are well understood and less expensive. On the other, research groups and governments in Asia and Europe are heavily investing in both quantum communications approaches. Their investment might be driven by the realization that while large-scale quantum computing is not currently achievable, quantum communications may be an interim step that primes a nation's technical capacity in the future. Or perhaps China and the EU see the metadata-shielding advantages of quantum communications as an opportunity to shrink the surveillance aperture of the US government.

In any case, we believe that it is prudent to move to post-quantum cryptography algorithms as soon as possible, rather than waiting for an announced quantum breakthrough.

Part 10, "Shaping the Quantum Future," turns to the social and policy issues raised by quantum technologies.

There are mechanisms that underlie quantum technologies that will result in a similar development cycle to predecessor classical technologies. We resist heroic innovation narratives that promote quantum technologies as unique and entirely new, because these narratives tend to charm the public, leading to the mistaken impression that existing tools of analysis and comparison are inadequate. Historical comparisons and previous technological revolutions can be used to help understand the implications of quantum technologies. Comparing classical technologies with their quantum counterparts is indeed like comparing dynamite to nuclear weapons: quantum technology is vastly more powerful, but also more specialized: in most cases, quantum technologies will complement, not replace, tools that are in use today.

We anticipate the arcs that quantum sensing, computing, and communications could take in Chapter 8. This portion seeds a policy discussion by modeling four possible futures for quantum technolo-

Gov't Dominant Scenario	A government enjoys enhanced and new sensing, computing, and communication powers and can deny them to others.
Public/ Private Scenario	Quantum technologies emerge in both public and private sectors; there is broad commercialization. Collaboration is relatively international and open.
East/ West Bloc Scenario	The US and Europe's public and private sectors develop quantum technologies in competition with China; secrecy and export control used aggressively.
Quantum Winter Scenario	Basic science challenges prevent large-scale, general purpose quantum computing from being realized; advances in quantum simulation, sensing, and communications proceed.

Scenarios for how quantum technologies could evolve are presented in Chapter 8.

gies. In the first, a government becomes dominant and superior in quantum technologies, enabling it to enjoy the powers of quantum technologies while denying those capabilities to others.

The government dominance scenario is foreseeable because quantum technologies are likely to be expensive and complicated for some time. The expense and complication mean that only large, moneyed institutions will have quantum technologies. Actors with access to outer space will be able to deploy quantum technologies in more powerful ways. Quantum technologies thus have the double whammy of being both institution-empowering and expensive, attributes that mean that masters of quantum technologies are likely to have asymmetric advantages over ordinary people.

In a government-dominant scenario, states and perhaps state-affiliated companies have more power to sense, more power to comprehend sensed data, and more ability to communicate secretly –

and be able to deny these powers to others. To make this explicit, those without quantum technologies will have less sensing, less sense-making, and less privacy from those with quantum technologies. Quantum technologies may result in *strategic surprise*, situations where a nation gains a substantial advantage over competitors, for instance, by using remote sensing to discover hidden facilities or critical infrastructures. The asymmetric advantage is, in a nutshell, why nation states see quantum technologies as a strategic issue much like advances in artificial intelligence.

The second scenario, where public/private partnerships blossom into an innovative landscape that uses quantum technologies broadly, is more likely. We recount the reasons why quantum research is similar to and different from previous technology efforts such as the Manhattan Project and the Apollo Space Program – the most important being that barriers to entry in quantum technologies are lower. Prototype quantum computers can be made for tens of millions of dollars, instead of the billions required by atomic bomb and space research. That price differential means that even startup companies can be strategically relevant in quantum technologies. Strategic surprise in a public/private scenario looks different. Surprise may take the form of a company proposing to eliminate public governance with private governance, perhaps with a smart city that is optimized by a quantum computer.

The third scenario is a variation on the public/private partnership, where such partnerships exist but follow East/West bloc divisions, for instance, separate, quantum technology programs in the US and allied nations primarily competing with China. In both public/private scenarios, innovation blossoms for industrial and consumer applications of quantum technologies. Surprise in a block division scenario might include a different nation taking a fundamentally different approach to quantum computing than other actors and succeeding, causing the other nation to advance in ways others cannot.

Finally, we consider a "quantum winter," a scenario where scalable and general purpose quantum computing cannot be realized in the next 10 to 15 years, leaving just quantum sensing and communications as the most vibrant form of quantum technologies. In this scenario, governments must contemplate surprise coming from other big technology bets. Perhaps one nation squanders billions developing small, ineffectual quantum computers while another becomes technologically superior by focusing on traditional machine learning and automation.

For each of these scenarios, understanding the complex relationships among companies, the market for quantum technologies, and the state is critical for norm development and regulatory capacity. With an understanding of the technology and its possible paths, we turn to the political economy of quantum technologies and policy options in Chapter 9.

We do not need to draw on a blank canvas when discussing the implications of quantum technologies: many of the questions facing us today faced scientists, engineers, and policymakers during the first half of the twentieth century. This means that we can look to the history of computing and sensing and make reasonable predictions about quantum technologies. The highest-level policy issues include:

Innovation Policy. Although a German inventor had an innovative computer years earlier than the Americans, the German government failed to fund the project. Meanwhile, the US and UK incubated computing in pursuit of military and intelligence needs. Government patronage overcame the initial, high costs of developing computers. Particularly in the US, continued government needs for computing – an *industrial policy* that seeks national technological superiority in computing to this day – kept the industry alive and innovating and eventually created a consumer marketplace. Silicon Valley benefited from decades of Department of Defense patronage, seeding the region for high-technology innovation.

Like classical computing, quantum technologies also require large, multidisciplinary teams to properly develop them. We should cast off romantic narratives about individual, heroic inventors, and see that the path to success will be a group one. Similarly, we must recognize popular libertarian technology innovation narratives that malign or minimize governments' role in technology as specious. If technology development were left to the private sector, America's technical achievements in the twentieth century almost certainly would have happened elsewhere. History suggests instead that governments will be key to the realization of quantum computing, as governments have also been the driver of innovations like global positioning systems and the Internet.

And yet at the same time, the private sector has an important role to play. Barriers to entry in quantum technologies are much lower than in aerospace or nuclear weapons, making private companies strategically relevant in the field. Private sector investment in

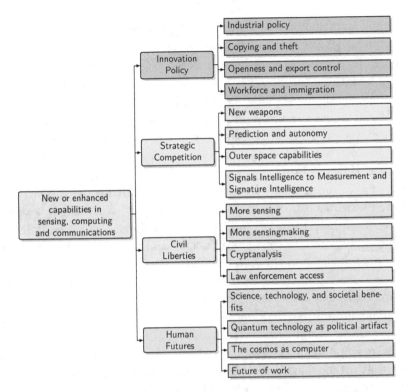

The highest-level policy issues implicated by new capabilities and improvements on classical methods from quantum sensing, computing, and communication.

quantum research is substantial, sometimes in parallel with government funding and sometimes separate from it. The balance of public and private funding shapes economic incentives and ultimately what applications will get developed first.

Openness. Sometimes, important technologies are developed by researchers in secret government organizations and then re-invented in public at universities or corporations. One well-known example is public key cryptography, which was first discovered, then discarded, by the UK communications intelligence agency GCHQ. Public key cryptography was then re-invented by a group of US university professors in 1976 and 1977. As a result, US companies commercialized the technology and made billions; UK companies didn't.

Several precursor developments to quantum technologies, such as the transistor and the laser, played important roles in Cold War weapons systems. Fortunately these technologies were developed at organizations interested in commercializing them, rather than keeping them bottled up. We can easily imagine an alternative history where the transistor was tightly controlled and the computing revolution was delayed by decades, or was centered in Japan rather than the US. Similarly, the relative openness of quantum technologies will affect how these technologies are used but also who can develop further enhancements to these technologies.

While nations develop quantum technologies, governments must make innovation policy tradeoffs. A policy of openness might grease the wheels of innovation and democratize quantum technologies, leading to innovations that are unpredictable and wonderful. Openness might just as well allow nations to free ride on the investments made by others, and even come into parity with the powers developed by China, Europe, and the US. Nations have several levers including export controls, patent secrecy, and classification to shape who can get access to the leading-edge technologies. Nations that fear strategic destabilization, for instance those that fear that quantum technologies will allow detection of stealth jets and silent submarines or compromise legacy communications systems, might pursue something akin to a non-proliferation strategy.

The Value of Basic Research. Many of the breakthrough ideas in QIS that are now attracting billions of dollars in investment started off as fringe ideas in academic and corporate research organizations. This shows once again the value of allowing – and funding – basic research that has no obvious near-term payoff. For policymakers this presents a quandary, because of the challenges posed in distinguishing solid basic research proposals, that deserve funding, from wayward or even crackpot ideas that suck resources but never produce anything of value.

One way to minimize the risk of funding basic research is by increasing the size of research funding in general and earmarking a percentage for basic research, so that funding managers can pursue innovative ideas without risking their own professional reputation, and by giving more leeway to redirect or repurpose funds with minimal administrative overhead. The current path is concerning, because in the five decades since the birth of quantum information science, the amount of US government funding spent on basic research has steadily declined, while the administrative restrictions associated with using those funds have steadily increased.

Immigration Policy. Just as quantum physics and early computing in the US and other liberal nations benefited tremendously from the bright lights from around the world, today's quantum technology companies assemble experts from all over the world to solve fundamental challenges. Nations that make it easier for skilled scientists to emigrate and to gain access to sensitive new inventions will have advantages in quantum technology development.

The future of the US as a quantum technology power depends on our immigration policy. Many students and researchers working within the US on QIS are foreign nationals. If individuals are unable to remain in the US at the completion of their studies, US universities today will train the nation's competitors of tomorrow.

Virtuous Cycles and Winner-take-all Risks. Computers can be used to build faster computers, allowing computers over time to grow in speed, capacity, and efficiency more quickly than other kinds of technologies. This is known as a *virtuous cycle* and it is not present in most technological endeavors. For instance, faster aircraft do not permit aircraft manufacturers to build significantly faster aircraft.

Classical computing enjoyed several kinds of virtuous cycles, where advances in computing justified investments that produced even faster computing. Quantum computing will likely enjoy such a virtuous cycle once computers have reached the scale that they can be used for simulating basic physics. Quantum sensing may enjoy such a cycle; quantum key distribution almost certainly will not.

A strong virtuous cycle also raises the risk that the first group to make a stable quantum computer enjoys a virtuous cycle that is unachievable by competitors. We have to anticipate the risk that quantum computing may be a winner-take-all technology.

The Risk of Hype. The policy discussion also highlights concerns that the private sector and investors have about the technology. Quantum technology, as a field, is particularly vulnerable to unfounded claims of capabilities and unlikely paths to profit. The precursors for fraud are all present: privately held companies with fewer transparency requirements than others, technology optimism, boosterism, limited availability of independent expertise, complexity, and a class of employees and investors who could make a fortune if a company merely enjoys speculative success. Decisionmakers need to understand whether the quantum market is "frothy." Answering this question requires knowing the difference between quantum foam (a real quantum phenomenon) and quantum fluff (a classical phenomenon as old as markets). Beyond investor losses, one risk of hype is that it could lengthen a quantum winter, making it more difficult to recognize a thaw where investment in quantum computing becomes fruitful again.

Strategic Competition. Nations are spending lavishly on quantum technologies because of the risk of strategic surprise, the notion that a nation will somehow gain a fundamental, decisive advantage over others. Here too, the history of conflict, military and intelligence investments in technology, and norms of conflict all help predict how quantum technologies might be used. Parallels can also be drawn from existing logistical limits on conflict, such as how nations decide to use limited, valuable resources in situations of uncertainty.

Strategic competition shares space with innovation policy concerns. Nations' strategic goals may rest uneasily with companies' desire for profit from their quantum inventions. Companies will want to sell their products and services for many purposes, and will be concerned with a different kind of secrecy: the protection of their engineering secrets.

New Weapons. At the same time, strategic concerns may motivate greater controls on quantum technologies, especially as quantum technologies' dual-use nature is realized. While use of nuclear weapons comes with a taboo, governments have been willing to use conventional devices that create nuclear-like effects. Quantum simulations intended to improve processes in peaceful contexts could be re-purposed to create new, more powerful, or more discriminate conventional weapons. We have to contemplate use of quantum simulation to create biological, chemical, and even genetic weapons.

SIGINT and MASINT. Even without simulation, militaries will find the intelligence, surveillance, and reconnaissance uses of quantum sensing irresistible. The last half century has been characterized by intelligence power gained by signals intelligence (SIGINT) prowess, but quantum communications might limit that power. The next century may be defined by greater measurement and signature intelligence (MASINT), brought about by electromagnetic and gravimetric quantum sensors. Militaries might soon find it impossible to hide matériel and their current secrecy strategies, such as using underground facilities, may be rendered ineffective.

Complementary Technologies and Space Programs. As with other innovations, the future of quantum technologies will be shaped by the availability of complementary technologies that make adoption of quantum technologies easier or implementations more powerful. In the former category, improvement of precursor technologies such as lasers and single-photon detectors lower barriers to entry for those who wish to develop quantum technologies. In the latter, nations that have outer space launch capabilities can do more with quantum technologies than nations limited to terrestrial applications.

Civil Liberties. Privacy and fairness tussles loom large as quantum sensing devices become less expensive and smaller so that they can be used in more environments, including mounted on unmanned aerial vehicles. With the power to see through roofs and walls, or as sensing peers into the body and possibly the human mind, society will have to reconsider boundaries and rules on what may be observed.

Not Just Sensing, More Sensemaking. As quantum computing enables more complex *sensemaking* through link analysis and other techniques, those who possess quantum computers will be able to understand more about the world than those who do not. That is, even if two parties possess the same "facts" about the world, the party with quantum technologies might know more about the world.

Cryptanalysis. The most common risk articulated about quantum computers is their potential to undo the most popularly used encryption systems in the world. This risk is real, but as we explain in detail, also greatly overstated. Cryptanalysis will require a large quantum computer, time to perform the analysis, and of course access to the underlying secrets being discovered. The greater near-term risk to civil liberties comes from quantum sensing advances.

Devolution to Law Enforcement Agencies. Powerful tools developed in intelligence and military contexts tend to find their way into the hands of law enforcement agencies, even on the local level, and often without political oversight. How can policymakers prepare intelligence, military, and law enforcement agencies to contemplate the implications of quantum technologies? For many kinds of surveillance enabled by quantum technologies, ordinary people are unlikely to ever develop countermeasures. Window coverings and fences are effective countermeasures against classical privacy intrusions, but to keep up in the quantum age, homeowners would have to install electromagnetic shielding. Norms and laws will have to suffice to protect privacy.

Human Futures. Quantum technologies present tremendous potential for societal benefits, particularly if the Feynman vision for quantum computing is realized. Understanding quantum-level phenomena may make it possible to support more human life and at a higher quality of living while mitigating damage to the environment.

Is our Reality Just a Computer? Existential crises might lurk in the shadows of a bright quantum technology future. As people realize that the basis of these benefits is the random interactions of quantum events, what will this mean for how people conceive of meaning and their place in the universe? Seeing the world as random might unmoor us from ideals of free will, undermine individual responsibility, and spoil the notion of humans' special place in the universe.

Future of Work. In practical terms, quantum sensing and computing might erode the barriers

Quantum Communication
Uses quantum states to transfer data or to ensure confidentiality and integrity.
Capabilities
Creates fundamentally stronger encryption keys, may enable end-to-end data transmission with quantum states.
Near-term applications
Key distribution systems already realized, works in progress to create more ambitious quantum internet that could block metadata surveillance and even interconnect small quantum computers to create a grid system.
Outlook
Mixed: some applications are less challenging than computing, and implemented in small systems. More ambitious achievements require basic science breakthroughs to store quantum states. Prospects brightened because of massive investment in China and the EU, as well as precursor advances in sensing devices.

to creating more capable systems. As computers become more capable, humans' range of useful work may shrink, undermining our value as economic actors. If computers also become more creative than people, the technology will present a challenge to human meaning and value far worse than privacy invasions. *An inevitable downside* of quantum sensing and computing is interference with privacy norms. *The downside* is a future where humans make themselves irrelevant with an invention that outshines our creativity and ability to take action.

Quantum Technologies as Political Artifacts. Before those existential questions are realized, we should contemplate the political norms that may come with quantum technologies. Quantum technologies, like the atom bomb – a quantum weapon – are associated with specific forms of power, authority, and secrecy. Today, elites from educational, government, and (mostly) defense-industrial base companies can understand and employ quantum technologies. For the foreseeable future, much like the history of early computing, powerful institutions will be the exclusive adopters of quantum technologies. Who can understand and adopt quantum technology matters, because their uses of the technology will dominate for some time.

As with early computing, quantum technologies will at first be used to solve the kinds of problems that powerful institutions are concerned about. Quantum technologies could thus be politicized, and a quantum taboo could emerge.

Finally, Chapter 10 ends our exploration of quantum information science. We are at the cusp of a quantum revolution, yet we have not countenanced the social challenges presented by the technology. We have the opportunity to set normative goals for how the technology is applied. The choices will have to be taken and we hope this book helps elucidate our options.

Part 01

Quantum Technologies

In this first part of the book we show why quantum technologies offer new opportunities and bring with them new challenges.

After first introducing our topic in Chapter 1, we explore the world of quantum sensing in Chapter 2. Quantum sensing is an aspect of the quantum science revolution that is already here and promises to be increasingly important in the coming years.

We then have four chapters devoted to computation. We explore the history and a bit of the math of traditional mechanical and electronic computation in Chapter 3, followed by the history of quantum computing in Chapter 4. While the roots of quantum computing go back to the 1960s, the field got its great push forward in the 1990s with the discovery of two quantum algorithms; we discuss these algorithms and their importance, as well as the more likely near-term use of quantum computers for simulating physics and chemistry, in Chapter 5. We end our tour of quantum computing in Chapter 6.

We close this part of the book with a discussion of quantum communications in Chapter 7.

Small Phenomena, Big Implications

Q UANTUM Information Science (QIS) is the merger of quantum mechanics and information science. These are rich fields that few study in great detail. The three most important QIS concepts that underlie quantum technologies are: *uncertainty, entanglement,* and *superposition.* After introducing the three technical concepts, we outline the highest-level policy challenges in quantum technologies.

This chapter is written for people who neither have nor want a background in quantum physics. It is written at a high level, and thus necessarily omits nuance in favor of basic comprehension. After reading this short summary, the reader then has a choice: continue on where we present quantum technologies from a functional perspective, or you can turn to the appendixes of this book (p. 471), where the three concepts receive a much deeper treatment.[1]

1.1 Uncertainty

The concept of *uncertainty* is the core concept of quantum mechanics. Simply put, uncertainty means that it is physically impossible to know everything about anything. More specifically, uncertainty

---

[1]The Appendixes explore the how and sometimes the why of quantum technologies. Readers who have to make key decisions surrounding quantum technologies, such as the decision to invest money or to make predictions surrounding the technology, should first invest in understanding the basics as presented in the last part of this book.

means that it is impossible to know specific things about the physical world with total accuracy.

We all manage to get by with significant uncertainty in society. For example, it is uncertain how many dollars there were in the US economy at the stroke of midnight in Washington, D.C. on January 1, 2021. But this is a different kind of uncertainty than we deal with in quantum physics. In principle we could obtain the bank records of every US corporation and individual, go house-to-house and count all of the cash, go through every hotel and look at all of the loose change that had fallen into every sofa, and with all that information come up with the size of the money supply. Making that measurement with high accuracy would depend upon having a precise time at which the measurements were being made and an army of auditors to make it. With enough information, one could be certain about the state of the economy.

Quantum uncertainty is different than economic uncertainty because it is typically described in terms of two quantities that are antagonistic: the more accurately one is measured, the less accurate the other. Physicists use the word *complementarity* to describe this antagonistic quality. One explanation for this is that measurement is a physical act, and thus measuring an object requires interacting with that object, which influences its quantum state.

Physicists will not like this simple explanation of uncertainty, but it is good enough for our purposes.

1.2 Entanglement

Entangled particles are somehow linked on the quantum level, even though they are separated in physical space and have no known way of communicating with each other. When two particles are entangled, a measurement made on the first may be correlated with a measurement made on the second. Albert Einstein called entanglement "spooky action at a distance" because it means the measurement of one particle somehow effects another distant one. Yet there is no transmission of information from particle to the other: the two particles are simply linked in some "spooky" way.

Entanglement is a quantum phenomenon with no classical analogue. One way to think about entanglement is that particles that are entangled are part of the same system. When one measures one part of the system, for instance the *polarization* of a photon (the orientation of a light wave), the measurement of that photon's polarization reveals something about the other, entangled photon.

Two entangled photons can be produced by shining an infrared laser on a lithium niobate crystal: sometimes a single photon will appear to be "split" into two photons, each with half the frequency of the first, traveling in different directions.[2] The two resulting photons are entangled: if the polarization of one is measured to be horizontal, the polarization of the second will be measured to be vertical.

Entanglement as a technique is used in all three classes of quantum technology discussed in this book. In metrology and sensing, an entangled photon pair can be used such that one photon illuminates an object while the linked photon is measured. An example comes from still-in-development radar and navigation systems, where the illuminating photon is focused on airplanes in the sky or on underwater hazards. By comparing measurements of the reflected photon with the photon that was retained, it may be possible to detect an adversary's jet or an underwater mine that could not be sensed with a single photon. In quantum computing, entanglement is used to coordinate ensembles of "qubits," the quantum version of classical computing bits. In quantum communications, entanglement can be used to ensure the distribution of secure encryption keys.

1.3 Superposition

Because of uncertainty, quantum mechanics is fundamentally a probabilistic view of reality: some outcomes are more probable, and some are less probable. *Superposition* is the word that quantum physicists use to describe the state of a quantum system before we measure it and learn the outcome of a specific experiment or manipulation.

One way to think about superposition is to consider the state of a coin at an athletic event when a referee flips it up into the air and catches it – but before the coin's state is revealed (see Figure 1.1). The coin could be heads-up or it could be tails-up. Until the referee and the players know, either outcome is equally possible.

A coin toss isn't *actually* quantum superposition, however, because there are ways that the outside universe could know about the coin's state even before it is revealed by the referee. For example, a spectator with a telescope and a high-speed camera might have recorded the position of the coin at the exact instant that it was

---

[2]Prabhakar et al., "Two-Photon Quantum Interference and Entanglement at 2.1 Mm" (2020).

Figure 1.1. Superposition is like the state of a coin in an official coin toss at an athletic event after the coin has stopped spinning but before it has been revealed. Coin toss photo by Keith Johnston from Pixabay.

covered by the referee's hand. Or the referee may be able to feel the position of the coin, and somehow telegraph that knowledge.

A quantum-random coin toss would be truly random and invulnerable to the observation tricks of the spectator or corrupt referee. But the process would impose constraints on the referee. The referee would have to isolate the quantum coin toss from the noise and energy of the universe, lest inference affect the randomness. Instead of using a coin, hands, and eyes, the referee might use a particle of uranium and two Geiger counters entombed in a special, sealed room. As the uranium naturally and randomly emits single quanta of radiation, a Geiger counter clicks. One Geiger counter is labeled "heads," the other "tails." The referee turns on the two Geiger counters at precisely the same time and then notes which counter clicked first.

In this example, the uranium, the referee, and the Geiger counters *are* in a superposition of two states: one where the heads Geiger counter clicks first, the other where tails clicks first. Until the referee leaves the room and interacts with the rest of the universe, either outcome is equally possible, even many seconds (or even hours) after the Geiger counters were first switched on.

Of course, we would have a hard time building a room that would truly isolate the referee from the rest of the universe, and so in practice we do not experience superposition in our daily lives. Nevertheless, supposition is a critical component of many QIS-based instruments. Superposition allows quantum computers to compute directly with quantum probabilities. QIS engineers use the term *coherence* to describe such a system that is in the probabilistic superposition state, before it has interacted with the rest of the universe.

Most QIS systems today require cooling the active components to near absolute zero, in order to shield the quantum state from thermal noise and maintain coherence. QIS devices may also be surrounded by a curious material called mu-metal which shields from magnetic fields.

Because the quantum systems that we use are typically based on the interaction of photons, electrons, and occasionally entire atoms, they don't require shielding from gravity. This is a good thing, because we (currently) don't know how to shield something from gravity. At the same time, there are some quantum sensing devices that use our inability to shield from gravity as a way of measuring minor changes in gravity, which can be used to detect underground mineral deposits and even objects. With even more sensitive devices we can detect gravity waves, although such waves are detected indirectly by their impact on the fabric of space–time, as we shall see in the next chapter.

The exquisite sensitivity of quantum states is both the source of quantum technologies' utility and a challenge to technology development. Quantum states' fragility make them sensitive to small perturbations, a fantastic quality for measuring subtle phenomena, such as the precise contours of the Earth's magnetic field. Yet, that same fragility is a barrier to quantum computing, where information processing requires maintaining quantum states free of environmental perturbations.

1.4 Conclusion

From this intentionally brief introduction to quantum effects, we turn to covering the most exciting developments in quantum sensing, quantum computing, and quantum communication.

Quantum Sensing and Metrology

Q UANTUM sensing is the most exciting quantum technology and it has the most potential to change our lives in the next decade and beyond. Quantum sensors will offer new capabilities with benefits for medicine, defense, intelligence, extractive industries, and many others. Quantum sensing is a precursor technology to quantum computing and communications. In quantum computing, quantum sensors are the literal devices that get information out of a quantum computer, while in quantum communications, quantum sensors are the devices that recover the stream of encryption bits. Thus, quantum sensing will advance as governments pour money into quantum computing and communications.

Quantum sensors use quantum properties and effects to measure or sense physical things.[1] Sensors based on quantum properties must be constructed such that they are sensitive to the smallest perturbations. This is because the smallest perturbations of the universe necessarily take place at the atomic and subatomic levels – and as such, the only way to measure them is with quantum devices.

In a functioning quantum computer, those perturbations can cause *decoherence* and thus limit the complexity of programs that the computer can run. In quantum communications, decoherence means that photons traveling down a long fiber-optic cable or through the atmosphere interact with the surrounding medium, losing their in-

---

[1] This book refers to metrology (that is, measurement, not the study of weather) and sensing under the common label *quantum sensing*.

tegrity. This integrity loss imposes limits on the length of a quantum link and the size of quantum networks.

Quantum sensing flips our vantage point. In the quantum sensing context, the exquisite sensitivity of quantum systems is a strength.[2] Quantum sensors harness this sensitivity of individual quantum particles to measure extraordinarily subtle phenomena.

Atomic clocks, nuclear magnetic resonance (NMR), and magnetic resonance imaging (MRI) are all decades-old forms of quantum sensing. These *first-generation quantum sensing* systems used classical physics and electronics to make precise measurements of quantum phenomena. Newer *second-generation* forms of quantum sensing rely on the quantum effects discussed in Appendix B.4, particularly quantum entanglement and superposition. We examine two specific applications of second-generation sensors: *signals intelligence* (SIGINT), which focuses on communications systems, and in *measurement and signature intelligence* (MASINT), which focuses on the physical attributes of targets.

We believe quantum sensing is the most consequential technology for our lifetimes because:

- Quantum sensing is the most mature quantum technology and some quantum sensors are already commercialized. The market is likely to grow. That's because sensing is simpler than quantum computing, and because many large, mature industries, such as healthcare, mining, and construction, can directly benefit from measurements that are both more accurate and more precise.

- Quantum sensing has applications in military, intelligence, and law enforcement. Nations with outer space programs have a wider range of quantum sensing options than nations limited to terrestrial applications.

- Some applications of quantum sensing are "stealthy," that is, one may be able to use quantum sensing without being detected.

- Advances in lasers – such as increased power, efficiency, and stability – make some kinds of quantum sensors more powerful

---

[2]Degen, Reinhard, and Cappellaro, "Quantum Sensing" (2017).

Figure 2.1. Improvement in laser technology is one factor contributing to more exquisite quantum metrology and sensing. Laser stability sets limits on precision, and is affected by the most subtle factors, such as the vibrations caused by photons striking mirrors inside the laser. To address instability, scientists at JILA, a joint University of Colorado/NIST research center, developed a "superradiant" laser. Based on a 1953 hypothesis by R. H. Dicke, the JILA laser traps rubidium atoms between mirrors separated by 2 cm – the small disks in the center of the photograph. By manipulating the rubidium transitions, the atoms themselves emit a dim, yet coherent laser. In doing so, the rubidium atoms produce light while avoiding the normal, noisy process of synchronizing large numbers of photons. Image public domain by Burrus/NIST.

by increasing their range, their resolution, or the speed with which a measurement can be taken.

- Advances in measuring time precisely using quantum technologies have knock-on effects for the precision of all other kinds of sensing, including location and image resolution.

- Finally, some quantum sensing methods do not require supercooling down to liquid helium temperatures, making them easier to work with and miniaturize. For example, such sensors might be readily made portable – or even used inside the body.

Precision, Accuracy, and Repeatability

Precision, accuracy, and repeatability are three complementary ways to evaluate the performance of a measuring device. For example, imagine you are tracking the height of a child by marking a door-frame with a pencil, and measuring the height of the mark with a yardstick:

precision The word *precision* means the ability of the measuring device to distinguish two numbers: for example, it may be difficult to tell in the example above if the child is 48 inches or 48.05 inches tall. That's because the precision of a yardstick is roughly $\frac{1}{16}$ inch. With most digital meters, the precision is typically the significance of the rightmost digit.

accuracy This word refers to the difference between the result that you might get using the calculation method described above and the true number. For example, you might report the child as being between 0 inches and 1000 inches tall, which is accurate but not very precise. On the other hand, you might say that the child is 978.01 inches tall. This is precise, but it is unlikely to be accurate. In our example most yardsticks are reasonably accurate unless they are damaged: if your yardstick is missing its first inch, it will be just as precise, but it will be significantly less accurate.

repeatability Something is *repeatable* if the same answer is obtained by following the same sequence of operations. If the child is fidgeting, it might be very difficult to get a repeatable measurement.

It is possible for a measuring method to be repeatable and precise without being accurate: we might consistently measure the child as being 978.01 inches tall. It is also possible to be accurate without being repeatable or precise: we might measure the child as being between 0 inches and 500 inches, and then as being between 50 and 600 inches. However, it is generally not enough for a measurement to be accurate and precise without it also being repeatable.

Defining and Redefining The Second

The second was traditionally defined as being $\frac{1}{86\,400}$ of a day, each day being divided into 24 hours, each hour divided into 60 minutes, each minute into 60 seconds ($24 \times 60 \times 60 = 86\,400$). But how should it actually be measured?

Several proposals for using the swing of precise pendulums as the standard measure for time started in the seventeenth century. Measurements of such pendulums resulted in the discovery that the earth's gravity is not constant, a result of the planet's bulge and the unequal distribution of minerals beneath the surface.

In the 1930s astronomers discovered that the Earth's rotation is not constant either, because the Earth's atmosphere and water do not turn lockstep with the planet. Instead, the Earth is slowing down at a rate of roughly 2.5 milliseconds per century, which means that each day is imperceptibly longer than the previous.

For the next three decades physicists and astronomers argued which discipline should standardize time. Physicists proposed using the vibration or resonance of a crystal, molecule, or atom, while astronomers proposed using a readily observable periodic motion, such as the rotation of the Moon around the Earth.

The physicists ultimately won, and the second was redefined on October 13, 1967 by the General Conference on Weights and Measures (Conférence Générale des Poids et Mesures, CGPM) to be exactly "the duration of $9\,192\,631\,770$ periods of the radiation corresponding to the transition between the two hyperfine levels of the ground state of cesium-133 atom."[a]

It may be more correct to say that the result of the discussions was a truce, however. (These days *both* physicists and astronomers keep track of the time and the two are synchronized by the addition or subtraction of *leap seconds.*)

---

[a]Bureau International des Poids et Mesures, "50th Anniversary of The Adoption of The Atomic Definition of The Second" (2017); Weyers, "Unit of Time Working Group 4.41" (2020).

2.1 First-Generation Quantum Sensing

"First-generation quantum sensors" use classical physics and electronics to observe quantum phenomena.

A familiar technology, the atomic clock, invented in 1959, is based on a quantum hyperfine transition that occurs within a cesium-133 atom. The transition happens when the atom absorbs and then re-emits a photon with a frequency of exactly 9.192 631 770 GHz, which is in the microwave frequency range. Modern atomic clocks use a tube of cesium atoms suspended in a vacuum and cooled to nearly absolute zero, to minimize the impact of external forces on each atom's electrons. The clock then adjusts the frequency of the microwave beam until it resonates with the cesium atoms. Once the resonance is achieved, the circuit keeps the frequency locked in place. In 1967 the second was defined in terms of the atomic clock (see the sidebar "Defining and Redefining the Second" on page 35), which measures time by simply counting the number of cycles that elapse: every 9 192 631 770 cycles is precisely one second. Arias and Petit note that the official definition "refers, without saying[,] to 'unperturbed' atoms, that is, those at rest, at zero magnetic and electric fields," reminding us that it is one thing to define a standard in terms of a quantum property and another thing to measure that property with accuracy and precision.[3]

In 1997 physicists clarified that the 1967 definition also required that the cesium atom be at the temperature of absolute zero. Cooling to near absolute zero is performed using lasers. This technique was developed by Steven Chu, Claude Cohen-Tannoudji, and William D. Philips, who received the Nobel Prize in Physics that same year, "for development of methods to cool and trap atoms with laser light."

Quartz watches and computers use a similar approach, although they typically measure the vibrations electronically stimulated in quartz crystals using the piezoelectric effect. Such crystals are easily packaged and are inside most computers, cell phones, and digital watches. Unlike individual atoms of cesium near absolute zero, which absorb energy at a precise frequency dictated by quantum physics, the vibrational frequency of a stimulated quartz crystal can be tuned by altering the crystal's thickness and shape when it is cut. To make

---

[3]Arias and Petit, "The Hyperfine Transition for The Definition of The Second" (2019).

Ethanol

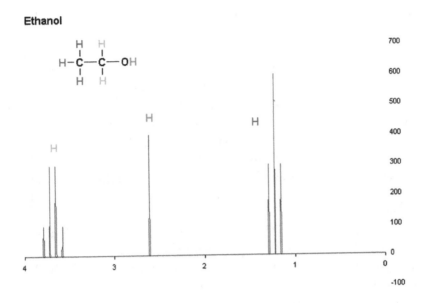

Figure 2.2. The nuclear magnetic resonance spectrum of an ethanol molecule shows that hydrogen atoms exist in one of three configurations. From Wikipedia, by T. Vanschaik. Used under CC-A-SA 3.0.

the engineering easier (and cheaper), the quartz crystals in most digital watches are tuned to vibrate at roughly 32 768 times per second.

Another familiar technology that leverages principles of quantum phenomena is Magnetic Resonance Imaging (MRI). MRI creates images of body parts by detecting the magnetic spin of hydrogen.[4] MRI is based on a molecular property called Nuclear Magnetic Resonance (NMR), in which molecules resonate with a radio frequency (RF) field when placed in a strong magnetic field. Because magnetic resonance depends on each atom's local environment, chemists have long used NMR for clues to figure out the molecular structure of organic chemicals. For example, the ethanol molecule (CH_3CH_2OH) has three kinds of hydrogens: the three hydrogens attached to the terminal carbon, the two hydrogens attached to the middle carbon, and the one hydrogen attached to the terminal oxygen. In an NMR machine, the complex resonance pattern for each set of hydrogens appears in a slightly different part of the RF spectrum. The pattern

---

[4]Berger, "Magnetic Resonance Imaging" (2002).

also reveals that the groups contain three, two, and one hydrogen atoms, respectively.

MRI applies this principle to a section of the human body: the body is placed into a large magnetic field, and then electricity pulsed through coils both to make systematic changes to the magnetic field and generate radio waves, systematically scanning through the three-dimensional space. This is why MRI machines are so loud – the pulses also cause the coils to vibrate with significant force.

The atomic clock, NMR, and MRI all measure quantum effects. However, the technique that they use for detecting that effect is resonance with a radio frequency signal, which is based entirely on classical electronics.

Another first-generation quantum sensing technique is Positron Emission Tomography (PET), which uses small amounts of radioactive material to image metabolic processes in the body.[5] Two-photon microscopy uses lasers to fluoresce tissues[6] including those in live animals.[7] Many nations use gamma-ray and neutron detecting devices, known as *radiation portal monitor systems*, at their borders to detect attempts to smuggle radiological materials or radioactive weapons. The devices can also detect radioactive waste in hospitals and landfills.[8]

Such passive detectors are commercially available from firms such as Bertin Technologies SAS.

Recent developments in the mastery of other quantum effects enable new advancements in quantum sensing. The next section turns to these approaches.

---

[5]M. A. Taylor and Bowen, "Quantum Metrology and Its Application in Biology" (2016).

[6]Svoboda and Yasuda, "Principles of Two-Photon Excitation Microscopy and Its Applications to Neuroscience" (2006).

[7]Holtmaat et al., "Long-Term, High-Resolution Imaging in The Mouse Neocortex through a Chronic Cranial Window" (2009).

[8]In December 1983, the cobalt-60 source from a radiation therapy device broke open on the way to a junkyard in Juarez, Mexico, just across the Rio Grande from El Paso, Texas. The capsule contained 6010 tiny silvery pellets which contaminated the bed of the truck. The truck was scrapped and its steel recycled, which contaminated 5000 metric tons of steel. This steel was used in appliances and construction materials in Mexico, the US, and Canada. Ultimately 109 houses had to be condemned and several people were exposed to radiation as high as 200 rads, a life-threatening amount. See Blakeslee, "Nuclear Spill at Juarez Looms As One of The Worst" (1984). A similar incident happened in Taiwan in 1982. See Hwang, J. B. Chang, and W. P. Chang, "Spread of 60Co Contaminated Steel and Its Legal Consequences in Taiwan" (2001).

2.2 Modern Quantum Sensing Approaches

"Second-generation" quantum sensing technologies advance on the first generation because they directly use quantum effects, such as entanglement, superposition, spins of subatomic particles, or superconductivity. An example would be a device that illuminates a remote object with one of two entangled photons, and then inspects the retained photon to make a measurement that would not have been otherwise possible.

The challenge in quantum sensing is to create a system that can be sufficiently controlled and monitored so that the changes to the system's quantum state are the result of the target object and not intrinsic noise from the device itself. External noise causes decoherence in the sensor, shortening the time in which the sensing can take place. In some cases, sensing requires electromagnetic shielding and cryogenic cooling, adding expense and limiting the environments in which quantum sensors can be used. Because coherence times are short and sensors in a superposition ultimately report a binary outcome, second-generation quantum sensors typically require many repeated measurements. Some approaches use ensembles of sensors so that these measurements can be performed in parallel.

Scientists are experimenting with more than a dozen kinds of quantum sensors that attempt to measure magnetic fields, electric fields, gravity, temperature, pressure, rotation, acceleration, frequency and time.[9] Key approaches include:

Superconducting Quantum Interference Devices (SQUIDs)

These are magnetometers based on "Josephson junctions," circuits that sandwich a small insulating material between superconducting loops.[10] The SQUID is connected to a detection coil, the shape of which can be matched to sensing needs. SQUIDs can detect the strength and gradient of magnetic fields, and since electrical current creates a magnetic field, SQUIDs can be used for non-invasive imaging of the human body.[11] In the medical field, SQUIDs have been

---

[9] Degen, Reinhard, and Cappellaro, "Quantum Sensing" (2017).

[10] Josephson junctions are named after their inventor, Brian David Josephson, who won the 1973 Nobel Prize "for his theoretical predictions of the properties of a supercurrent through a tunnel barrier, in particular those phenomena which are generally known as the Josephson effects." Josephson predicted the effect as a 22-year-old PhD candidate at the University of Cambridge.

[11] Fagaly, "SQUID Magnetometers" (2014).

used for analysis of the heart (magnetocardiography), the lungs, and to record brain activity (magnetoencephalography).[12] SQUIDs also have myriads of uses outside healthcare. For example, SQUIDs can be used for detecting corrosion rates as small as 70-millionths of an inch per year in aluminum,[13] as well as performing other kinds of non-destructive evaluation of materials.[14]

Among the oldest and most sensitive quantum sensors, SQUIDs are commercially available from firms such as US-based Quantum Design Incorporated and UK-based Cryogenic Limited. The SQUID sensor and coil are enclosed in a supercooled, vacuum-insulated container, so they are physically separated from sensed objects and, as such, currently cannot be used in living subjects. Superconducting circuits are the basis for many companies' quantum computing efforts, including Google, IBM, Intel, BBN (Raytheon), and Rigetti.[15]

SQUIDs may eventually be replaced by Optically Pumped Magnetometers (OPMs), devices that do not require cryogenic cooling.[16] Today individual SQUID sensors can be purchased for a few thousand dollars, while fully functioning SQUID magnetometers easily cost hundreds of thousands of dollars.

Atomic vapor technologies These sense electric and magnetic fields with atoms suspended in a resonant medium. Atomic vapor technologies are promising because they can be initialized and read optically, and they operate at room temperature. Two promising variants of this technology are Electromagnetically Induced Transparency (EIT) and Spin Exchange Relaxation Free (SERF) magnetometry. In EIT, an otherwise opaque medium exhibits transparency when two lasers of different frequencies are pumped into it. Measurement of the transparency can detect subtle magnetic fields.[17] In

---

[12]Heidari and Nabaei, "SQUID Sensors" (2019).

[13]Juzeliunas, Y. P. Ma, and Wikswo, "Remote Sensing of Aluminum Alloy Corrosion by SQUID Magnetometry" (2004).

[14]Faley et al., "Superconducting Quantum Interferometers for Nondestructive Evaluation" (2017); Jenks, Sadeghi, and Wilkswo Jr., "Review Article: SQUIDs for Nondestructive Evaluation" (1997).

[15]Buchner et al., "Tutorial: Basic Principles, Limits of Detection, and Pitfalls of Highly Sensitive SQUID Magnetometry for Nanomagnetism and Spintronics" (2018).

[16]Tierney et al., "Optically Pumped Magnetometers: From Quantum Origins to Multi-Channel Magnetoencephalography" (2019).

[17]EIT is exciting because it can produce what is known as a "slow light" effect, when optical pulses travel through a medium with a low group velocity. This

SERF, a high-density vapor of alkali atoms is polarized with a laser to an initial state. After being exposed to the magnetic field, a second probe light is used to detect changes in the atoms' polarization from the magnetic field.[18]

Despite active research on these physical phenomena, there is no evidence of an emerging commercial marketplace for atomic vapor technologies.

Nitrogen vacancy   These approaches exploit imperfections in diamond crystals, that is, where a single nitrogen atom is trapped by the strong bonds of neighboring carbon atoms, and remains relatively insulated from the outside world. A laser is used to initialize the state of the nitrogen atom, and based on photons emitted from the crystal, one can measure magnetic fields at room temperature. Although artists sometimes illustrate articles on nitrogen vacancy with images of large diamonds, the size found on the ring fingers of the rich, in reality the "diamonds" are nanoscale thin membranes. Nitrogen vacancy diamonds are synthetic diamonds made by chemical vapor deposition (CVD), a process involving irradiation and annealing.[19]

Nitrogen vacancy devices are small enough to measure magnetic fields *in vivo*.[20] To speed measurement, they can be arranged in an ensemble, but controlling these ensembles remains a key technical challenge.

Nitrogen vacancy is entering the commercial market. Swiss-based QZabre LLC offering a microscope integrating the approach, and precursor materials, such as CVD diamond films sold by Delaware-based Applied Diamond Inc. Nitrogen vacancy is also considered a promising medium for quantum computing because it operates at room temperature, and it is being pursued by Australia-based Quantum Brilliance, Japan's Nippon Telegraph and Telephone Corporation (NTT), and research groups at Tu-Delft's QuTech, MIT Lincoln Labs, and at Oxford University.

---

effect makes EIT a candidate for quantum memory and for optical transistors. (L. Ma et al., "EIT Quantum Memory with Cs Atomic Vapor for Quantum Communication" (2015))

[18]Budker and Romalis, "Optical Magnetometry" (2007).

[19]Ruf et al., "Optically Coherent Nitrogen-Vacancy Centers in Micrometer-Thin Etched Diamond Membranes" (2019).

[20]Fujiwara et al., "Real-Time Nanodiamond Thermometry Probing in Vivo Thermogenic Responses" (2020).

Photonic approaches   These approaches to quantum sensing have several advantages over other approaches owing to the relative resilience of photons. Photons can be sent out into free space and still retain their critically important quantum phenomena – their "spin" (see Appendix B.3). Photonic approaches sometimes require cooling, but often not the *super*cooling used to maintain quantum states in other media.

Photonic sensing uses techniques such as *light squeezing*, entanglement, single-photon detection, optical interferometry, and quantum "dots."[21] Light squeezing involves limiting the uncertainty of a light wave for some portion of its phase (and thus increasing the uncertainty in other portions of its phase) in order to reduce errors. Photonic entanglement approaches use a pair of photons which have been correlated in some specific way. The pair of photons are split and go in different paths. One of the photons is aimed at something, either to detect it or to illuminate it. The other photon is simply measured directly. In quantum illumination, entanglement enables one to discern reflected light from noise, making it possible to filter and produce a cleaner image. In theory, this would be useful for sensing objects with extraordinarily low reflectivity, such as aircraft designed to have minimal radar cross-sections – sometimes called "stealth" aircraft.[22] These techniques require development of devices that can emit and detect single photons (see Figure 2.3).

In optical interferometry, a beam of light is split and then superimposed upon itself.[23] A detector compares the phases of the superimposed beams, and the patterns reveal evidence of other phenomena. For instance, the Laser Interferometer Gravitational-Wave Observatory (LIGO) uses optical interferometry to detect minute changes in the fabric of space–time that result from the passage of gravity waves.[24]

---

[21]Pirandola et al., "Advances in Photonic Quantum Sensing" (2018); Flamini, Spagnolo, and Sciarrino, "Photonic Quantum Information Processing: a Review" (2018).

[22]Guha and Erkmen, "Gaussian-State Quantum-Illumination Receivers for Target Detection" (2009).

[23]We discuss in Section B.1.4 (p. 493) how Michelson and Morley famously used an interferometer in 1887 to show that there is no aether.

[24]The 2017 Nobel Price was awarded to Rainer Weiss, Barry C. Barish, and Kip S. Thorne "for decisive contributions to the LIGO detector and the observation of gravitational waves." See Abbott et al., "Observation of Gravitational Waves From a Binary Black Hole Merger" (2016).

A range of sensing applications are being explored for quantum dots, including chemical detection (for instance, the presences of extremely small amounts of heavy metals, pollutants, and pesticides) and for biological purposes, such as monitoring drugs and DNA.[25] Quantum dots are crystals fabricated so that they control the movement of electrons; the restriction of movement enables quantized emission of energy. Quantum dots can even be grown to absorb or emit certain wavelengths of light. In effect, quantum dots act as large, artificial atoms. Grown in lattices, quantum dots range from 10 nanometers to a single micrometer.

Like other sensing substrates, quantum dots are a candidate for building quantum computers. Such computers would use the spin of the outermost electron as the qubit.[26] Quantum dots are also being considered for improving existing systems such as solar panels, and for quantum communications, as they can be tuned to emit single photons.

Lasers have been used in many scientific contexts for decades, and quantum photonics has reached significant commercial maturity as a result. Today hundreds of vendors sell various kinds of photon-based components and whole systems. One can readily find sellers of single-photon detectors, single-photon emitters, bucket and Charge Coupled Device (CCD) photonic detectors, beta barium borate crystals to generate entangled photons, lenses to manipulate light polarity, and of course lasers of all varieties. New Jersey-based ThorLabs even sells demonstration kits for colleges that illustrate how polarization in 3D movies works and a small tabletop Michelson Interferometer for just a few thousand dollars. One cannot buy a fully assembled photonic sensor for the advanced applications discussed in this chapter, such as ghost imaging (see Section 2.3.4, p. 68), but one could purchase commercially the necessary components and assemble a ghost-imaging rig in a garage if one was so inclined (and had sufficient financial resources).

Several research groups are pursuing photonics as a medium for quantum computing, including Paris-based Alice&Bob, the UK's Orca Computing, Swiss-based ID Quantqiue, California's PsiQuantum, and Canada-based Xanadu.

---

[25]M. Li et al., "Review of Carbon and Graphene Quantum Dots for Sensing" (2019).
[26]Loss and DiVincenzo, "Quantum Computation with Quantum Dots" (1998).

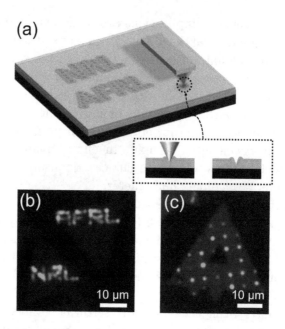

Figure 2.3. Scientists at the Navy and Air Force Research Laboratories placed 1–2 micrometer-sized light sources on semiconductors that can emit single photons on demand. Single-photon emitters and detectors are among the technologies that lower entry barriers for innovation in quantum technologies. Photo by Daniel Parry, courtesy of the US Naval Research Laboratory.

As quantum sensing improves, the world's measurement standards are getting upgraded as well.

Consider *Le Grand K*, the century-old piece of platinum iridium alloy in a secure underground vault in Paris. This metallic cylinder is 90 percent platinum by weight; in those 900 grams of platinum there are roughly 2.78×10^{24} individual atoms.[27] At least, that was the cylinder's weight when it was manufactured; measurements made in 1988 found that Le Grand K had lost roughly five-hundredths of a milligram – perhaps the result of improper handling, or perhaps the result of the material somehow outgassing.[28]

The problem with the cylinder's weight change mentioned in the preceding paragraph is that it is logically inconsistent. In 1988 Le Grand K *was the world's reference standard.* If Le Grand K had

---

[27]Platinum has an atomic mass of 195.078 amu (atomic mass units); to convert kilograms to amu, divide by 9.223×10^{18}. For a discussion of quantum sizes, see Appendix A.

[28]Keats, "The Search for a More Perfect Kilogram" (2011).

The Origins of The Metric System

Article 1, Section 8 of the US Constitution grants Congress the power "To coin Money, regulate the Value thereof, and of foreign Coin, and fix the Standard of Weights and Measures." That is, Congress has the power to determine how things in the US are weighed, how lengths are measured, how time will be kept, and so on. Other world governments claim similar powers. Fair and consistent taxation required a single system for money and a consistent set of measurements, given that many taxes were tariffs collected on a measure of a thing being imported.

By 1787 there was also wide realization within the scientific community that the exchange of scientific knowledge also required a consistent set of weights and measures, and efforts to create a standardized set of measures had been underway for some time. In 1790, the French Academy of Sciences was charged by the National Assembly to devise a new system of weights and measures. Over the next five years the Academy devised the Metric System.

The French Academy originally defined the meter as one 10-millionth the distance at the surface of the Earth from the North Pole through Paris to the equator; this is why the diameter of the Earth is 6371 km: the circumference was defined to be 10 000 000 m (10 000 km) and circumference = $2 \times \pi$, so $40\,000 \div (2 \times \pi) = 6\,366$. The survey was completed in 1798, at which point a bar of platinum was created to be the primary reference standard meter from which all others would be measured. (Platinum was used so that the length would not be affected by oxidation.) In 1959 the foot was redefined by the international yard and pound agreement to be exactly 0.3048 meters.

The kilogram was defined based on the meter and pure water: it is the mass of 1 liter of water (a liter is the volume of a cube that is 10 cm on a side). But as with the meter, this formal definition gave way to a platinum reference standard, the *Kilogramme des Archives*, which was cast in 1799.

somehow lost a fraction of the atoms that made up its mass, in principle all of the world's scales should have required recalibration to the cylinder's new mass. That didn't happen because even though officially Le Grand K was the reference standard, in practice scientists had created near-perfect replicas of the standard, used them to develop extraordinarily accurate scales, and then used the consensus of all of those physical objects to reason about Le Grand K's changed composition. It is as if the scale at Le Grand K's gym, its doctor's office, and its best friend's house all said that it had lost a little weight, even though K's bathroom scale said that it weighed the same as it did back in 1889. Who are you going to believe?

The situation is even worse: as scales became increasingly accurate in the twentieth century, they became increasingly able to detect minor variations in the weight of nearly all of the kilogram's official copies.[29] In part this is because the metallic surfaces absorb or release molecular impurities in the air – and even the air itself. It may also be a result of the wear that results from the need to physically handle these objects in order to measure them. So over the past fifty years, metrologists have worked diligently to redefine all of the standards of measurement in terms of quantum processes or measurements, just as the second was redefined in 1967 (see the sidebar "Defining and Redefining the Second" on page 35).

From a public policy perspective, moving the world's scientific standards from the measurement of specific physical objects to measurements of plentiful and identical quantum objects is democratizing. The movement means that any government, organization or individual with sufficient technical capabilities can make measurements as repeatable, as precise, *and as accurate* as they are able, without stopping to calibrate their measuring devices against some reference standard in Paris, France or Gaithersburg, Maryland. Previously, those groups could make measurements that were repeatable and precise, but accuracy depended upon performing that repeatable and precise measurement on a national standard. Thus it was a great step forward not just for science, but for the *practice of science*, when in 1960 the meter was redefined to be the length equal to 1 650 763.73 wavelengths of radiation in vacuum for a specific transition of krypton-86.

---

[29]Gibney, "New Definitions of Scientific Units Are on The Horizon" (2017).

Surprisingly, the kilogram itself wasn't redefined until 2019 as part of the 2019 redefinition of the International System of Units (abbreviated SI from the French-language name Système International d'Unités).[30] According to the English version of the standard:

> the kilogram will continue to be the unit of mass, but its magnitude will be set by fixing the numerical value of the Planck constant to be equal to exactly $6.626\,06X \times 10^{-34}$ when it is expressed in the SI unit $m^2\,kg\,s^{-1}$, which is equal to J s.

(The symbol X represents additional digits that were added in a technical memorandum. The SI value of Planck's constant is currently $6.626\,070\,15 \times 10^{-34}$.)

As in other areas of quantum information science, the improved metrology creates both the tools and the economic incentives to further improve metrology: this is another example of a virtuous circle.

2.3 Quantum Sensing Applications

At first it may seem to a non-scientist that there are few compelling commercial needs to be able to measure objects to within a nanometer (one-billionth of a meter) or time events to the nearest picosecond (one-trillionth of a second). After all, yardsticks and stopwatches seem like they are good enough for most day-to-day measurements. While it is true that the original motivation for making precise measurements was that of scientists seeking to have a better understanding of the natural world, many of the foundations of our modern technological society depend upon the ability to make measurements that are precise, repeatable, and accurate according to established international standards.

The wide range of quantum sensing technologies currently under development promise a new generation of measurement technology that is not only more precise and accurate, but also widely available and (eventually) low-cost. Critical to all these applications is more precise measurement of time and location, which is discussed next.

2.3.1 Measuring Time

Measuring time with more absolute accuracy and higher precision are the first benefits of quantum sensing; they are also requirements for

---

[30]Bureau International des Poids et Mesures, *The International System of Units* (n.d.).

breakthroughs in sensing other kinds of modalities. This is because precise measurement of time contributes to increases in precision for all other forms of sensing.

The Global Positioning System (GPS) is an example of the power that precise time measurement confers. Mathematically, the accuracy of GPS comes from the stability of orbital mechanics, Einstein's equations of relativity, and precise timekeeping. Each GPS satellite is individually numbered and orbits the Earth along a predetermined path. The satellite sends a radio signal down to the planet below consisting of the satellite's number, coefficients for various mathematical equations that allow computing the satellite's position at any given time, and the precise time that the radio wave left the satellite's antenna, as defined by the satellite's on-board atomic clocks.

GPS receivers listen for these signals from the satellites. If a receiver can "hear" and resolve signals from three satellites, it can solve a series of equations and determine its latitude, longitude, and the precise time. This is possible because all of the satellite clocks are synchronized, but because the distance between the receiver and each satellite is different, the timestamp on each received signal is slightly divergent. Thus, the distance to a specific satellite is simply the receiver's computed time minus the time that a specific satellite is reporting.

Light moves at $299\,792\,458\,\mathrm{m\,s^{-1}}$ – roughly $30\,\mathrm{cm}$ every nanosecond. This means that if the clock on the satellite were to lose or gain just 10 nanoseconds, the satellite's computed position would be off by 3 meters. In practice, such precision in the rigors of outer space requires more than just an atomic clock: it also requires compensating for the impact of time dilation caused by each satellite's orbital speed of roughly $3.9\,\mathrm{km\,s^{-1}}$,[31] which causes the satellite's atomic clocks to tick slightly slower than they would on Earth. The GPS receiver computes your speed from the Doppler Shift of each incoming radio signal: if it is at a slightly higher frequency than expected, then the distance between you and the satellite is decreasing; if it is lower than expected, the distance is increasing. In practice, GPS receivers are so sensitive that they are able to measure the speed of a person walking from its impact on the Doppler Shift.

GPS clocks are pretty accurate, but they do drift. Twice each day, each satellite synchronizes its internal atomic clocks with a ground

---

[31]Caro et al., "GPS Space Segment" (2011).

Figure 2.4. US Army Research Laboratory researchers tested Rydberg atoms' sensitivity, finding them to be sensitive to radio waves across the entire spectrum. Rydberg atom sensors may enable new ways to detect stealthy communication and without the inconvenience of multiple antennae, because the laser that excites the atoms can tune the sensor to detect desired frequencies. Image courtesy of US Army.

station. NASA has also developed deep space atomic clocks based on ion traps and mercury ions for applications where the clocks cannot be readily updated.[32]

Another application for precise timing is to increase the resolution of individual sensors by making repeated exposures and combining them. This approach is sometimes called "super resolution." Photographers can do it today by taking four or eight photos of the same scene and then combining the images using software: this technique requires having the same exposure with each photo. But the same approach can be applied in principle to all kinds of measurements.

Metrologists are now developing approaches for creating so-called *optical clocks* that measure vibrations of atoms in the optical region of the radio spectrum, where light cycles roughly a million times faster than it does in the microwave region used by today's atomic clocks. Using ion trap technology the National Institute of Standards and Technology (NIST) built optical clocks based on the vibration of a single mercury ion (in 2006) and a single aluminum ion (in 2010).

Highly accurate clocks can measure subtle changes in gravity, thanks to the way that the flow of time changes depending on the clocks' position in a gravity well, as predicted by Einstein's general

---

[32]Samuelson, "What Is an Atomic Clock?" (2019).

Figure 2.5. The National Institute of Standards and Technology developed this atomic clock based on an ytterbium lattice in 2013. Image public domain by Burrus/NIST.

theory of relativity. As NIST scientists explained, "if two identical clocks are separated vertically by 1 km near the surface of Earth, the higher clock emits about three more second-ticks than the lower one in a million years."[33] In that same 2010 paper, NIST reported that its atomic clock was sensitive enough to detect an up-or-down movement of just 33 cm.

In 2018 NIST announced a breakthrough for their atomic clock based on a lattice of ytterbium atoms (going back to the mid-1950s, previous clocks were based on cesium).[34] This clock will contribute to *geodesy*, the study of the shape, orientation in space and gravitational field of the Earth. The ytterbium atomic clock can make geodesic observations to within a centimeter accuracy.[35] This clock, and another clock based on strontium ions, are so accurate that one of these clocks would neither gain nor lose a second if it were left running for more than *10 billion years* (assuming, of course, that

---

[33]Chou et al., "Optical Clocks and Relativity" (2010).

[34]McGrew et al., "Atomic Clock Performance Enabling Geodesy Below The Centimetre Level" (2018).

[35]Gaithersburg, MD: National Institute of Standards and Technology, "Second: The Future" (2019).

the rest of the clock could be engineered with the required level of reliability).[36]

The next stage goal is to move beyond measuring the oscillation of electrons as is performed in today's atomic clocks to the "nuclear clock," that measures time by focusing on the states of an atomic nucleus.

The Defense Advanced Research Projects Agency (DARPA) and private-sector investments in quantum sensing have resulted in more accurate atomic clocks,[37] and smaller devices that are now commercially available. For instance, Microsemi Corporation sells a "chip scale atomic clock" that is only 35 grams. "Today's microwave-based atomic clocks on GPS satellites provide 10-nanosecond (billionth of a second) timing, whereas optical clocks could provide 10-picosecond (trillionth of a second) precision," explains a DARPA brochure on the quantum technology projects section of the agency's website.[38]

2.3.2 Sensing Location

Since the 1990s the primary source of positioning (determining where one is and orientation), navigation (determining one's desired position and routes to it), and timing (determining accurate and precise time) information are complex systems built from satellites that orbit the planet, ground stations that service those satellites, and billions of handheld receivers that sense the extraordinarily faint radio signals from the satellites and use them for the basis of complex mathematical operations. Although these systems are typically called GPS, after the US Global Positioning System, there are actually four competing positioning, navigation, and timing (PNT) systems in the world today.

Satellite navigation has become such a critical part of both the modern economy and the modern military that the Russian Federation, Europe, and China have all spent billions of dollars developing and fielding their own systems (see Table 2.1), so that they will not be dependent upon the continued diplomatic goodwill of the United States.

---

[36]Chou et al., "Optical Clocks and Relativity" (2010).

[37]Nicholson et al., "Systematic Evaluation of an Atomic Clock at 2×10^{-18} Total Uncertainty" (2015).

[38]Defense Advanced Research Projects Agency, "Quantum Sensing and Computing" (2020).

Table 2.1. Satellite navigation systems

Year operational	Sponsor	System
1990	US	Global Positioning System (GPS)
1993	Russia	GLONASS
2016	EU	Global Navigation Satellite System (GNSS or Galileo)
2020	China	Beidou (Běidǒu Wèixīng Dǎoháng Xìtǒng)

GPS History

The US Global Positioning System (GPS) satellites were the first satellite-based navigation system; GPS became operational in 1990, 12 years after the launch of the first GPS satellite in 1978. Although the system was designed and funded and intended for use by the US military, provisions were made for incidental use by civilians as well.

Each GPS satellite has a synchronized atomic clock and sends information about its location and current time that can be received anywhere on Earth's surface or in the air (see Section 2.3.1, p. 47). Although it is commonly believed that the satellites track the receivers, this is not the case. Just as a boat at sea determines its position from observing a lighthouse, each GPS satellite tracks itself, and each receiver determines its position by finding and tracking the satellites.

Each US GPS satellite broadcasts on multiple frequencies, including a set of civilian frequencies that are open for public use and one or more military frequencies that are protected by various technologies. In March 1990, shortly after the GPS system became available, the US government intentionally made the civilian signals less accurate through a system called *selective availability*, which increased the uncertainty of civilian receivers from 20 m to 100 m.[39] This was done so that foreign militaries and terrorist organizations could not use the high-resolution GPS signal against the interests of the US.

But shortly after selective availability was switched on, the US found itself at war in the Persian Gulf with Iraq. There was a shortage of military-grade GPS receivers, and civilian-grade GPS receivers

---

[39]Thorton, "Selective Availability: A Bad Memory for GPS Developers and Users" (2018).

were sent from the US to warfighters in the theater.[40] The US responded by disabling selective availability until the conclusion of the war in July 1991, when it was promptly turned back on.

Even with selective availability, GPS found increasing uses in the civilian economy. Although a handheld GPS receiver cost a few hundred dollars, GPS became wildly successful in the civilian marketplace. While the designers of GPS had expected that it would be used in boating, aviation and by hikers, in-car navigation systems soon appeared on the market, a result of the navigation revolution working synergistically with the computer revolution. Selective availability was a constant annoyance for these systems, so approaches were found to get around it.

Differential GPS (DGPS) was one approach for addressing the error introduced by selective availability. DGPS uses a second set of ground stations that "listened" to the GPS signal, figured out how much error was being introduced by selective availability at that very moment, and sent out a correction. Two proponents of DGPS in the US government were the Federal Aviation Administration and the US Coast Guard, both of which had stakeholders that required high-precision PNT to allow for instrument navigation at night and during inclement weather. The fact that two different parts of the US government couldn't agree on how GPS should be controlled, and that each was willing to spend money and engineering effort to deploy a system that advanced its interests in a manner that was antagonistic to another government agency, should be carefully noted.

As commercial use grew, so did pressure on the US government to permanently switch off selective availability. This finally happened on May 1st, 2000. The newest GPS "Block III" satellites, first launched in December 2018, do not even have selective availability capability. In part this may be because the multiplicity of satellite-based PNT makes a satellite-based system such as selective availability less relevant: if the US switched on selective availability, a US adversary could simply use the Russian system instead.

Today you can purchase a 72-channel satellite navigation receiver from China for just $4; the package is just 34 by 28 by 9 mm and includes an embedded antenna (alas, battery not included). The chip

---

[40]"GPS Navigation: From The Gulf War to Civvy Street" (2018).

works with both the US and the Russian systems, presumably allowing users to align their PNT supply chains with their geopolitics.

GPS Spoofing and Jamming

The signals from navigation satellites are quite weak by the time they reach the Earth, leaving all satellite navigation systems vulnerable to spoofing or jamming by stronger signals. Attacks on GPS are motivating investment in quantum PNT approaches.

GPS spoofing is the act of generating radio signals that confuse GPS receivers into thinking that they are in one place when they are really someplace else. In December 2011, the *Christian Science Monitor* reported that the Iranian military had stolen a US bat-wing RQ-170 Sentinel unmanned aerial vehicle by spoofing the GPS signals that the drone received.[41] Likewise, Russia may use GPS spoofing to mask the whereabouts of its high-ranking officials.[42] The Russian armed forces are renowned for their electronic warfare prowess on the battlefield,[43] meaning that local jamming may disrupt equipment in a specific conflict. Increasingly, there is anxiety that satellites themselves will be attacked,[44] leading to a regional or global outage.

GPS jamming is a simpler attack in which the faint GPS signals are simply overrun by other signals on the same frequency. Today GPS jammers that plug into an automobile's cigarette lighter can be purchased for as little as $10.79 from the Walmart website.[45]

---

[41] Peterson and Faramarzi, "Exclusive: Iran Hijacked US Drone, Says Iranian Engineer" (2011).

[42] C4ADS, "Above Us Only Stars: Exposing GPS Spoofing in Russia and Syria" (2019).

[43] Creery, "The Russian Edge in Electronic Warfare" (2019); McDermott, *Russia's Electronic Warfare Capabilities to 2025* (2017).

[44] Kan, *China's Anti-Satellite Weapon Test* (2007).

[45] Such devices are popular with truckers, as they defeat GPS vehicle trackers, allowing the truckers to take unauthorized routes or drive over the speed limit without the rig's owner taking notice. In 2013 a man in New Jersey was fined $31 875 by the Federal Communications Commission for operating such a device near Newark Liberty International Airport, where it interfered with aircraft operations. See Strunsky, "N.J. Man Fined $32K for Illegal GPS Device That Disrupted Newark Airport System" (2013). Indeed, notes the FCC, "The use of a phone jammer, GPS blocker, or other signal jamming device designed to intentionally block, jam, or interfere with authorized radio communications is a violation of federal law" and "It is also unlawful to advertise, sell, distribute, import, or otherwise market jamming devices to consumers in the United States." US Federal Communications Commission, "Jammer Enforcement" (2020).

The US government uses the term "denial" to describe technologies such as spoofing and jamming that can deny the use of GPS to an adversary. Developing these technologies has been part of US strategy since selective availability was turned off. This is clearly signaled on the government's GPS website: "The United States has no intent to ever use SA again. To ensure that potential adversaries do not use GPS, the military is dedicated to the development and deployment of regional denial capabilities in lieu of global degradation."[46]

Of course, other governments are developing similar technology. The US Navy is responding, in part, by training midshipmen and navigators to use charts and sextants so that they will have a low-tech navigational fallback (see Figure 2.6).[47] Concerns about attacks on satellites were one of the factors behind the elevation of the US Space Command to the status of being a unified combatant command.[48]

Inertial Navigation

Inertial navigation is an alternative to both GPS and celestial navigation. These systems use a combination of on-board accelerometers and gyroscopes to continuously track changes in a vehicle's motion and orientation. This approach to navigation, called dead reckoning, is sort of like closing your eyes while you are walking down a sidewalk with the goal of walking another 50 feet and then stopping at the traffic light to press the "walk" button.

The first inertial navigation system (INS) was designed and built by Robert Goddard at his research facility in Roswell, New Mexico, with its first successful demonstration in September 1931. Goddard's work was largely ignored by the US Government but was replicated and extended by Wernher von Braun in Nazi Germany, who perfected the system and used it to guide Germany's V2 rockets to their targets. "More than 9000 civilians and soldiers were killed in total in V2 attacks on the Allies. That excludes the estimated 12 000 labourers and concentration camp prisoners killed while making the

---

[46]National Coordination Office for Space-Based Positioning, Navigation, and Timing, "Frequently Asked Questions About Selective Availability" (2001).

[47]Brumfiel, "US Navy Brings Back Navigation by The Stars for Officers" (2016).

[48]The Space Force, originally formed and housed in the Air Force out of Vandenberg Air Force Base, now has the leadership of a four-star general and a seat at the Joint Chiefs of Staff. Space Command now has authority over all military actions in space, defined as 100 km above sea level.

Figure 2.6. Quartermaster Seaman Delaney Bodine, from Elkton, Maryland, uses a sextant on the bridge wing while standing watch as the Boatswain's Mate of the Watch (BMOW) aboard the guided-missile destroyer USS *Spruance* (DDG 111) in the Arabian Gulf, March 20, 2019. In 2011 the Navy's Surface Warfare Officers School in Newport, RI, resumed training in celestial navigation for navigators and assistant navigators. Navy photo by Mass Communication Specialist 1st Class Ryan D. McLearnon/Released.

missiles."[49,50] After the war, von Braun was invited by the US government's Operation Paperclip to continue his research in the United States. A parallel effort to develop inertial guidance systems was initiated at the MIT Instrumentation Laboratory in the 1950s, which separated from MIT in 1973 to become The Charles Stark Draper Laboratory.

INS is another example of a dual use technology. Pan Am, once a great US airline, installed its first INS on a Boeing 707 in 1964. That early system had problems: because it was based on a gyroscope with mechanical bearings, it tended to drift over time.[51] An improved

---

[49] Arkell, "Death From above without Warning: 70 Years after The First One Fell, Interactive Map Reveals Just Where Hitler's V2 Rockets Killed Thousands of British Civilians in Final Months of WW2" (2014).

[50] The BBC estimates that 20 000 prisoners pulled from concentration camps died constructing the V2s. See Hollingham, "V2: The Nazi Rocket That Launched The Space Age" (2014).

[51] Morser, "Inertial Navigation" (2020).

inertial guidance system that used gas bearings was deployed in the Boeing 747. Even so, the Federal Aviation Administration required that *three* INS units be installed in each cockpit, so that the pilots could rely on two systems at all times, with the third standing by as a hot spare. (The 747 was also equipped with a sextant port, as was the Vickers VC10, a British long-range jetliner produced in the 1960s.)

Today's mobile phones can also perform inertial guidance, thanks to their MEMS (micro-electromechanical system) accelerometers and gyroscopes. Although the sensors are not accurate enough for extended dead reckoning, they provide sufficient accuracy that a person's precise location can be determined by fitting the patterns of acceleration, movement, and deceleration to a street map.[52] This demonstrates one of the ways that external information can be used to increase the effective sensitivity of measuring instruments.

Quantum sensors offer the promise of dramatically improved inertial navigation. Just as the lower friction of gas bearings made gyroscopes less subject to drift than mechanical bearings, gyroscopes based on ring lasers and eventually cold atoms promise even more improvements.

As the name implies, ring laser gyroscopes are gyroscopes based on the principle of sending laser light around a ring: the light remains on its current path even if the ring moves while the photons are in flight, allowing the movement of the ring to be precisely measured. In 2017 the ring laser market was estimated to be at $720 million, with a projected annual growth rate of 3.5 percent. "Ring laser gyroscopes are primarily implemented in defense applications owing to their excellent measurement accuracy and [the absence] of moving parts that are in mechanical gyroscopes."[53] Today a single ring laser inertial navigation system can be purchased from China for between one thousand and ten thousand dollars.

Dramatically more accurate gyrometers based on "cold atom" technology are now being developed. "For inertial navigation, atom interferometers are particularly important because they provide an absolute measurement of the physical quantity of interest, be it acceleration or rotation...In geophysics, a gyrometer can be used for local

---

[52] Jun Han et al., "ACComplice: Location Inference Using Accelerometers on Smartphones" (2012).

[53] Research, "Ring Laser Gyroscope Market – Snapshot" (2017).

monitoring of the variations in Earth's rotation rate due to seismic or tectonic-plate displacements."[54]

The researchers developing inertial guidance for mobile phones matched acceleration patterns against street maps of Pittsburgh and Mountain View to determine a mobile phone's location.[55] Similarly, an advanced INS could in principle match changes in the Earth's magnetic field against a map to dramatically improve its accuracy. Such systems would work equally well underground or underwater. Today's submarines navigate using a variety of strategies, including GPS antennae attached to a long tether that can be sent to the surface and then either reeled back down or cut as necessary.

One could imagine vessels of all types being equipped with GPS, inertial guidance, and quantum magnetometers. Properly equipped, comparisons between GPS and the quantum sensor should reveal when GPS is being jammed or degraded, and tell the operator where the vehicle is located with certainty. An unclassified summary of a 2015 Air Force quantum technologies study concluded that quantum navigation sensors would be ready for demonstration between 2020 and 2025.[56] Efforts are underway to miniaturize the devices so they can be used in all kinds of applications. For example, in 2019, MIT scientists created a microchip-sized nitrogen vacancy magnetometer using standard complementary meta-oxide-semiconductor (CMOS) technology, paving the way to small devices.[57]

Quantum Sonar

Because SQUID-based magnetometers and gravimeters can sense exceedingly minute changes in the Earth's magnetic and gravitational fields, these devices can be used to create three-dimensional models of the underground mineral deposits or man-made structures that are responsible for those changes. Such models are made by moving the sensor in three-dimensional space while precisely recording the location and orientation of the sensor, and then using a computer to fit a mathematical model of the presumed underground object

---

[54]Alzar, "Compact Chip-Scale Guided Cold Atom Gyrometers for Inertial Navigation: Enabling Technologies and Design Study" (2019).

[55]Jun Han et al., "ACComplice: Location Inference Using Accelerometers on Smartphones" (2012).

[56]US Air Force Scientific Advisory Board, *Utility of Quantum Systems for The Air Force Study Abstract* (2016).

[57]Kim et al., "A CMOS-Integrated Quantum Sensor Based on Nitrogen–vacancy Centres" (2019).

Sensing and The Fundamental Forces

There are four fundamental forces in nature: electromagnetism, gravity, and the strong and weak nuclear "forces." Of these, electromagnetism is the base of nearly all remote sensing today, as it is easy to generate, control, and measure. This ability to control also means that it is possible to shield from electromagnetic waves.

We currently lack the ability to generate, control, or shield from gravity waves. (Such an ability would presumably enable the anti-gravity and artificial gravity devices commonly seen in science fiction.) But we can detect gravity waves based on their interaction with other masses and, thus, with the fabric of space–time. Increasingly precise quantum sensors create the opportunity for high-resolution gravity sensing. Today such techniques are creating new possibilities for both astronomy and geology, although increased resolution might create possibilities for even more precise measurements of human artifacts in the future.

The two remaining forces are mostly confined to the nucleus of the atom. The strong force, more properly called the strong interaction, is responsible for holding the atomic nucleus together: without it, the protons would repulse and matter as we know it would not exist. The weak interaction, meanwhile, is responsible for radioactive decay.

Electromagnetism and gravity can be used for remote sensing because they follow the inverse square law, which is to say, because the force that they exhibit between two objects is proportional to $\frac{1}{r^2}$ where r is the distance between the two objects. The strong and weak forces do not follow the inverse square law. They are much stronger than the electromagnetism and gravity within and in the immediate vicinity of the atomic nucleus, but they appear to play no role at larger scales of measurement.

responsible for the disturbance in the magnetic force to the observations. In principle, such methods are no different from techniques that geologists have used for decades to explore for mineral wealth and oil. In practice, the exquisite sensitivity of SQUID-based sensors creates new opportunities for observing the hidden world.

It is likely that there will be many applications outside of the extractive industries for such underground sensing technology. For example, the Chinese military has reportedly developed next-generation, sonar-like systems that can detect submarines and other underground objects based on their mass and shape.[58] Other publications describe how Chinese scientists flew a SQUID-based magnetometer over a field to detect buried iron balls of various sizes based on how the balls changed the Earth's magnetic fields.[59,60] There are obvious applications for landmine detection.

The iron-ball-detecting device is fascinating. The scientists created a proof-of-concept by hand-carrying an array of six SQUIDs. Each SQUID in the device contained a 24-bit analog-to-digital converter (which means that its precision is 2^{-24} or one part in 16 million). Illustrating the importance of super-precise timing, the papers emphasized that the device could make 2000 measurements a second, with a time synchronization of 1 microsecond. The device included a sensitive inertial navigation system to know the precise location and orientation of the detector ensemble. By knowing the device's location and orientation, it is then possible to know the precise direction and magnitude of the Earth's magnetic field at a series of measurement locations – and remember, this device is making 2000 measurements each second.

The papers report that the scientists detected all of the iron balls. A follow-up experiment replicates the procedure, but the SQUID array is dangled from a helicopter with a towrope. Given the speed of measurement, the ability to know location, and orientation of magnetic fields, these devices should be able to detect the existence of underground tunnels or structures, and even the movement of military matériel or even drugs through such tunnels.

Many details of the helicopter experiment are vague. The experiment suggests that the helicopter approach worked, yet one paper says that the data were still being processed at time of publication. Other details, such as the altitude of the helicopter, whether it was modified to avoid interference with the device, the size of the balls, and so on, were either vague or omitted. Yet, a photograph in Qiu et al. gives some hints: it reports that the rope suspending the SQUID

---

[58]Hambling, "China's Quantum Submarine Detector Could Seal South China Sea" (2017).

[59]Wu et al., "The Study of Several Key Parameters in The Design of Airborne Superconducting Full Tensor Magnetic Gradient Measurement System" (2016).

[60]Qiu et al., "Development of a Squid-Based Airborne Full Tensor Gradiometer for Geophysical Exploration" (2016).

array is 35 meters long. Based on the length of the towrope, people's presence in the photograph, and its angle, the helicopter appears to be flying rather low, with the sensor perhaps 5–10 meters off the ground. The iron balls appear to vary in size from a golf-ball-sized object to one the size of a melon. But without a scale, it is hard to be more precise.

Such approaches for quantum sensing have obvious implications in counter-terrorism and counter-smuggling operations. But quantum sensing extended to the ocean might have implications for submarine tracking and anti-submarine warfare, which would have significant geopolitical repercussions for nuclear deterrence (see Figure 2.7). Since the 1960s, the US nuclear strategy has been based on the so-called *nuclear triad* and the serviceability of some US nuclear forces in the event of a massive first strike by another power. The most survivable of the nuclear forces are those in submarines, since the subs can remain underwater for months at a time and deliver a massive retaliatory second strike, with the intent of utterly destroying an attacking nation (and also ending all remaining human life on the planet in the process). Without survivable forces, game theory says that there is an incentive for a nuclear nation to strike first and wipe out its adversary's forces before they can get off the ground.

Quantum sensors that would allow an adversary to accurately pinpoint and track the location of an adversary's nuclear forces would appear to impact the survivalability of those weapons. If all such weapons can be tracked, it might be able to destroy them all in a single surprise attack. Thus, high-precision quantum sensing that could recognize mass distributions such as those that appear in Figure 2.8 might be destabilizing.

Aside from nuclear attack and counter-terrorism, there are certain hot zones where even a limited-range quantum sonar might change how countries posture. SQUIDs might improve the effectiveness or decrease the cost of minesweeping, for example, perhaps to the point of allowing for low-cost detection systems that could detect mines and even pirate vessels (significant research has already been devoted to landmine detection[61]). In areas such as the South China Sea, a magnetometer-based surveillance system might tip off a nation to the presence of another nation's underwater vessels and lead to their exclusion.

---

[61]Garroway et al., "Remote Sensing by Nuclear Quadrupole Resonance" (2001).

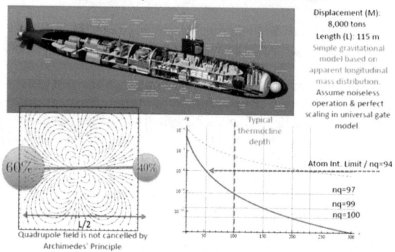

Detecting the Gravitational Quadrupole of a Simple Submarine Model

Displacement (M):
8,000 tons
Length (L): 115 m
Simple gravitational model based on apparent longitudinal mass distribution. Assume noiseless operation & perfect scaling in universal gate model

Typical thermocline depth

Atom Int. Limit / nq=94

nq=97
nq=99
nq=100

60%   40%

L/2

Quadrupole field is not cancelled by Archimedes' Principle

Figure 2.7. In a 2018 address to the National Academies, Dr. Marco Lanzagorta, explained how quantum gravimeters might detect a submarine. Image courtesy US Naval Research Laboratory.

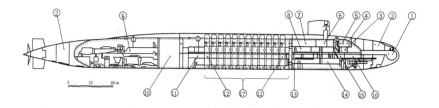

Figure 2.8. The engine room (9), reactor compartment (10), torpedo room (14), and missile compartment (17) in this Ohio class nuclear submarine provide several highly dense areas – arranged in a distinctive pattern unlikely to occur naturally in the ocean – that quantum sensing devices could detect. Image CC-By Wikimedia Commons user Voytek S.

Magnetic Field Sensing Futures

Quantum sonar is an application of magnetic field sensing. As explained in Chapter 9, today it is militaries that are the principal patrons of quantum technology research, and perhaps that funding shapes our imagination for applications of quantum sensing. As quantum sensing matures, more entrepreneurs will understand the potential for the technology, and be able to actually deploy it. Sensing magnetic fields has enormous application that we sketch below. A wealth of applications could flow from the SQUID magnetometer approach, especially as the technology is miniaturized and becomes usable higher in the air or even in outer space.

- Extractive industries, ranging from oil and gas to technology companies desperate for the rare-earth minerals necessary for mobile phones, will benefit from sensing below-ground magnetic fields. A subsidiary owned by De Beers, in an effort similar to the Chinese iron-ball detecting experiment, suspended an array of six low-temperature SQUIDs 20 meters below a helicopter, and used the array to map the magnetic gradients over a 7 km square area in South Africa.[62] The implications for extraction industries are clear. By mapping magnetic and gravimetric waves, these industries should be able to locate valuable minerals, allowing for exploration and mining that is dramatically more effective than before – and thus lowering the net price of extracted materials. The implications for the environment are less clear: will such technologies allow for highly targeted extraction, or more extraction overall?

- Scientists interested in brain–machine interfaces could use quantum sensing to detect subtle electrical signaling in the brain. Signaling in the brain is fainter than the electrical field created by the heart, requiring more sensitive instrumentation and advances in locating and isolating brain signals. It might also allow for improved lie detectors and even *brain wiretapping*.[63]

- Medical treatment and research centers, as evidenced by broad adoption of MRI and PET, were quantum sensing early adopters.

---

[62]Chwala et al., "Full Tensor SQUID Gradiometer for Airborne Exploration" (2012).

[63]Garfinkel, *Database Nation: The Death of Privacy in The 21st Century* (2000).

As quantum sensors improve in resolution, there will be corresponding benefits to diagnosis and treatment.

- Today conventional magnetometers are widely used to detect weapons. In the future, machine learning might be used to allow the identification of popular rifles and handguns, in all their possible orientations. With such a model, a magnetometer might be able to detect all guns within a specific range while ignoring other items. Using an airborne magnetometer, firearms might be easily detected in a crowd, or even in homes and vehicles because of the lack of shielding. This might be used to find weapons that are unregistered or that are possessed by people ineligible to own firearms; courts would need to decide if flying a magnetometer over a neighborhood or crowd constituted a search that required a warrant.

- Quantum sensing might allow for dramatically smaller antennae in consumer and professional electronics (see Figure 2.4). For example, a paper from the Delft University of Technology in Delft, Netherlands, reports that a supercooled sensor was able to detect single quanta of radio waves.[64] A 2020 paper by the Army Research Laboratory demonstrated that Rydberg atoms (atoms with excited valence electrons) were sensitive to the entire radio band spectrum.[65] The future of radio communications may not rely on bigger antennae or stronger transmission but rather on more sensitive sensing and narrower allocations of frequency spectrum. This is in part because the lasers controlling the sensors can tune focus to specific frequencies without reliance on multiple, different antennae. More sensitive radio would augur more efficient use of communications spectra. Intelligence agencies too might be interested in the interception capabilities of single-quanta devices, as multiple such devices working together should be readily able to determine the source of signals. One can also imagine the possibility of stealthy communication capabilities.

---

[64]Gely et al., "Observation and Stabilization of Photonic Fock States in a Hot Radio-Frequency Resonator" (2019).

[65]Meyer et al., "Assessment of Rydberg Atoms for Wideband Electric Field Sensing" (2020); Cox et al., "Quantum-Limited Atomic Receiver in The Electrically Small Regime" (2018).

- Manufacturers in many fields will benefit from advances in materials science from quantum technologies. While quantum computing will help manufacturers design new materials, quantum sensing will allow inspection and characterization of them. Someday, super-precise fabrication may be possible where objects are crafted at the atomic level, making them perfectly matched in size and composition.

2.3.3 Sensing Gravitational Fields

Albert Einstein predicted the existence of gravitational waves, but thought them too weak to be sensed. They were, for a time.

Gravitational waves are caused by the acceleration or change of mass through space. This means that you create gravity waves every time your heart beats (or your quartz crystal vibrates). This is similar to the way that a moving charge creates electromagnetic waves, which is the basis of how a radio transmitter works. Gravity is much weaker than electromagnetism, however, and so gravity waves are correspondingly much smaller.

To date, the only gravity waves that we have managed to detect are the waves created as a result of cosmic events that released tremendous amounts of energy – such as the collision of two black holes. Once formed, gravity waves travel at the light speed and pass through our planet (along with eventually everything else in the universe) without much interaction. Almost a hundred years after Einstein's prediction, researchers at the Laser Interferometer Gravitational-Wave Observatory (LIGO), an enormous, ambitious scientific project, made the first direct observation of gravitational waves and of two black holes merging to form a single black hole.[66] The detection of gravitational waves was accomplished with an interferometer that was able to detect the ever-so-slight compression of space–time in one direction compared with another as a result of the passing gravity wave (see Appendix B for more information).

An interferometer uses a source of light and a beam splitter to send the light in different directions. Mirrors at the end of the arms of the interferometer reflect the light back, where it is superimposed on a sensor. Turn it on and the two beams of light form an interference pattern. If the interference pattern changes, then either the distance between one of the mirrors and the beam splitter must have changed,

---

[66] Abbott et al., "Observation of Gravitational Waves From a Binary Black Hole Merger" (2016).

or else something between the mirrors and the beam splitters must have changed the phase of one of the light beams. One application of interferometers is thus making precise measurements of distance and making sure that physical systems stay in calibration.

In the case of the LIGO, the system is designed so that *nothing* should be able to change the distance between the beam splitter and the mirrors. For starters, all of the optical components are in a vacuum chamber. The devices are built in a region that is not seismically active, and far away from equipment that might cause the ground to vibrate. The idea of the system is that a passing gravity wave literally changes the distance between the splitter and the mirrors. Because gravity waves are directionally aligned, the distance for each mirror changes by a different amount, and the diffraction pattern changes. The longer the arms of the interferometer, the more sensitive the device will be to distortions in the fabric of space–time.

The LIGO interferometer has arms 4 km long. The beam of light in LIGO is prepared so that by default, if the distance between mirrors does not change, the photodetector senses no light. That is, LIGO harnesses destructive interference as a tool to detect waves. If a gravitational wave is sensed, an arm expands or contracts, thus eliminating the destructive inference and revealing a pattern on the photodetector.

Of course, the interferometer is not perfectly isolated from the ground on which it is built. Trucks drive around, planes pass overhead, and there is always a risk that some stray vibration will also move one of the mirrors. So the LIGO consists of *two* 4 km interferometers separated by a great distance. If one senses a change in distance and the other doesn't, that vibration was no gravity wave. But if both sense the same change at the same time, a gravity wave has been detected.

The curvature of the Earth, and other challenges such as the need for seismic stability and a vacuum, limit the size of a terrestrial optical interferometer and thus its sensitivity. A collaboration between the European Space Agency and NASA seeks to build the Laser Interferometer Space Antenna (LISA), which will be formed by three space vehicles separated by 2.5 million km. LISA is expected to be completed in the 2030s. The Chinese Academy of Sciences has a similar project, the TianQin observatory, on a similar timeline, but with a target of 3 million km-distant interferometer arms.[67] Separately, some are proposing the construction of "space-borne gravitational wave detectors based on atom interferometry" that would detect gravity waves acting on collections of perhaps a hundred million atoms falling in the vibration-free environment of space.[68]

LIGO, LISA, and TianQin are all focused on gravitational sensing of the cosmos. What if similar highly accurate sensing technologies based on interferometry were focused on Earth? In the 1990s, the first experiments were conducted that used interferometric synthetic aperture radar (InSAR), an approach that enhances the sensitivity of downward-pointing space-based radar systems by comparing carefully timed radar imagery (interferograms) of the Earth.

In 2002 the GRACE (Gravity Recovery and Climate Experiment) employed twin satellites that orbited roughly 220 km apart. The GRACE satellites used a microwave ranging system to detect minute changes in the distance between them that are the result of variation in the Earth's gravity field.[69] The GRACE system is designed to be especially sensitive to changes that result from the collection of water (liquid or solid) on the Earth's surface. As the satellites approach stronger gravity fields, signaling greater concentrations of water, gravity pulls the lead vehicle a little faster and thus increases the distance to the trailing satellite, which itself speeds up a short time later. As the water recedes into the distance, the lead satellite slows down a bit, followed by the second. The GRACE mission produced a monthly, whole planet survey of water, tracking millimeter-level changes in density. Originally planned for a 5-year mission, the GRACE mission was decommissioned in 2017.[70] The Gravity Recovery and Climate Experiment Follow-On (GRACE-FO) mission, launched in 2018, uses laser interferometry for increased precision. Also focused on water movement, GRACE-FO will help forecast rising seas and the development of drought.

Optical interferometry has occupied a central place in this discussion, but scientists have also developed cold-atom interferometers to

---

[67]H.-T. Wang et al., "Science with The TianQin Observatory: Preliminary Results on Massive Black Hole Binaries" (2019).

[68]Loriani et al., "Atomic Source Selection in Space-Borne Gravitational Wave Detection" (2019).

[69]Tapley et al., "The Gravity Recovery and Climate Experiment: Mission Overview and Early Results" (2004).

[70]The European Space Agency operated a similar mission called the Gravity Field and Steady-State Ocean Circulation Explorer (GOCE) from 2009 to 2013.

sense gravity.[71] In this approach, two different collections of atoms are initialized in a superposition using lasers; the atoms then interact with gravitational signals, the characteristics of which are revealed in the differences between the two ensembles of atoms. Because the ensembles of atoms themselves are being compared, the ensembles need not be separated by great distances. Indeed, atom interferometers are now miniaturized, with devices resilient enough to be mounted on aircraft and even on small unmanned aerial vehicles (UAVs, also known as drones).[72] These devices are not sensitive enough for detecting gravity waves, but they are just fine for measuring the Earth's gravity. Miniaturized gravimeters were still research curiosities just a decade ago; today they are available from companies such as AO Sense, Inc.

As gravimetric detection improves in resolution, one might imagine strategic uses of the data collected. With GRACE-FO, the US is sharing with the world information about water and drought. Such predictions inherently have strategic implications, given the likelihood of conflicts resulting from climate change. (For example, the so-called Arab Spring of 2010 and 2011 was driven in part by high food prices attributed to that year's poor crop yields, a likely result of climate change.)

2.3.4 *Quantum Illumination*

In quantum illumination,[73] entanglement can discern between reflected light and noise, or be used as a kind of object detector. In experimental systems, entangled photons are generated. One of the pair is sent out to the environment while the other is measured. As photons are received in a detector, the measured, entangled photon is compared to received ones to see if it is thermal noise or a reflected photon.

Ghost Imaging

In ghost imaging, which has both classical and quantum methods, entangled photons are used to sense objects that are not "in view" of a camera.

---

[71]Bongs et al., "Taking Atom Interferometric Quantum Sensors From The Laboratory to Real-World Applications" (2019).

[72]Weiner et al., "A Flight Capable Atomic Gravity Gradiometer With a Single Laser" (2020).

[73]Quantum illumination, as defined here, goes by several names, including correlated-photon imaging and two-photon approaches.

Figure 2.9. Carefully counting scattered and reflected photons from this toy soldier created a "ghost image" of it – an image constructed of an object that was outside the view of a camera. Courtesy of Office of the Secretary of Defense Public Affairs.

In an exciting demonstration of this approach, researchers at the Army Research Laboratory (ARL) published a paper that indirectly imaged a toy soldier (see Figure 2.9). The image was generated using a split beam of light. One beam was directed to the toy soldier, illuminating it. The soldier reflected and scattered photons from the first beam, some of which were collected by a nearby "bucket" detector, a special type of single-photon sensor that, like a water pail, collects photons without mapping out their specific location. The second beam was directed into a CCD camera. A separate system correlated the photons between the bucket detector and CCD to reveal which light was reflected and which was scattered. The resulting image is clearly of a toy soldier. The approach works on all wavelengths, meaning that shining different frequency light could reveal chemical composition of an object (perhaps revealing it to be real or a decoy).[74]

The ARL scientists built on this achievement with a demonstration of how ghost imaging could be applied to challenges in satellite-based sensing, and sensing in other difficult conditions, including underwater. In a follow-up study, the ARL team introduced "turbulence" to the setup by adding a 550 °C heater. Despite the turbu-

---

[74]Meyers, Deacon, and Shih, "Ghost-Imaging Experiment by Measuring Reflected Photons" (2008).

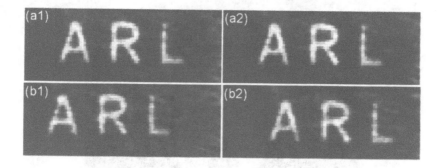

Figure 2.10. The ghost images in panels a1 and a2 were captured under 550 °C turbulence. In frames b1 and b2, the experimenters correlated photons captured at different times, also under turbulence. In b1, the image is based on photons five frames before the other detector; in b2, five frames after. Figure CC-BY Meyers and Deacon (2015).

lence, one can make out the letters A R L in their demonstration. The ARL's advances in ghost imaging could make it possible to see clearly on chaotic, turbulent, hot, and smoky battlefields. Another demonstration used cloudy water as the "turbulence." Nevertheless, the ARL's "A" is relatively readable, elucidating implications for underwater detection (light is absorbed by water, thus limiting sensing distance). Subsequently, ARL showed that it could image objects with photons measured at different times (see Figure 2.10). This demonstration is important, because it signals the potential to use ghost imaging for moving objects.[75]

The ARL techniques would be useful for many civilian contexts. Vehicle safety systems might use indirect evidence from "unseen" vehicles around corners or difficult-to-see pedestrians based upon how light reflects and scatters around them. Scientists are also excited about ghost imaging's potential to contribute to image compression and to multi-spectral analysis. Scientists at the Brookhaven National Laboratory announced in 2020 that they intended to use ghost imaging in an X-ray microscope to take advantage of radiation avoidance from a split beam of light. That is, X-rays damage many study samples, thus the scientists will try to use ghost imaging to reduce the amount of X-ray exposure to the sample while using correlated photons to maintain high resolution (see Figure 2.11).

---

[75]Meyers and Deacon, "Space-Time Quantum Imaging" (2015).

Figure 2.11. Scientists at Brookhaven National Laboratory are constructing a micro-scope that will use ghost imaging to minimize the harm to samples under examination. The idea is to split and entangle the X-ray photons, and send just a portion of the harmful X-rays to the sample. Because the photon beams are correlated, it should be possible to infer data from the photons that never interact with the sample. The goal is to achieve high resolution without exposing the sample to the full, damaging effects of X-rays. Courtesy of Brookhaven National Laboratory.

2.3.5 Quantum Radar

Quantum illumination is a candidate sensing approach for expanding the sensitivity of military radar systems[76] and to make radar systems themselves more difficult to detect.[77]

Quantum radar involves generating billions of entangled photon pairs to illuminate targets. One photon from the pair, the *signal* photon, is sent to the environment in hopes it will hit a target and be reflected back to the radar array. The other photon, the *idler* or *ancilla* photon, is retained in memory (see Figure 2.13). Photons received by the array are then compared to the retained idler photons, where the operator can determine whether those received photons are correlated or not. Non-correlated photons are noise from the atmosphere, but correlated ones reveal information about the reflective object.

---

[76]Barzanjeh et al., "Microwave Quantum Illumination" (2015).
[77]Marco Lanzagorta, *Quantum Radar* (2011).

Figure 2.12. Lockheed Skunk Works' HAVE BLUE brought about a revolution in low-observable "stealth" aircraft, and minted billions for Lockheed, Silicon Valley's largest employer until the internet revolution in the 1990s. Quantum sensing imperils stealth technologies. Photo public domain DARPA.

To an adversary, those billions of photons are simply atmospheric noise. Thus quantum sensing is "stealthy"; the idler photon allows the operator to distinguish between background noise while focusing attention on the quantum radar signal, those reflected signal photons that correlate with the idler photons.

The military applications of such quantum illumination for radar are many. A photonic approach should detect low-observable objects, such as vehicles that use "stealth" technology (see Figure 2.12) or even forms of electromagnetic jamming. For instance, one use foreseen by the Air Force is to use quantum technology to counter "digital radio frequency memory jamming," a technique where an enemy fighter captures emitted radar pulses and replays them at a different speed in order to confuse air defense systems. Adversaries might try to jam quantum radar by sending billions of noise photons into the array. However, if quantum radar works properly, the operator can simply filter out those noise photons based on correlations between the desired signal and idler photons. Militaries might also use these techniques for navigation. A submarine, for instance, could use reflected photons to sense undersea dangers, such as mines.[78]

---

[78]Marco Lanzagorta, Jeffrey Uhlmann, and Salvador E. Venegas-Andraca, "Quantum Sensing in The Maritime Environment" (2015).

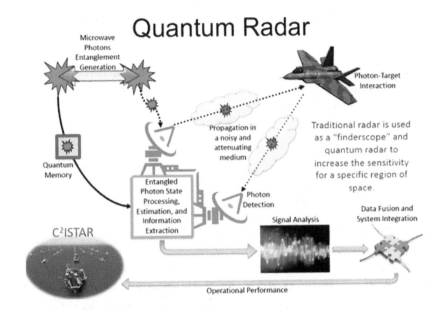

Figure 2.13. In a 2018 address to the National Academies, Dr. Marco Lanzagorta, explained how quantum radar might detect low-observable aircraft. Image courtesy US Naval Research Laboratory.

There are significant engineering challenges to quantum radar;[79] however, theoreticians believe these are surmountable, and recent developments suggest alternative approaches that could produce working quantum radar. The main challenges surround generation and entanglement of billions of photons and the need to have some form of quantum memory to retain idler photons for comparison to signal photons. But in both challenges, there are reasons to believe that innovations or different approaches could make quantum radar possible.[80] With respect to entangled photons, great strides have been made in recent decades in lasers, a key complementary technology for quantum innovation. Investment in laser technology is bound to continue and expand, because of lasers' importance to quantum sensing, computing, and communications (see Figure 2.1 and Figure 2.3). Because radar systems typically operate in the microwave band, research funding for photonic generation at microwave frequencies may

---

[79]Cho, "The Short, Strange Life of Quantum Radar" (2020).

[80]Marco Lanzagorta and Jeffrey Uhlmann, "Opportunities and Challenges of Quantum Radar" (2020).

be a signpost of quantum radar programs. Separately, there have also been advances in quantum memory; however, an alternative approach to quantum radar may eliminate the memory requirement entirely. Researchers at University of Waterloo proposed a protocol that measures the idler photon immediately, thus allowing the signal photon to be compared at some later time using classical memory.[81]

Quantum radar has applications in outer space, which makes sense because there is less photonic attenuation in space than in a planet's atmosphere. A satellite equipped with a quantum radar system might be used for a range of applications: for detecting ballistic missiles, discovering adversaries' secret satellites, and even finding dangerous space junk.[82]

2.4   From SIGINT to MASINT

Some quantum technologies discussed in this chapter raise few unmanageable policy issues, in part because with some of these technologies, the subject would know they were being measured. The individual would have to be in a Faraday-caged room so their body could be isolated from sensor-befuddling power lines, radio waves and the like. The individual would also have to remain extremely still until technologies catch up to track moving objects. However, other quantum metrology and sensing approaches have characteristics of remote sensing. That is, like many surveillance technologies, they can be used against unwilling or unknowing subjects, raising policy issues ranging from individual privacy to national security concerns. The primary dividing line is between magnetic field sensing and gravitational sensing. Gravitational sensing can be made extremely sensitive, and because gravity cannot be shielded, countermeasures are limited.

The emergence of deployable and highly precise gravitational sensors could cause a shift in intelligence gathering. In recent decades, the power of signals intelligence (SIGINT) (Section 7.2 (p. 264)) has astonished many. Signals intelligence focuses on communications and radar systems and is the primary responsibility of the National Security Agency (NSA) in the United States. By monitoring fiber-optic and other forms of communications, many nations have surprising powers to track people, identify them, and to listen in to their con-

---

[81]C. W. Chang et al., "Quantum-Enhanced Noise Radar" (2019).
[82]Marco Lanzagorta and Jeffrey Uhlmann, "Space-Based Quantum Sensing for Low-Power Detection of Small Targets" (2015).

versations. In recent decades, the NSA has attracted respect and resentment as its surprising and strong capabilities have been brought to bear in conflicts. The NSA is thought to be the largest employer of mathematicians in the world;[83] it has emerged as a central asset with the rise of computing and the need to both secure computers and to attack them in surprising ways.

Quantum technologies, for reasons explained in Chapter 7, may secure more communications and make metadata surveillance impossible, thus frustrating SIGINT efforts. But at the same time, quantum sensing technologies will give governments more power to engage in *MASINT*, measurement and signature intelligence. MASINT approaches focus on the measurement of objects and their "signatures." This includes what objects are, whether they are moving, and whether they have been used recently. For instance, by sensing attributes of an armored fighting vehicle, one might identify it, understand whether it is vibrating or moving, and by studying heat dissipation, whether and how recently the tank gun has been fired. As quantum sensing comes into use, a trio of different agencies will become more important: the Defense Intelligence Agency (DIA), the National Reconnaissance Office (NRO), and the National Geospatial-Intelligence Agency (NGA). Companies that foresee the shift to MASINT, such as ColdQuanta, are already collecting strategic board members with experience at these agencies.

2.5 Quantum Sensing: Conclusion

This long chapter foreshadows a key finding of this book: that quantum sensing is likely to be the most consequential of quantum technologies. As the technologies discussed in this chapter transition from the laboratory to the marketplace, quantum metrology and sensing have the potential to alter how nations monitor and engage in conflict. Since the deployment of GPS in the 1990s, quantum measuring technologies have provided the US military with incremental advantage in conflict. Such capabilities are both increasing and becoming more widely available. Some quantum sensors and components are commercially available.

As impressive as the military and intelligence applications are, quantum sensing could also contribute to drug development, medical diagnostics, medical devices including prosthetics linked to the brain,

---

[83]M. Wagner, "The Inside Scoop on Mathematics at The NSA" (2006).

and more efficient and targeted mineral extraction. Quantum sensors are also a precursor for quantum computers and communications.

The next three chapters build on quantum sensing by explaining the history of quantum computing, its likely uses, and the current landscape of the field.

Understanding Computation

W HAT is computation? How is computation different from calculation? What kinds of tasks can computers perform? To provide
a foundation for what makes quantum computing special and to
ground our policy analysis, this chapter visits the history of computing, starting with the ancients and their concepts of mathematical
concepts, and proceeds to discuss modern classical computing.

Humans have been using numbers since at least ancient Sumer
and Babylonia, five thousand years ago. The Babylonians were fas-
cinated with the number 60; they thought that the number 60 was
mystical, since it could be divided into two, three four, five, six, 10, 15,
20 or 30 pieces. But for the majority of human history, manipulating
numbers was something done by people, not machines, sometimes
with tools such as the abacus, but more capacity was needed.

When machines took over the task of manipulating numbers, it
was often because of war or military efforts. Designing and building these machines took government funding, often supplemented
with support from private companies and brainpower from academia.
These facts are stressed here because just as early analog computers were electromechanical engineering marvels, building quantum
computers requires state-of-the-science engineering at particle-level
scales, with experts from several disciplines, and the funding to
match. Also emphasized is how computers can be miniaturized, be
reproduced for a fraction of their initial costs, and find their way into
everything, including even doorknobs. Computing can enjoy a virtu-

ous cycle where simple devices can reveal efficient design for even larger, faster computers. This insight will be key for the trajectory of quantum computers.

This chapter also introduces complexity theory to explain the kinds of problems that are hard for computers to solve. This lays the groundwork for understanding the different capabilities and potential advantages of quantum computers. This background is crucial to understanding quantum computers for two reasons. First, it dispels the common notion that quantum computers would be a kind of magical device that can ponder all possible solutions to a problem. Instead, quantum computers, like any other kind of tool, are good for some tasks but no better than ordinary computers for others. Second, complexity theory helps illuminate what is truly exciting about quantum computers (hint: it is not whether encryption can be cracked). Instead, if quantum computers can solve problems out of reach for classical ones, quantum computers will help solve some of the difficult, costly challenges in life. Complexity theory helps elucidate the kinds of efficiencies that could come about, from finding ways to optimize energy-intensive processes to finding valuable information in enormous datasets.

This chapter should be read by those who need to make investment decisions or otherwise understand the underlying technology and assumptions. This chapter lays the groundwork for understanding what quantum technologies are likely to do and, conversely, helps identify the specious claims so often made about the capabilities of quantum computers.

3.1 Mechanical Calculation

Machines are systems that use multiple parts and some kind of power for performing some kind of task. "Shovels are tools; bulldozers are machines," we are informed by Merriam-Webster.[1] Machines are different from tools in their complexity and their power. The earliest known calculating machine is the Antikythera Mechanism, a device with more than 30 interlocking bronze gears that was found in a shipwreck off the small island of Antikythera, Greece. Although the user's manual for the mechanism did not survive, this 2000-year-old mechanism has now been thoroughly reverse-engineered and is be-

---

[1] Merriam-Webster Incorporated, ""Machine."" (2020).

lieved to be a means to predict the movements of the planets and the occurrences of eclipses.[2] You can even download a simulator.[3]

The Antikythera Mechanism used differently sized wheels with teeth to account for the differing speeds of the planets; a peg that cycles back and forth in a slot accurately represents elliptical motion of the Moon, which is attributed to the Greek astronomer Hipparchus of Nicaea (*c.* 190–*c.* 120 BCE). The mechanism thus implements a kind of multiplication, but the ratios were set and unchangeable, like the motions of the planets themselves.

It took another 1700 years before the basic building blocks of flexible mechanical calculation were put into place. In the early 1600s, the Scottish mathematician John Napier invented two approaches for multiplying and dividing numbers using addition and subtraction. The first, called "Napier's bones," embedded numeric tables on wooden rods. The second and more powerful approach used logarithms, which Napier also invented. Napier published the first book of logarithms in 1614. Sixty years later, the German mathematician Gottfried Wilhelm Leibniz (1646–1716) started working on a mechanical calculator that could add, subtract, multiply and divide when the user set dials to various positions and turned a crank. Critical to this invention was what is now called the Leibniz wheel, which causes the dial that shows the tens' place to advance from "0" to "1" when the dial showing ones advances from "9" to "0." In 1820, the French inventor Charles Xavier Thomas de Colmar (1785–1870) introduced the Arithmometer, the first commercially produced mechanical calculator: his factory built a thousand of them before his death in 1870. Meanwhile in England, Charles Babbage (1791–1871) designed the world's first automatic calculator in 1822 for the purpose of calculating and printing tables of logarithms, trigonometric functions, and artillery tables. Babbage called his invention the "difference engine," and obtained funding from the British government to build it in 1832.

Although all of these devices proved to be helpful aids to humans performing tasks involving numbers, none of them could *compute* in the modern sense. That's because they all lacked the ability to alter their computations based on the results of a specific calculation. This is what distinguishes a machine that *calculates* from one that *com-*

---

[2]Spinellis, "The Antikythera Mechanism: A Computer Science Perspective" (2008).

[3]Goucher, "Antikythera Mechanism" (2012).

putes. Babbage realized that the difference engine was limited, and designed an improved system he called the analytical engine. Alas, Babbage never built his invention, although a group of enthusiasts in England called Plan28 are now working to do so. You can follow their efforts at plan28.org.

3.2 The Birth of Machine Computation

Babbage may have seen the future, but there is no clear evolutionary descent from his machines to the computers of today. Instead, the first computers of the 1940s descended from the invention of punch cards and card-sorting machines that were developed for the 1890 US Census. The invention of teleprinters and punched paper tape was a way of making more efficient use of telegraph lines, and to manage the growing demands of science, engineering, and various militaries to perform increasingly complex numerical calculations.[4]

World War II saw two significant efforts aimed at using automated calculation for the war effort. There were two radically different applications for automated calculators, with the United Kingdom leading the development of machines to solve combinatorial problems, and the Americans largely developing machines to solve numerical ones.

3.2.1 *Combinatorial Problems*

In the United Kingdom, a project headquartered at Bletchley Park developed a series of hard-wired special purpose devices for cracking the German military codes. Cracking those codes is a "combinatorial" problem because the encrypted text was created with a "key" represented by the complex (for its time) initial settings of German encryption devices. The goal of the project was to determine which combination of those settings produced the encrypted text sent by the Germans. This is the project on which Alan Turing worked, and which is featured in the somewhat factual Hollywood film *The Imitation Game.* Initially this project used electromechanical devices called "The Bombe" to search the possible settings for the Germans' Enigma encryption device. In the movie there is a single Bombe, but in reality there were hundreds of them, each one working on a different part of the problem, or a different encrypted message.

---

[4]While there are many histories of computing, we recommend the eminently entertaining coffee table book by Garfinkel and Grunspan, *The Computer Book* (2018), as well as the more scholarly book by Dasgupta, *It Began with Babbage: The Genesis of Computer Science* (2014).

As Copeland explained it:

> "The Bombe was a 'computing machine' – a term for any machine able to do work that could be done by a human computer – but one with a very narrow and specialized purpose, namely searching through the wheel-positions of the Enigma machine, at super-human speed, in order to find the positions at which a German message had been encrypted. The Bombe produced likely candidates, which were tested by hand on an Enigma machine (or a replica of one) – if German emerged (even a few words followed by nonsense), the candidate settings were the right ones."[5]

The second code-breaking project at Bletchley Park – one that was shrouded in considerably more security – used vacuum tubes to crack the military codes used by the German High Command. Tubes can switch electrical circuits 500 times faster than relays. This complexity was essential, as the encryption machine developed by C. Lorenz AG had 12 encryption wheels, compared with the three or four used by the Enigma. The system was called Colossus, and the UK only built ten of them. The engineering on these systems was fantastic. For example, input data was on punched paper tape, and the computers were so fast that the paper tape had to move at 35 miles per hour. The Colossus computers did their job so effectively that all were destroyed or dismantled at the end of the war in order to protect the secret of the UK's code-breaking capabilities – a secret that it kept until 1974, when F. W. Winterbotham published his book *The Ultra Secret*.[6] A similar code-breaking effort in the US called Magic was under the direction of William F. Friedman, at the US Army's Signal Intelligence Service, the precursor to the US National Security Agency. The US story of how early punch card tabulators from International Business Machines were modified to perform cryptanalysis has also been told,[7] but it is not as well known as the story of Bletchley Park.

---

[5] B. J. Copeland, *Alan Turing's Automatic Computing Engine* (2005).
[6] Winterbotham, *The Ultra Secret* (1974).
[7] Rowlett, *The Story of Magic: Memoirs of an American Cryptologic Pioneer* (1999).

3.2.2 Numerical Analysis

Digital computers were also under development by the US military, but on the western side of the Atlantic the generals wanted to solve numerical problems, rather than combinatorial ones. Specifically, the military was seeking solutions to differential equations.

The military's interest in calculus was a direct result of improvements in firepower.[8] In 1800 the range of a big gun on a naval vessel was only 20 to 50 yards, making artillery pretty much a load, point and shoot affair. By 1900 naval guns could reach 10 000 yards: scoring a hit on an enemy ship, or a target on land, required accounting for the speed of the firing platform; the speed, direction, *and temperature* of the wind; the weight of the shot and the amount of propellant; and even the rifling of the gun's barrel. Spotters looked for splashes with precision optics, measuring (to the best of their ability) the distance and direction of the misses. All of these factors were used to calculate the azimuth, elevation, and amount of propellant used in the next shot. Artillery had become highly mathematical.

In 1927, an MIT professor named Vannevar Bush began work on a mechanical device that could evaluate calculus integrals and other kinds of mathematical function using a combination of spinning rods, gears, wheels, and several metal spheres. Bush, who became MIT's Vice President and Dean of the School of Engineering in 1932,[9] knew that the machine had both scientific and military applications. Specifically, the machine could be used to simulate many slight variations of the trajectory of an artillery shell, making it possible to produce numeric tables that could be used at sea (or in the field) by gunners to target their artillery faster and with more deadly precision. Bush originally called the machine *a contin-*

---

[8] Clymer, "The Mechanical Analog Computers of Hannibal Ford and William Newell" (1993).

[9] Vannevar Bush went on to become president of the Carnegie Institution of Washington, a philanthropic research funding organization in 1938. He soon became chairman of two US government agencies: the National Advisory Committee for Aeronautics and the National Defense Research Committee, effectively making him the US government's chief scientist. Bush initiated the Manhattan Project and convinced President Harry S. Truman to create the National Science Foundation (NSF), which was signed into law in 1950. Today he is frequently celebrated for his 1945 essay in *The Atlantic*, "As We May Think," which forecast the development of machines that could help people access vast amounts of information, and his July 1945 report "Science The Endless Frontier," which provided the intellectual justification for creating the NSF.

uous intergraph,[10] renaming it the *differential analyzer* later that year.[11] Within a few years versions of the machine had been built and impressed into service in both the US and England. For example, differential analyzers were constructed at the Ballistic Research Laboratory in Maryland and in the basement of the Moore School of Electrical Engineering at the University of Pennsylvania, where the machines were used to compute artillery tables.[12]

3.3 Numeric Coding

Analog mechanical calculating devices like the differential analyzer (and like slide rules) take a fundamentally different approach to solving numeric equations than the digital calculators, desktop computers, laptops and cell phones with which readers of this book probably grew up. Analog machines use physical quantities like distance, speed, and the accumulation of electronic charge to directly represent numeric quantities. This approach is simplistic and straightforward, but it has many disadvantages.

For example, you can use a ruler, a pencil, and a piece of paper to add together the numbers 2 and 3: just draw a line on the paper that is 2 cm long, draw a second, connecting line that is 3 cm long, and measure the length of the resulting line:

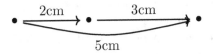

This is the basic principle behind the slide rule, except the rules on a slide rule are drawn using a logarithmic scale, so that adding the distances results in multiplication and subtracting them results in division (Figure 3.1).

The fundamental problem with analog mechanical calculating devices is that they are limited in *precision*, the ability to distinguish two numbers; *accuracy*, the difference between the true number and the one obtained by the calculation; and *repeatability*, whether the

---

[10]Bush, Gage, and Stewart, "A Continuous Integraph" (1927).

[11]Bush, "The Differential Analyzer. A New Machine for Solving Differential Equations" (1931).

[12]Bunch, *The History of Science and Technology: A Browser's Guide to The Great Discoveries, Inventions, and The People Who Made Them, From The Dawn of Time to Today* (2004), p. 535.

Figure 3.1. Using a slide rule to compute 2×3 = 6. The value of 2 is specified because the 1 on the C scale lines up with the 2 on the D scale. The cursor is then moved so that its center hairline is aligned with the 3 on the C scale, and the value of 6 on the D scale is the product of 2 and 3. Notice that the slide rule is simultaneously displaying that 2 × 4 = 8, 2 × 5 = 10, and many other values. it is you, the observer, who is actually doing the computation. (Slide rule simulation from www.sliderules.org/.)

same answer is obtained when following the same sequence of operations. These concepts are described in the sidebar "Precision, Accuracy, and Repeatability" on page 34.

Digital calculating systems use specific symbols – *digits* – to represent numbers and then perform math symbolically using these symbols. The mechanical computers developed by Charles Babbage in the nineteenth century used the position of wheels, rods and levers to represent decimal digits; modern computers use electric charge on a wire. Digital systems overcome many of the repeatability problems that plague analog systems by forcing intermediate physical measurements to a specific digit and then re-generating the signal. As a result, small variations in computations that result from wear or manufacturing defects can be detected and eliminated.

For example, an electronic circuit might store 5 volts (5 V) in an electronic storage device called a *capacitor* to represent a 1 , and 0 V to represent a 0 . A short while later the circuit might try to read the value: if it reads a 5 V, that's a 1 . But if a large amount of time has passed and some of the electricity has leaked out, the circuit may only read 4 V or even 3 V. As long as more than 2.5 V is read, the circuit still treats the value read as a 1 . As part of the reading operation, the circuit can then "top off" the electricity in the capacitor back to 5 V. On the other hand, if the circuit read 0.5 V, it would treat that as a 0 and not top it off—instead, it would drain the capacitor down to 0 V.

This forced choice between two values is called *digital discipline*, and it is the basis of how dynamic memory inside a modern computer works: a typical dynamic memory chip in 2020 might have 2 billion to

64 billion individual bits,[13] each one read and refreshed many times every second.

Like the differential analyzer, the first digital computing devices in the US were created to solve equations for scientific and military applications. The first was the Atanasoff Berry Computer (ABC),[14] built at Iowa State University by physics professor Dr. John Vincent Atanasoff and his graduate student Clifford Berry. Designed to solve systems of linear equations,[15] the ABC stored data on a pair of drums that rotated once a second. Each drum could store 32 sets of 50-bit binary numbers in 1600 capacitors: using binary numbers made the arithmetic circuits easy to design and construct. Although the basic system was functional, the input and output systems were not completed before Atanasoff was assigned by the War Department to the Naval Ordnance Laboratory in Washington, DC in September 1942. The ABC was eventually disassembled.

The second digital computing system in the US was built at the University of Pennsylvania's Moore School of Engineering by John Mauchly and J. Presper Eckert. Mauchly met Atanasoff at a scientific meeting in December 1940 where Mauchly was demonstrating an analog computer. Mauchly became interested in the promise of digital computation, and ended up traveling to Iowa and staying with Atanasoff for four days. In August 1942, Mauchly wrote a memo entitled "The Use of High-Speed Vacuum Tube Devices for Calculating," which proposed creating a fully electronic computing machine that could perform an estimated 1000 multiplications per second. The following year Mauchley was hired by Eckert, a profes-

---

[13]The word *bit* is short for "binary digit." Bits are the small unit of information. In normal usage we say that a bit can be either a 0 or a 1, but they could just as well be a black or a white, or an empty or a filled. Claude E. Shannon (1916–2001), the "father" of information theory, provided a mathematical definition for the *bit* in 1948, and attributed the coinage of the word to the American mathematician John Tukey (1915–2000), although the word was in use before that time. See Garfinkel and Grunspan, *The Computer Book* (2018). They're sort of like the Greek conception of atoms, but for information. The only problem with this analogy is that in the twentieth century we learned how to split atoms; bits, in contrast, cannot be split.

[14]Using the nomenclature adopted in this chapter, the ABC is not a computer because it is not Turing Complete, a concept that we explain later in this chapter.

[15]A system of linear questions describes one or more lines in two-dimensional space, planes in three-dimensional space, or hyperplanes in multi-dimensional space. Solving the set of equations finds the place where the lines or planes intersect. Rate/time problems from first-year algebra are examples of such problems.

sor at the University of Pennsylvania, and construction started on the Electronic Numerical Integrator and Computer (ENIAC) in secret during the summer of 1943. The project was funded by the US Army's Ordnance Corps for the purpose of creating a computer that could create artillery tables, which at the time were being created nearby in Philadelphia by a group of female "computers." Several of these women, Kay McNulty, Betty Jennings, Betty Snyder, Marlyn Meltzer, Fran Bilas, and Ruth Lichterman, became ENIAC's first programmers.

Two other early computer systems are worth mentioning. At Harvard University, Professor Howard Aiken conceived of a computer powered by relays that could perform computations and print numeric tables. Aiken partnered with IBM to design and build the computer; it was delivered to Harvard in February 1944 and started operations that summer. Called the Mark I, the machine was massive: 51 feet long, 8 feet high, and 2 feet deep. It had 500 miles of wire, 3500 relays, and 1464 10-position switches for entering numbers. Like the ENIAC the Mark I operated on decimal numbers, but because it computed with mechanical relays, rather than electronic tubes, it required 3 seconds to perform an addition and 6 for a multiplication – a thousand times slower than the machine in Philadelphia. The Mark I was built for the US Navy.

In Germany, Konrad Zuse built a series of computers: the Z1 (1936–1938), Z2 (1940), Z3 (1941), and Z4 (1945). Like the UK's Bombe and Harvard's Mark I, these computers were all built using relays. Unlike the others, none of them received significant funding from the host country's military. Zuse had to borrow money from his family and friends to construct the Z1, and he built the machine in his parents' living room! It wasn't until 1940 that Zuse received funding from the German government, and that was only partial funding. By failing to recognize the military applications of computing, the Germans squandered the significant lead in both computer theory and engineering that they had over the Allies.

3.3.1 Encoding Digital Information

Today many people tend to confuse the words *digital* and *binary*, but they are different. What makes digital computers *digital* is the use of specific, discrete values to represent information. We call these discrete values *digits*. Binary systems are digital, but they use just two mutually exclusive binary digits, typically `0` and `1`. The word "bit" is actually a contraction of the words "binary" and "digit."

One of the first binary systems was the Jacquard Loom (1801), which used holes punched into wooden slats to control the pattern woven into the fabric. Each hole determined whether an individual weft would pass over or under a wrap on each pass of the shuttle through the shed. The Jacquard Loom is frequently taken as the first use of punch cards to control a piece of machinery.

It is also possible to have digital systems that use more than two values: the early ENIAC at the University of Pennsylvania (1943) used a voltage moving down 1 of 10 wires to represent the digits 0 through 9, while today's multi-level cell (MLC) flash memory uses four discrete voltage levels within each flash cell, allowing them to store two bits per cell.[16] Not surprisingly, MLC flash costs less than single-level cell (SLC) flash memory, but it is more prone to errors.[17]

Digital computers need a way to store information and to read back the information that they have stored. The Jacquard Loom wasn't a computer because it had no way of writing to its punch cards: the same was true of the card sorters and tabulators that Herman Hollerith created for the 1890 US Census. Without such memory, these devices lacked the ability to alter computations based on an earlier calculation, thus failing the definition for computing. In contrast, the flash memory (1980) in a modern cell phone can be both read and written.

Computers can store all kinds of information beyond simple binary bits: even in the 1940s, computers were computing on integers,

---

[16] High-dimension storage and communication are active research areas in quantum technology. Some are investigating qutrits, quantum bits that have three states. Separately, one group has demonstrated that it can use modulators and mirrors to encode information in photons along seven dimensions, exploiting the photon's "orbital angular momentum" and "angular position" instead of polarization, which is the typical approach. See Mirhosseini et al., "High-Dimensional Quantum Cryptography with Twisted Light" (2015).

[17] Analog computers, in contrast, might use a specific voltage to represent the value of 1, half that voltage to represent the value of 0.5, twice that voltage to represent the value of 2, and so on. Although you might think that this approach provides for more flexibility, the problem is that there is no good way for such computers to distinguish values that are close together, like 1.001 and 1.002. As a result, analog computers tend to lack both accuracy and repeatability, as discussed in the sidebar "Precision, Accuracy, and Repeatability" on page 34. This is also the fundamental problem of proposals to use analog computers as an alternative to quantum computers.

floating point numbers, and text. Today's computers can store virtually any kind of information that can be contemplated, including pictures, sound, and movies.[18] Fundamentally, all of these things are ultimately transformed into a series of bits and recorded in the computer's memory, and then reconstructed on output.

Representation is a word that computer scientists use to describe how information is broken down and stored. One of the simplest representations uses different combinations of binary digits to represent different integers. For example, if you have three binary digits, you can represent eight different values, typically taken to be the numbers 0 through 7:

Bits			Value
A	B	C	
0	0	0	0
0	0	1	1
0	1	0	2
0	1	1	3
1	0	0	4
1	0	1	5
1	1	0	6
1	1	1	7

In modern computers, data is arranged in groups of eight bits called *bytes*. A byte can represent $2 \times 2 \times 2 \times 2 \times 2 \times 2 \times 2 \times 2 = 2^8 = 256$ different values. This is typically scaled from 0 to 255, but it can also be scaled from -128 to 127. The first case is sometimes called an *unsigned 8-bit integer*, the second a *signed 8-bit integer*.

It is common to group four bytes together to form a 32-bit *word* that can represent numbers from $-2\,147\,483\,648$ to $2\,147\,483\,647$. Rational numbers can be represented with two numbers, one for the numerator, one for the denominator. Alternatively, there are *floating point* representations; the IEEE single-precision floating point format uses 32 bits to represent floating point numbers: 1 bit for the number's sign, 8 bits for the exponent, and 23 bits as a binary

---

[18]Some things that modern computers can't store are complex physical objects, thoughts, space, time, or entanglement states.

fraction.[19] However, today most computations are done with 64-bit IEEE *double-precision* floating point numbers, since the additional four bytes of storage is typically inconsequential while the increase in precision is dramatic.

Computers also use combinations of bits to represent individual letters, like the letters typed into a computer that eventually became the sentence you are reading. Using combinations of bits to represent letters dates back to 1874, when the French inventor Émile Baudot devised a more effective way to send text down a telegraph line. Instead of using the dots and dashes of Morse code, Baudot designed a device with five keys and a rotating "distributor" that electronically connected a switch at the end of each key, in rapid succession, to the line. The device sent down the telegraph line a rapid succession of electric pulses corresponding to whether each key was up or down. Today this approach is called *time-division multiplexing*. Five bits allowed the operator to send one of 32 possible combinations down the line with each rotation of the distributor. Baudot used 27 of these codes for letters (E and É were represented with different codes) and another two for the space character and a marker for the end of the message. A device at the other end recorded the marks on paper: it didn't take long to invent devices that actually printed letters that corresponded to the codes that the operator was sending. And thus was born the printing telegraph, soon to be known as the teletype.

Just as the way that numbers are stored inside computers has been standardized, so too has the way that letters are stored. In the 1960s much of the industry adopted the American Standard Code for Information Interchange – ASCII – which dictates that letter "A" will have the binary code 0100001, the letter "B" will be 0100010, "C" will be 0100011, and so on. Lower case letters start with "a" at 0110001. These numbers correspond to the values 65, 66, 67 and 97 in decimal (base 10). In the 1990s ASCII was expanded to include the complex glyphs of Japanese, Chinese, Korean and all of the world's other languages. The new system is called UNICODE and has since been expanded to include dead languages like Cuneiform and even made-up languages like Klingon.

---

[19] Because numbers like 0.1 cannot be perfectly represented as a binary fraction, when floating point numbers like 0.1 are repeatedly added together, the result might end up as 0.9999999999999999 instead of 1.0. This is called *roundoff error*.

3.3.2 Digital Computation

Computers need to have a way to change their behavior based on the information that they read – that is, they need a way to *compute*. Computer engineers use the term *logic* to describe both the internal rules that a computer follows and the mechanism that implements those rules. Once again, logic can be built from many different technologies: from the point of view of a *computer scientist* the details of how the logic is actually implemented doesn't matter much.[20] In an electronic computer, the logic is assembled from fundamental building blocks called *gates*.

Gates can have 1 or more inputs and 1 or more outputs. These inputs and outputs are typically wires, but in a diagram you will see them drawn as lines that carry digital information. The simplest gates replicate the basic logic operations of Boolean algebra:

- The AND gate (Figure 3.2) combines its inputs and produces a 1 if both of its inputs are 1, otherwise its output is 0.

- The NOT gate (Figure 3.3) has an output that is the reverse of its input.

Any logic circuit can be created using combinations of just these two gates and the appropriate connecting wires.

For example, you can make an AND gate that has three inputs (A, B, and C) by taking the output of a single AND gate that computes (A AND B) and connecting it along with C to the input of a second AND gate, creating a circuit that computes ((A AND B) AND C). More generally, it is possible to use AND and NOT gates to build complex circuits that add, subtract, multiply, or divide numbers. For example, Figure 3.4 shows how such circuits are put together to create a one-bit "full-adder," while Figure 3.5 shows how four full-adders can be combined to form a four-bit adder.

It is also possible to create circuits that interface with memory units to load and store information. It is even possible to use a combination of AND and NOT gates to create memory units – such memory is called *static memory* and is much faster than other kinds

---

[20]In the 1970s, Danny Hillis and Mitch Kapor (who later went on to found the Lotus Development Corporation) created a computer out of Tinkertoy that played Tic Tac Toe. The computer is now part of the permanent collection at The Computer History Museum. See D. Hillis and Silverman, "Original Tinkertoy Computer" (1978).

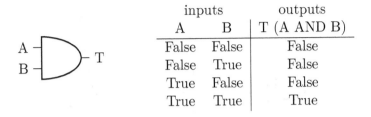

inputs		outputs
A	B	T (A AND B)
False	False	False
False	True	False
True	False	False
True	True	True

Figure 3.2. A simple AND gate and its "truth-table."

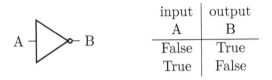

input	output
A	B
False	True
True	False

Figure 3.3. A simple NOT gate and its truth table.

of memory used inside a computer. In fact, *any* digital circuit can be built if you can combine sufficient numbers of AND and NOT gates with the correct wiring pattern. For this reason, the combination of these gates is said to be *universal.*

But one can do even better: the AND and the NOT gate can be combined into a single universal gate called the NAND – *not AND* – gate, from which *every* digital circuit can be built.

In practice, digital designers use all kinds of gates, safe in the knowledge that their designs can always be transformed in a series of universal NAND gates if needed. In fact, the process for doing this is so straightforward and automatic that such transformations can happen when a design is turned into silicon without the designer even knowing it.

3.4 Computing, Computability and Turing Complete

There are many questions to ask in comparing these computers and trying to assess the role that they played in World War II. What sort of monetary and human resources were required to build each machine? How hard was it to find skilled scientists to work on these projects? How much original research had to be done? Did these devices actually contribute to the war effort, as the machines at Bletchley Park clearly did, or were they merely fascinating historical footnotes, like the Zuse machines? One might consider their contribution to military efforts after the war: ENIAC's first official calcu-

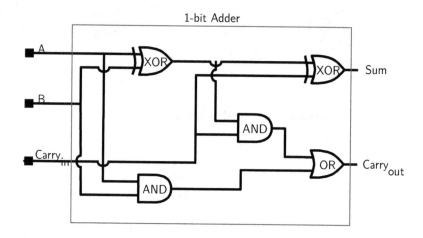

Figure 3.4. Circuit diagram of a "full adder." The inputs are A, B, and C (carry). The outputs is S (the sum) and Cout (carry out). S is true if either A, B, or C are true. If two of them are true then Cout is true and S is false. If all three inputs are true, then both S and Cout are true. Multiple full adders can be chained together to add any number of binary bits.

$$A_3A_2A_1A_0 + B_3B_2B_1B_0 = S_3S_2S_1S_0$$

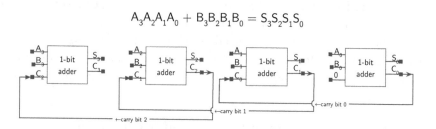

Figure 3.5. Four one-bit full-adders can be combined to form a four-bit adder. Each bit adds the input bits A_n and B_n and the carry bit C_{n-1}. (Note: This four-bit adder ignores the carry bit C_3. As a result, adding 1111 and 0001 will produce 0000, a condition known as an overflow.) This circuit is "clock-free," meaning that it runs without reference to an external clock, although it may take a few hundred picoseconds for the transistors that make up the gates to stabilize when the logic inputs change. Compare this with Figure 6.3, a 4-bit quantum adder.

lations were not for artillery tables, but for the development of the hydrogen bomb, and the team went on to create the Universal Automatic Computer (UNIVAC), while the UK's obsession with secrecy and its persecution of Alan Turing were major setbacks for the early UK computer industry.

Computer scientists evaluating these early machines tend to focus on two questions: how fast could the machines calculate and were they *Turing Complete?*

Speed. For the pioneers of the 1940s, faster calculations were the *only* reason that justified spending the time and money that it took to create calculating machines. It was clear that mechanical calculation had a much higher *initial cost* than human computers but a much lower *incremental cost*. Within the world of mechanical computation, electromechanical systems built with relays had a lower initial cost than electronic systems built with tubes, as the technology was better understood and more readily available. It was also a thousand times slower.

Turing Complete. Modern computers are said to be *general purpose machines*, in that they can be programmed to perform any calculation or any programmable function. This is sometimes called *Turing Completeness*, meaning that the computer implements the computational model described by Alan Turing.[21] Being Turing Complete is what differentiates a machine that calculates from one that computes. The easiest way to make a machine that is Turing Complete is to have it store the program in some kind of memory and for there to be some way to change the program's order of execution, either a mechanism that allows the program to modify itself, or to have the program's execution determined by a computed data value.

---

[21] Turing developed his model to solve a challenge posed by the mathematicians David Hilbert and Wilhelm Ackermann in 1928 called the Entscheidungsproblem (German for "decision problem"). The problem was to develop a procedure or algorithm for evaluating any mathematical statement to determine if it is true or false. Turing developed his model of computation to show that this was not possible; the American mathematician Alonzo Church also showed the impossibility of the Entscheidungsproblem, although using a completely different approach. Church published his solution (Church, "An Unsolvable Problem of Elementary Number Theory" (1936)) a few months before Turing (Turing, "On Computable Numbers, with an Application to The Entscheidungsproblem" (1936)); today these solutions are called the Church–Turing thesis or the Church–Turing hypothesis.

Surprisingly by today's standards, the pioneers were not concerned with storage the way we are today. Modern computers have storage systems that can both store data and load it back: such storage is used for both programs and data. But storage that could support such "load and store" operations on the early computers was minuscule. Code-breaking the ENIGMA required rapidly iterating through many possible encryption keys, but the intermediate results did not have to be archived. Cracking each Lorenz cipher required a lot of input data, which was provided on paper tape, but there was very little in the way of output. Creating artillery tables required a computer-controlled teleprinter, but such devices were write-once, read-never. Moreover, such printers were widely available in the 1940s, as they had been developed for printing telegraphs in the early 1900s.[22]

After the war, the pioneers turned their attention to building machines that could be easily reprogrammed to different tasks. This created the need for some sort of system that could be used to store the programs. Three main technologies emerged: first, acoustic delay lines, in which bits were stored as pulses of sound traveling down a tube of mercury (although Alan Turing suggested using gin instead); second, drum memory, in which bits were stored by changing the magnetization of a small region of a rotating magnetic drum; third, core memory, in which bits were stored by changing the magnetization of a tiny iron torus. Of these three, magnetic core became the dominant form of memory until the emergence of semiconductor memory in the late 1960s, and was widely used until the late 1970s.

3.4.1 Introducing The Halting Problem

In 1936, Alan Turing invented a modern concept of computers when he proved that it is impossible to examine a computer program and determine if the program will halt or will run forever. Here we present Turing's idea by showing that such a program-analyzing program must sometimes be wrong. This is called a proof by *contradiction*.

---

[22]The Morkrum Company, established in 1906 by Charles Krum and the Morton family, developed the M10 printer in 1908. It was adopted by the Associated Press in 1915. The company merged with the Kleinschmidt Companies in 1925, and in 1929 the combined company changed its name to Teletype after the name of its most successful product. See D. R. House, "A Synopsis of Teletype Corporation History" (2001).

There are some programs that obviously halt:

```
PROGRAM A:
1: PRINT "Hello World."
2: HALT
```

Thus, HALT_CHECK(PROGRAM A) = HALTS.

Likewise, there are programs that obviously do not halt:

```
PROGRAM B:
1: PRINT "Hello World."
2: GOTO 1
```

Thus, HALT_CHECK(PROGRAM B) = DOES NOT HALT.

Here we use *functional notation* to denote a computer program called HALT_CHECK that examines a second computer program (variously PROGRAM A and PROGRAM B) and returns HALTS or DOES NOT HALT.

If only a program like HALT_CHECK could exist! With it, we could answer *any* mathematical question! For example, we could use it to determine the correctness of Fermat's Last Theorem, which holds that there is no solution to equation $A^n + B^n = C^n$ for $A > 0$, $B > 0$, $C > 0$ and $n > 2$. We would just code up a new program called FERMAT:

```
PROGRAM FERMAT:
 1: A ← 1
 2:   B ← 1
 3:     C ← 1
 4:       N ← 1
 5:         IF A^N + B^N = C^N THEN
                 PRINT "FERMAT'S LAST THEOREM DISPROVED!"
                 PRINT A,B,C,D
                 HALT
               N ← N + 1
 6:         IF N < C THEN GOTO 5
             C ← C + 1
 7:       IF C < B THEN GOTO 4
           B ← B + 1
 8:     IF B < A THEN GOTO 3
 9:   A ← A + 1
       GOTO 2
10: THIS LINE WILL NEVER BE REACHED
```

We would then compute HALT_CHECK(FERMAT). If the result was DOES NOT HALT, we know that FERMAT never halts, and thus Fermat's Last Theorem is true!

3.4.2 The Halting Problem Cannot Be Solved

Sadly, the Halting Problem cannot be solved. Computer scientists say that the function HALT_CHECK is *undecidable* or *uncomputable*.

To see why we cannot create a HALT_CHECK program that works reliably, in all cases, we simply construct a second program, which we will call H2:

```
PROGRAM H2:
  1: IF HALT_CHECK(H2) = HALTS, GOTO 1
  2: PRINT "H2 HALTS!"
  3: HALT
```

Program H2 asks HALT_CHECK if H2 itself halts. If HALT_CHECK reports that H2 halts, then H2 runs forever. But if HALT_CHECK reports that H2 runs forever, then it must not halt, so HALT_CHECK(H2)=False. But then H2 halts! Clearly, HALT_CHECK cannot correctly report if H2 halts or runs forever.

Program H2 is the logical equivalent of what's called the Liars Paradox. The paradox is that when a person says "I am lying," they are speaking a contradiction. If the person is telling the truth, then they are lying. But if they are lying, then they are telling the truth. So HALT_CHECK can't exist, and finding out if Fermat's Last Theorem is true or not requires years of mathematical research, rather than simply coding up the question and giving it to a computer.[23]

The theory of computation is a lot of fun intellectually, and it is closely related to Gödel's theorem of incompleteness, which holds that in any system of mathematics there are statements – an infinite number, in fact – that are true but unprovable. In fact, it is possible to use the theory of computation to prove Gödel's theorem. But the core ideas of Turing's theory give us more than a simple parlor game that lets us show that some functions are not computable: it gives us a theory that allows us to prove that the only difference between

---

[23]The British mathematician Sir Andrew Wiles published two papers proving Fermat's Last Theorem in 1995; combined, the papers totaled 129 pages and required more than seven years of research. Wiles was knighted as a result of his accomplishment and received the Abel Prize, which is generally regarded as the Nobel Prize of mathematics.

different computers is the size of a problem that they can process, and the speed with which they can arrive at a correct answer. That is, computability concerns *whether a computer can* perform some task, and not how long that task will take or how much memory and storage is necessary. Unfortunately, we are limited by time and memory. The time and other practical limits on computation are the domain of "complexity theory," which we discuss in Section 3.5 (p. 98).

3.4.3 Using The Halting Problem

To recap, the theory of computation tells us that even given a computer that is *infinitely powerful,* has an *infinite amount of storage,* and an *unlimited amount of time,* there are still problems that cannot be solved. The Halting Problem is one such problem.

One of the best uses that you can make of the Halting Problem is as a kind of snake oil detector. For example, upon close examination, many disreputable computer security companies are effectively claiming to have solved the Halting Problem.

Consider a (hypothetical) company that claims to have an anti-virus program called `WIPE_CHECK` that can determine with perfect accuracy if a cell phone app can wipe your cell phone. If such a program existed, we could use it to solve Fermat's Last Theorem! All we would have to do is write a new program and test it with `WIPE_CHECK`:

```
PROGRAM FERMAT-WIPER:
1: A ← 1
2:    B ← 1
3:       C ← 1
4:          N ← 1
5:             IF (A^N) + (B^N) = C^N THEN
                  PRINT "FERMAT'S LAST THEOREM DISPROVED!"
                  PRINT A,B,C,D
                  PRINT "NOW WIPING YOUR PHONE"
                  WIPE_CELL_PHONE
                  N ← N + 1
6:             IF N < C THEN GOTO 5
               C ← C + 1
7:          IF C < B THEN GOTO 4
            B ← B + 1
8:       IF B < A THEN GOTO 3
```

```
 9:   A ← A + 1
      GOTO 2
10: THIS LINE WILL NEVER BE REACHED
```

Something here must be wrong! If a program called WIPE_CHECK could really examine *any* program and always, reliably, determine if that program could wipe your phone, then the program-analyzing program would need to be at least as powerful as HALT_CHECK, because we could use it to solve the same problems.

As with HALT_CHECK, we can prove that WIPE_CHECK cannot exist by using contradiction:

```
PROGRAM W2:
 1: IF WIPE_CHECK(W2) = WILL_WIPE_PHONE THEN GOTO 1
 2: WIPE_CELL_PHONE
```

WIPE_CHECK(W2) *cannot* return the correct answer, for the same reason that HALT_CHECK(H2) cannot: if W2 wipes your phone, then it doesn't, but if it doesn't wipe your phone, then it does. Clearly, a perfectly accurate WIPE_CHECK program cannot exist.

3.5 Moore's Law, Exponential Growth, and Complexity Theory

Computing's pioneers realized that computers would get faster and that storage capacities would increase with every coming year – in principle, they realized, there is no limit to how fast computers could get or how much they could store.

For example, in his seminal 1951 article "Computing Machinery and Intelligence," Alan Turing wrote that in 50 years' time computers would have a storage capacity of 1×10^9 ($1\,000\,000\,000$) binary digits. As it turned out, he was right: Apple's PowerBook G4, a laptop introduced on January 9, 2001, came with 128 MiB of memory ($1\,073\,741\,824$ bits), expandable to 1GiB ($8\,589\,934\,592$ bits).

In his article, Turing hypothesized that a person chatting (by text!) simultaneously with such a computer and a second person would be unable to distinguish between the computer and the second person roughly 70 percent of the time. This challenge is the infamous "Turing Test." Yet here Turing over-estimated the powers of his fellow humans: communications from Joseph Weizenbaum's ELIZA program were regularly mistaken for those of a human just a few months after it was operational in 1964, and many so-called "chatterbot" programs have passed versions of the Turing Test since the 1990s. Today the Internet is awash with programs that not only imitate humans, but attempt to get them to take actions in the physical world, all without revealing that they are bots. And even when users know they are interacting with software, some treat them as people, fall in love with these computer personalities, and take major life decisions based on interactions with them.[24]

Turing's predictive powers were pretty amazing when you consider that the computer Turing built in 1950 – the Pilot ACE (Automatic Computing Engine) – had a main memory of just 4096 bits (arranged as 128 32-bit words). Turing was predicting that the storage capacity of computers would increase by a factor of a 250 000 in 50 years. He pretty much nailed it.

Other engineered systems have not enjoyed similar continued growth in speedup. Consider the passenger airplane:

- In 1903 the Wright Flyer reached an airspeed of 31 mph. It carried one person.

- In 1957 the Boeing 707-020 jet aircraft had a cruising speed of 600 mph;[25] it carried 140 passengers.

- Between 1976 and 2003, the Concorde supersonic jet ferried well-heeled passengers across the Atlantic at 1340 mph. The Concorde carried 92 to 128 passengers.

- The Boeing 787 Dreamliner made its debut 2011, with a maximum operating speed of 600 mph and a cruising speed of 560 mph. The Dreamliner carries 242 passengers.

Planes have certainly improved over the past 100 years. They can carry more passengers and do so more safely. But no technical metric over the past 100 years, from fuel efficiency to safety to cost, compares to the performance improvements that computers have experienced in just 50. Computers have experienced eye-popping increases in speed of computation, storage – and in the efficiency of their algorithms.

In part, planes are limited by the physics of sound: the speed of sound where jets fly is roughly 660 knots, and planes experience

---

[24]Olson, "My Girlfriend Is a Chatbot" (2020).
[25]Repantis, "Why Hasn't Commercial Air Travel Gotten Any Faster Since The 1960s?" (2014).

significant turbulence as they approach it, thus creating a real "barrier" that planes must be engineered to overcome. No similar barrier exists in the world of computing. Planes must overcome the physics of moving large objects: computers need move only electrons.

Turing's Pilot ACE computed with 800 vacuum tubes, but within a few years computers were being constructed with semiconductor transistors. In 1965 Gordon Moore, who at the time was director of research and development at Fairchild Semiconductor, wrote an article exploring the technology trends that the semiconductor industry was facing. Unlike aircraft, semiconductors are not made one at a time: they are made in batches on round disks of silicon called *wafers* and then cut up into individual chips and put into packages that we think of as integrated circuits:

> At present, packaging costs so far exceed the cost of the semiconductor structure itself that there is no incentive to improve yields, but they can be raised as high as is economically justified. No barrier exists comparable to the thermodynamic equilibrium considerations that often limit yields in chemical reactions; it is not even necessary to do any fundamental research or to replace present processes. Only the engineering effort is needed.[26]

What this meant, Moore wrote, is that the number of components on semiconductors was likely to rise exponentially over time "at a rate of roughly a factor of two per year." He added: "certainly over the short term this rate can be expected to continue, if not to increase." Eventually this prediction was named *Moore's Law* and the rate was scaled back to a doubling every 18 months.[27]

The increase in computing over the past 50 years has truly been incredible. In the 1940s the ENIAC could perform 350 multiplications per second; today one can purchase a high-end graphical co-processing card for under $6000 that can perform "100 teraflops," or 10^{14} floating point operations per second, an increase of roughly 3×10^{11}.[28] Iowa State's ABC stored 3200 bits in the size of an actual

---

[26]Moore, "Cramming More Components Onto Integrated Circuits" (1965).

[27]Moore, "Progress in Digital Integrated Electronics [Technical Literature, Copyright 1975 IEEE. Reprinted, with Permission. Technical Digest. International Electron Devices Meeting, IEEE, 1975, pp. 11–13.]" (2006).

[28]The "floating point" operations referred to in the measure "flops" are typically addition, subtraction, multiplication, or a multiplication paired with an addition.

desktop; today you can purchase a desktop disk array with six 16 TB drives for under $6000 that stores roughly 8×10^{14} bits, an increase of roughly 2×10^{11}.

Danny Hillis (b. 1956) is a beloved, accomplished, insightful computer scientist and innovator. He earned his Ph.D. at MIT (advised by Marvin Minksy and Claude Shannon), founded the supercomputer company Thinking Machines in the 1980s, and went on to be a Fellow at the Walt Disney Company. Hillis once gave a talk at the New York City Hilton in which he predicted that one day computers would be so inexpensive that they would be everywhere – in numbers exceeding the world's population. "What are you going to do with all of them?" a heckler in the audience shouted. "It's not as if you want one in every doorknob."

In the 1990s, Hillis returned to the hotel and noticed that each door had been equipped with an electronic lock. "You know what?" he told the audience at the tenth anniversary of the MIT Media Lab. "There is a computer in every doorknob!"[29]

Moore's Law held until roughly 2016, when the market leader in chip production, Intel, signaled that developments in chip-shrinking would slow.[30] In part, this was a reflection of economic realities: for many years Intel and other companies had moderated their technology investments to match the prediction of Moore's Law, bringing a breath of predictability to the topsy-turvy world of high-tech. But starting in the 2000s, other factors such as power consumption came to dominate semiconductor design requirements: no reasonable amount of technology investment could keep Intel on the technology curve that had been forecast in the 1960s. This slowdown was also a result of quantum effects – as gate sizes shrink, there's a greater chance for electrons to "tunnel" from one semiconductor tract to another, causing an error.

Moore's Law isn't really a law: it's really a prediction about the likely progress in semiconductors, given continued investment of dollars in research, engineering and production.

But it is not a precise measurement, because any given processor typically takes a different amount of time for each of these operations, and the amount of time that it takes can also depend on the input data. The ENIAC did not support floating point operations, but most of its contemporary systems did.

[29]Garfinkel, "1985–1995: Digital Decade. MIT's Computing Think Tank Chronicles The Electronic Age" (1995).

[30]Simonite, "Intel Puts The Brakes on Moore's Law" (2016).

3.5.1 Software Speedups

Computers operate from the interplay of hardware and software. In the last section we recounted the dramatic improvements in storage capacity and speed that hardware has experienced over the past 50 years. There have also been improvements in software, but those improvements are of a fundamentally different nature and harder to quantify. This is relevant for our exploration of quantum computing, as the performance that quantum promises is paradoxically much closer to performance improvements of the kind that software has experienced.

Software performance improvements are primarily the result of improvements in algorithms and data structures. An *algorithm* is a method, typically described by a sequence of steps, that performs some kind of computation. *Data structures* refer to the stylized ways that information is stored inside a computer's memory.

It is difficult to quantify changes to an algorithm or a data structure that can change the performance of a system, because performance depends on a dizzying number of specifics.

For example, consider a simple database of the first 17 US presidents (Table 3.1). Each president's information is stored in a *record*, and each record is put in a row, which are numbered 0–16. The records are sorted by the president's date of birth (normalized to the Gregorian calendar). This database is a data structure. Let's say that the computer's memory in which this data structure is stored allows *random access* – that is, it can immediately access any record by simply knowing the row number.

Now, let's say that we need two algorithms. The first is called BIRTHDATE_TO_PRESIDENT; given a president's birthdate, it returns the president's name. A simple algorithm would be:

```
ALGORITHM BIRTHDATE_TO_PRESIDENT(DATE):
BEGIN VARIABLES
N: 0
ROW: DATABASE Table 3.1

BEGIN CODE
 1: IF ROW[N].birthday = DATE:
      PRINT ROW[N]
      HALT
 2: N ← N + 1
 3: IF N < ROW.length:
```

```
        GOTO 1
4: PRINT "NO PRESIDENT WITH BIRTHDAY", DATE
5: HALT
```

Here we have introduced some notation. N is a variable that can hold any number. N is initialized to zero when the program is loaded. ROW is an array of records drawn from Table 3.1. ROW[0] is the first row in the database, and ROW.length is the total number of rows, in this case the number 17. The program starts by checking to see if the row referenced in the database has a date_of_birth equal to the DATE that is provided when the program starts running. If it does, the program prints the entire record and stops. If the record at ROW[N] does not have the requested birthdate, line 3 increments the value of N by 1. Line 3 causes the algorithm to jump back to line 1 if the N is less than the number of rows (17). If N is 17 then line 4 runs: the program prints that there is no president with that birthdate and stops.

The amount of time this program takes to run[31] depends on many factors, such as:

1. The amount of time it takes to load the program into memory and start execution.

2. The amount of time it takes to set variable N to zero.

3. The amount of time it takes to fetch the contents of ROW[N].

4. The amount of time it takes to compare two dates.

5. The amount of time it takes to increment N.

6. The amount of time it takes to compare N to the number 17.

7. The amount of time it takes to jump from line 3 to line 1.

8. Whether DATE is in the database or not.

Times 1 and 2 are constant for any database. Times 3 through 7 are the amount of time that it takes to check any given record. If

---

[31]When examining algorithms like this, it is common for computer scientists to consider both average and worst-case performance. In this example we only consider worst-case performance.

Table 3.1. The first 17 US presidents, sorted by date of birth.

row	Birthday	President	#	Tenure
[0]	1732-02-22	George Washington	1	Apr 30, 1789 – Mar 4, 1797
[1]	1735-10-30	John Adams	2	Mar 4, 1797 – Mar 4, 1801
[2]	1743-04-13	Thomas Jefferson	3	Mar 4, 1801 – Mar 4, 1809
[3]	1751-03-16	James Madison	4	Mar 4, 1809 – Mar 4, 1817
[4]	1758-04-28	James Monroe	5	Mar 4, 1817 – Mar 4, 1825
[5]	1767-03-15	Andrew Jackson	7	Mar 4, 1829 – Mar 4, 1837
[6]	1767-07-11	John Quincy Adams	6	Mar 4, 1825 – Mar 4, 1829
[7]	1773-02-09	William Harrison	9	Mar 4, 1841 – Apr 4, 1841
[8]	1782-12-05	Martin Van Buren	8	Mar 4, 1837 – Mar 4, 1841
[9]	1784-11-24	Zachary Taylor	12	Mar 4, 1849 – Jul 9, 1850
[10]	1790-03-29	John Tyler	10	Apr 4, 1841 – Mar 4, 1845
[11]	1791-04-23	James Buchanan	15	Mar 4, 1857 – Mar 4, 1861
[12]	1795-11-02	James K. Polk	11	Mar 4, 1845 – Mar 4, 1849
[13]	1800-01-07	Millard Fillmore	13	Jul 9, 1850 – Mar 4, 1853
[14]	1804-11-23	Franklin Pierce	14	Mar 4, 1853 – Mar 4, 1857
[15]	1808-12-29	Andrew Johnson	17	Apr 15, 1865 – Mar 4, 1869
[16]	1809-02-12	Abraham Lincoln	16	Mar 4, 1861 – Apr 15, 1865

DATE is not in the database, then the total amount of time will be proportional to the sum of times 3 through 7.

Because the birthdates are sorted, we could try to improve the algorithm by having it stop when DATE is larger than the date of birth of the president in ROW[N]:

```
ALGORITHM BIRTHDATE_TO_PRESIDENT2(DATE):
BEGIN VARIABLES
N: 0
ROW: DATABASE Table 3.1

BEGIN CODE
  1: IF ROW[N].birthday = DATE:
       PRINT ROW[N]
       HALT
  2: N ← N + 1
  3: IF (N < ROW.length) AND (ROW[N].birthday <= DATE):
       GOTO 1
  4: PRINT "NO PRESIDENT WITH BIRTHDAY", DATE
  5: HALT
```

Unfortunately, it isn't immediately clear if this change actually improves the performance of the algorithm. If the date being re-

quested is somewhere before the end of the list, the algorithm will
stop early, but if the date requested is after February 12, 1809, the
algorithm will still need to scan the entire database. And as an added
penalty, there are now two comparisons on line 1 each time the comparison each time through the loop.

A better approach is to use what's known as a binary search:

```
ALGORITHM BIRTHDATE_TO_PRESIDENT_BINARY_SEARCH(DATE):
BEGIN VARIABLES
GUESS: 0
MIN: 0
MAX: 16
ROW: DATABASE Table 3.1

BEGIN CODE
 1: IF MAX < MIN:
      PRINT "DATE NOT FOUND"
      HALT
 2: GUESS ← INTEGER (( MIN + MAX ) / 2 )
 3: IF ROW[GUESS].birthday is DATE:
      PRINT ROW[GUESS]
      HALT
 4: IF ROW[GUESS].birthday < DATE:
      MIN ← GUESS + 1
      GOTO 2
 5: MAX ← GUESS - 1
 6: GOTO 2
```

This program is more complicated than the first, but in the worst
case it only needs to check 5 of the rows, not 17. Mathematically, we
can say that its typical performance is going to be proportional to
the base-2 logarithm of the size of the table, rather than the length
of the table.

Computer scientists have a notation for describing this performance concept succinctly called *Big-O notation*. Using this notation,
we can describe the runtime of the first two algorithms as $O(n)$ because the runtime is proportional to the length of the table (n), while
the third algorithm has a runtime of $O(log\ n)$ because its runtime is
proportional to the natural log.

As a final thought, all of the examples in this section assume that
the records in the database were stored in sorted order. If they aren't

stored in some knowable order, then the only search that works is a sequential search from the beginning to the end. In a real application we would want to be able to search by not just birthdate, but by the other fields as well. A modern database management system would handle this by having additional tables called *index tables*, one sorted by name, one sorted by birthdate, and so on. These tables would consist of just the item being indexed and the row number.

3.5.2 Polynomial Complexity (P)

Programs to sort and search through databases were among the first to be written by computing's pioneers. John von Neumann's first computer program for the Electronic Discrete Variable Automatic Computer (EDVAC, the successor to the ENIAC) was a program to sort numbers, and von Neumann concluded that the EDVAC would be "definitely faster" at sorting than special purpose hardware that IBM had created for sorting punch cards, which could sort about 400 cards/minute.[32] Then von Neumann realized that he could improve the speed of his program by a factor of 80 simply by making changes to the EDVAC's hardware and corresponding changes to the program.

Early computer systems were extremely limited in their main memory, so sorting programs had to perform complex sequences in which data was read from one tape and written to others. A surviving article by Remington Rand describes how to sort data on its UNIVAC computer with six tape drives, and notes that it is possible to sort 12 000 10-word items (a full tape) in just 28 minutes.[33,34]

If all of the numbers to be sorted can fit into a computer's memory, the most obvious way to sort is something called an *exchange sort* or *bubble sort*. The algorithm is simple: start at the beginning of the list and see if the first two numbers are out of order. Now consider the second and third numbers, swapping them if they are out of order. Continue to the end of the list, then start again at the beginning. Repeat until the list is sorted. This approach never fails to produce a sorted list, but it requires n passes through the list to assure completion, where n is the number of elements in the list.

---

[32]Knuth, "Von Neumann's First Computer Program" (1970).

[33]Remington Rand, *Sorting Methods for UNIVAC Systems* (1954).

[34]The UNIVAC had a word size of 72 bits. For comparison, a 3GHz Intel Core i5 microprocessor can sort on array of 12 000 floating point values, each of which requires 192 bits, in 4.5 milliseconds.

Since each pass through also requires $n-1$ comparison and swap operations, the algorithm requires at most $(n)(n-1) = n^2 - n$ operations. As n gets large the value n^2 dominates the value $(-n)$, so we say that bubble sort requires "order n squared" time to solve, which is written $O(n^2)$. It is said that bubble sort "requires polynomial time" or that it has "polynomial complexity." Here, the polynomial is n^2.

There are a few obvious ways to improve on the bubble sort algorithm presented above, but it is hard to improve it by more than a factor of two. Then in 1959, Donald Shell came up with a fundamentally new sort algorithm that is now called Shell Sort. Although Shell Sort still has $O(n^2)$ performance in the worst case, it typically runs much faster. Two years later, Tony Hoare invented one of the best sorting algorithms we have today, known as Quicksort. It also has $O(n^2)$ worst-case performance, but its average performance is $O(n \log n)$.

All of these sort algorithms have performance in P, because they all take an amount of time to sort the array that is proportional to a geometric function of the function of the array's length. But in real world situations, some of these algorithms are faster than others. When sorting large datasets, such performance improvements can be dramatic.

3.5.3 Nondeterminism

Sorting turned out to be one of the easier problems for the pioneers to conquer: a harder one was scheduling, such as the classic traveling salesperson problem (TSP). Here we provide a simple variation of the problem:

> A sales representative needs to visit 20 cities by car and can only drive 350 miles on a single tank of gas: is it possible to reach all 20 cities in a single day without refueling?

If any two of the cities are more than 350 miles apart, then the answer is obviously no. But if the cities are scattered throughout Pennsylvania (which is 285 miles across), and some of the cities are directly connected by roads while others aren't, the answer to the question isn't obvious. If all of the cities are within a mile of the main branch of the Pennsylvania Turnpike, then the answer is clearly yes. But what if some of the cities are close to the Turnpike's Northeast

Extension? What if one of the cities is State College (not quite a city, but home of Penn State University), and far from both the Turnpike and the Northeast Extension?

With 20 cities there are actually $20 \times 19 \times 18 \ldots 2 \times 1 = 20! = 2.43 \times 10^{18}$ different ways of driving between them in theory, which is *way* too many to consider with even a modern computer.[35]

Complexity Theory, which is a part of *Theoretical Computer Science*, is the branch of computer science that is devoted to understanding the differences between problems like sorting and the TSP.[36] *Operations research* is the academic discipline that has taken on solving problems like this. Operations research emerged as a field during the Second World War for solving problems such as shipping supplies, deciding how much armor to put on aircraft, and searching for submarines. Problems like TSP arise on a daily basis for organizations that are trying to make optimal use of their fuel and vehicles. Today airlines and delivery companies solve versions of these problems when trying to decide where they should buy fuel and the routes that their vehicles should travel.

This version of TSP is called a decision problem: the answer is either yes or no, and it is the job of the algorithm to come up with the correct answer. The curious thing about the TSP decision problem is that, while it might be very hard to find a solution, it is easy to discover if the solution is correct: just add up the distance between the cities in the given order. If the distance is less than 350 miles, then you have a solution. Such a solution is called a *certificate*.

(We've seen decision problems before: the Halting Problem is also a decision problem. Specifically, it is a decision problem that is provably unsolvable.)

A more complex version of TSP is known as an *optimization problem*: find the *best* possible solution. If you have an efficient way to solve a decision problem, you can efficiently solve the optimization problem by increasing the time that it takes by another factor of *log(n)* by using binary search. Here, we could start by solving the decision problem for 300 miles. If the answer is yes, we try to solve the decision problem for 150 miles, if the answer is no, we try to solve the decision problem for 600 miles, and so on. Eventually we will find the optimal decision. (There are much more efficient ways

---

[35]If you could consider a billion (10^9) combinations every second, it would take 2.43 billion (2.43×10^9) seconds to find the answer. That's 77 years.

[36]Aaronson, *Quantum Computing since Democritus* (2013).

to solve the TSP optimization problem, but they are beyond what is needed here.)

In 1959 computer scientists Michael Rabin and Dana Scott proposed a model for a theoretical computer that made it easy to write algorithms for solving problems like TSP. They called it a *nondeterministic machine*;[37] today we call these creations-of-the-mind *nondeterministic Turing machines* (NTM).[38] The idea is that such a machine can explore all possible solutions simultaneously: when the right solution is found, the NTM recognizes that solution as the correct one.

Another way to conceptualize the NTM's theoretical module is to imagine that an NTM is just an ordinary computer that is equipped with a special module called `CORRECT_GUESS` that always guesses correctly.

In their paper, Rabin and Scott show that NTMs are no more powerful than conventional, deterministic Turing machines, but for many problems, the description of how to solve it is shorter when the write-up uses a NTM than the equivalent TM. That is, the two models are *mathematically* identical in the kinds of problems that they can and cannot solve.

To understand why TMs and NTMs are mathematically equivalent, but why it is easier to write up the program for a NTM, consider a program that factors a number N into two factors P and Q. The program on an NTM is simple:

```
ALGORITHM NTM_FACTOR(N):
    Z⁺ ← SET OF POSITIVE INTEGERS
    FOR ALL POSSIBLE P ∈ Z⁺, Q ∈ Z⁺:
        (P,Q) ← CORRECT_GUESS(P,Q, GIVEN (P × Q = N))
        RETURN (P, Q)
```

That is, the program tells the computer to correctly guess P and Q given that $P \times Q = N$ and that P and Q are integers.

If this looks like cheating, well ... it is! Nondeterminism is all about cheating. The breakthrough insight of the 1959 paper is that one is allowed to cheat and not design algorithms if one does not care how long those algorithms take to complete.

---

[37]Rabin and D. Scott, "Finite Automata and Their Decision Problems" (1959).

[38]Rabin and Scott's article variously refers to the machine that they created as *nondeterministic machines* and *nondeterministic automata*, but for our purposes, we can take the article as describing NTMs as well.

There are a lot of ways to find two factors of a number N. Here is one that is both naïve *and* inefficient:

```
PROGRAM NAIVE_FACTOR(N):
10 P ← 2
20 Q ← INTEGER( N ÷ P)
30 IF P × Q = N:
   RETURN (P, Q)
40 P ← P + 1
50 IF P > N ÷ 2:
   ABORT
60 GOTO 20
```

This program uses an approach called *trial division*. It tries to divide N by every number from 2 up to $\frac{N}{2}$. If it finds a number which evenly divides N, it returns that number and N divided by that number. If it doesn't find that number, it aborts.

Another way to describe this program is to say that it takes a *brute force* approach to the problem of factoring: it just tries every possible solution and stops when it finds one that works. This is the reason why TMs and NTMs are mathematically equivalent.

A common misconception about quantum computers is that they cheat in this way. They do not: *quantum computers are not NTMs*. Indeed, for a long time Scott Aaronson's blog had the tagline, "If you take just one piece of information from this blog: Quantum computers would not solve hard search problems instantaneously simply by trying all the possible solutions at once." Quantum computers can perform some functions dramatically faster than classical computers because of the algorithms discovered for certain problems. In some cases, these algorithms are just somewhat faster than classical counterparts. And yet in others, quantum computers will offer no real advantage over fast classical computers.

3.5.4 *NP-Complete and NP-Hard*

In 1971 Stephen Cook, a professor at the University of Toronto, presented a paper at the Third Annual ACM Symposium on the Theory of Computing that contained a startling discovery: any problem that could be solved by an NTM in polynominal time can be reduced to a specific *NP* problem called SATISFIABILITY.

SATISFIABILITY asks if there is an arrangement of Boolean variables that can solve a particular equation. Boolean variables can

have the value of TRUE or FALSE; a Boolean equation combines
these variables with the operators AND, OR and NOT. So if A and
B are Boolean variables, a simple instance of the SATISFIABILITY
problem is:

```
SATISFIABILITY PROBLEM 1:
CHALLENGE:  (A AND B) IS TRUE
```

In this case, it is satisfied if A is TRUE and B is true. Here is the
certificate:

```
SATISFIABILITY PROBLEM 1:
  CHALLENGE: (A AND B) IS TRUE
  SOLUTION:
    A: TRUE
    B: TRUE
```

Here is a problem that cannot be satisfied:

```
SATISFIABILITY PROBLEM 2:
  (A AND B) AND (NOT B) IS TRUE
```

This problem can't be satisfied, because the first clause can only
be TRUE if both A and B are TRUE, while the second clause can
only be TRUE if B is FALSE.

Cook's paper was astonishing, because it showed that any problem that can be solved in polynomial time on a nondeterministic Turing machine can be transformed into a SATISFIABILITY problem
that can be solved in polynomial time. The following year, Richard
Karp published a paper showing that 21 other problems have this
property, including the TSP decision problem. This means that any
given TSP decision problem can be quickly rewritten as a Boolean
SATISFIABILITY problem. Conversely, any SATISFIABILITY problem can be rewritten as a TSP decision problem. If you can come up
with a general solution for efficiently solving a SATISFIABILITY
problem, you can solve TSP. If you can efficiently solve any TSP,
you can efficiently solve SATISFIABILITY. Today this property is
called *NP-complete.*

Since 1971, computer scientists have proven that hundreds of
similar problems, including the traveling salesperson problem, are
also NP-complete. On the positive side, this means that a solution
to one of these problems could be easily repurposed to solve the
others: a good solution to TSP can be used to solve packing problems,
for example. But no such solution has ever been found, and many

111

researchers suspect that no such solution exists. Indeed, after the discovery of NP-completeness in 1971, many theoreticians thought that within five or ten years there would be a proof showing that problems in *P* (like sorting) are fundamentally easier than problems in *NP* (like TSP). But nobody could create such a proof.

Today, after 50 years of searching, computer scientists still lack proof that *P* and *NP* are fundamentally different kinds of problems. This is astonishing, because we have problems that are clearly easy, such as sorting a list of numbers into ascending order, and problems that are clearly hard, like solving complex Sudoku puzzles. Sorting is clearly in P, because there are algorithms of polynomial complexity that sort. Sudoku, meanwhile, is NP-complete. That is, there is no efficient algorithm for solving Sudoku, but there is an efficient algorithm for turning any other NP-complete problem *into* a Sudoku problem and vice versa. Perhaps there is some trick to solving Sudoku problems, just waiting there for someone to find it. Alternatively, there may be a proof that Sudoku is actually quite hard. And yet ... nothing, even after 50 years of trying.

Even more infuriating, there are a few problems that were thought to be hard, yet turned out to be easy. One such problem is *primality testing*. Primality testing means to take a number and determine if it is a prime number or a composite. For decades the computing world had a fast, probabilistic primality test that could determine with high probability if a number was prime or not, but there was no fast *deterministic* test that could determine in a reasonable amount of time if a number was prime or not.[39] Then in 2002, Manindra Agrawal, Neeraj Kayal, and Nitin Saxena at the Indian Institute of Technology Kanpur announced the discovery that "PRIMES is in P,"[40] which presented a polynomial time algorithm for primality test-

---

[39] Note that primality testing is fundamentally a different problem than factoring. With primality testing algorithms it is relatively straightforward to take a thousand-digit number and determine in seconds if the number is prime or not. However, these primality testing algorithms do *not* yield the factors of the number that is being tested. This is similar to the fact that you can tell quickly if a number is divisible by 3 – just add up all the digits, and then take the resulting number and add up all the digits, and keep going until you have a single digit. If that digit is 3, 6, or 9, the original number was divisible by 3.

[40] Although the preprint of the article was published on the Internet in 2002, the formal article wasn't published for two more years, finally appearing as (Agrawal, Kayal, and Saxena, "Primes Is in P" (2004)).

ing. This means that you can now take any number and determine quickly if the number is prime or not.

Primality testing is one thing, but what about factoring? Is that hard? Or is there some hidden algorithm waiting to be discovered? We don't know. Although factoring is clearly in the complexity class *NP* – there is a simple *NP* algorithm for factoring any number *N* – efforts to prove that it is or is not NP-complete have failed.[41] Perhaps a polynomial-time algorithm for factoring exists just out of reach, about to be discovered.[42] Today most computer scientists believe both that $P \neq NP$ and that factoring is not NP-complete, but this is a matter of faith, not of proof. For more information, see Section 3.5.6 (p. 116).

In addition to NP-complete problems, there is another complexity class called NP-hard. NP-hard problems are problems at least as hard as NP-complete problems, but possibly harder. One way to think of these problems is to consider the set of problems for which it is not obvious how to create a certificate. These problems might be fundamentally harder than NP-complete problems, or perhaps there is a way to efficiently create a certificate, and it just hasn't been discovered (yet).

Consider the game of chess. Assuming that it is white's turn to move, any given board position may be a winning position for white, meaning that there is a specific sequence of moves and counter-moves that white can play for which every possible response by black always leads to a victory for white or a draw. Likewise, any given board position may be a losing board for white, meaning that no matter what white does, black can always either win or achieve a draw. It is not clear what a certificate for Chess would look like. The most straightforward certificate would be a list of every possible move by white, followed by every possible response by black, and so on. But such a certificate would grow exponentially large with respect to the

---

[41] The simple *NP* algorithm for factoring any number *N* is to try all possible combinations of the numbers a and b such that $1 < a \leq b < N$ until you find a value of a and b such that $a \times b = N$. If you can find such a pair of numbers, then those are the factors.

[42] It turns out that factoring can be done in polynomial time on a quantum computer. Such algorithms are said to be in the complexity class *BQP* (bounded-error quantum polynominal time). Such algorithms are discussed in the next chapter. Perhaps one day someone who learns enough about quantum computing will come up with a fast factoring algorithm that runs quickly on conventional computers. If such an algorithm is found, that will prove that factoring is in *P*.

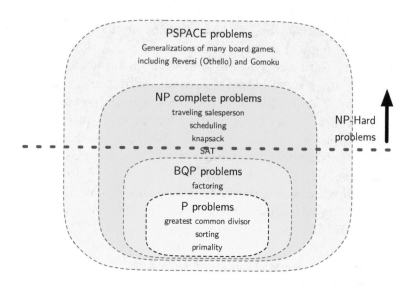

Figure 3.6. The *P, BQP, NP* and *PSPACE* complexity spaces as they are thought to be if $P \neq BQP \neq NP \neq PSPACE$. It is currently unproven if *PSPACE* and NP-complete problems are in the same complexity class, or if there is a partition between the two. Likewise, it is unproven if *NP* and *BQP* are in the same complexity class, and if *BQP* and *P* or in the same class. If *NP = P*, then *NP, BQP* and *P* are all in the same class. However, it is straightforward to prove that *P* and *PSPACE* are in different complexity classes.

number of pieces on the board, and it would therefore take exponentially long to check. In fact, the only way to check such a certificate would be to regenerate the certificate and prove it for yourself, and so this list-of-all-possible-moves-certificate doesn't actually accomplish its objective of being a certificate – that is, it doesn't save any time when you go to check it. Chess is said to be in the complexity class PSPACE, meaning that it requires polynomial space to solve – in this case, that space holds all of the possible chess games. In fact, Chess is said to be PSPACE-complete, actually meaning that all PSPACE problems can be reduced to the problem of finding a winning chess game (assuming you have an infinite number of chess pieces and an arbitrarily large chess board in which to express your problem). Perhaps there is a dramatically more efficient representation for all possible moves in a specific Chess instance; perhaps you will discover it.

3.5.5 NP-Complete Problems Are Solvable!

Just because a problem category is NP-complete doesn't mean that a specific instance of a problem in that category is impossible, or even hard, to solve. SATISFIABILITY is NP-complete, but problems 1 and 2 above are both trivial to solve. Indeed, TSP has been recognized as an important problem for more than 100 years,[43] and there are a growing number of approaches for solving the problem faster, such that in 2004 a challenge problem with 85 900 cities was solved in 136 years of computer time on a cluster of 2.4 GHz computers. (Because the program can be parallelized, it could be on a single computer for 136 years, or on 136 computers in 1 year, or on 1360 computers in 37 days.[44]) The actual computation was performed on a mix of computers between February 2005 and April 2006, because the TSP only ran when the computers were not being used for other purposes. In 2009 the group published a certificate proving that their solution was optimal: that certificate is 32.2 MB (uncompressed) and can be verified in just 569 hours.[45]

As mentioned above, the field of operations research really got going during World War II. One of the exciting early developments was the discovery of the simplex algorithm, an approach for optimizing a system of linear equations. Although simple problems can be solved exactly using symbolic mathematics, many optimization problems are solved in practice using iterative numerical methods – that is, the computer performs a series of computations, examines the results, and then repeats the computations many times in a row, with each iteration producing a more accurate result. Programs that can perform these kinds of optimizations are called, unsurprisingly, *optimizers.* Some optimizers are designed to solve a specific kind of problem, while others are general-purpose solvers, employing a broad range of algorithms and heuristics. The best optimizers today are commercial programs that cost thousands of dollars *per month* to run and save their users considerably more – according to one case study, Air France saves 1 percent of its fuel costs by using an optimizer to help assign planes to routes.[46]

---

[43] W. J. Cook, *In Pursuit of The Traveling Salesman* (2012).

[44] Applegate, Bixby, Chvátal, and W. J. Cook, *The Traveling Salesman Problem* (2006).

[45] Applegate, Bixby, Chvátal, W. Cook, et al., "Certification of an Optimal TSP Tour through 85,900 Cities" (2009).

[46] Gurobi Optimization, "Air France Tail Assignment Optimization" (2019).

3.5.6 BQP, BPP, and Beyond

So here's where things stand in the Summer of 2021, as we get ready to send this book to the printer:

- The class *P* contains problems that can be solved in polynomial time with respect to the problem's size. That is, they don't get dramatically harder as the problem gets larger. For example, determining if a word is in a book is a problem that's in P: just look at every page in the book to verify that the word is not there. If a second book has twice as many pages, it will take you twice as long to check that book. Many common computer problems are in P, such as sorting a list of numbers or taking the square root of a number. It turns out that determining if a number is prime or not is also in P.

- The class *NP* are the problems that take exponentially longer to solve as the problem gets larger, but can be *verified* in polynomial time. Factoring is a good example: there is no fast way to factor a large integer like N, but if somebody gives you two small integers a and b and claims that $a \times b = N$, you can verify this pretty fast. Factoring is in *NP*.

- Some *NP* problems have a property called NP-complete. It turns out that SATISFIABILITY, the Traveling Salesperson Problem, Sudoku played on an arbitrarily large $n \times n$ grid, and many other problems are all fundamentally the same problem. By this we mean that a SATISFIABILITY problem can be transformed (in polynomial time) into a TSP, and vice versa. Transforming the problem is fast, but solving the transformed problem is still hard. On the other hand, this means that if we find a fast way to solve *any* NP-complete problem, we've identified a fast way to solve them all.

- It's unknown whether or not *P* and *NP* are actually the same class. There might be some clever way to transform an *NP*-complete problem into a *P* problem – that is, to solve it in polynomial time. If we find that way, then $P = NP$. Most computer scientists think that this is highly unlikely, but even after decades of trying, nobody has been able to prove that $P \neq NP$. We write this confusion as: $P \overset{?}{=} NP$.

BQP is a complexity class that is conjectured to be between *P* and *NP*. As we will see in Chapter 5, there are a growing number of problems that take exponential time to solve on a classical computer, but which can be solved quickly, in polynomial time, on a quantum computer. This is the class *BQP*, short for bounded-error quantum polynomial time. The proofs that these algorithms are correct typically involve a combination of quantum mechanics and number theory, but they are irrefutable – that is, they are irrefutable if you believe in mathematics and quantum mechanics. We used the word *conjectured* at the start of this paragraph because if *P* = *NP*, then *P* = *BQP* = *NP*. But it might also be the case that *P* = *BQP* ⊆ *NP* or that *P* ⊆ *BQP* = *NP*. We just don't know!

A second complexity class that is conjectured to be between *P* and *NP* is *BPP*, the bounded-error probabilistic polynomial time complexity class. *BPP* is like *BQP*, but instead of using a quantum computer, the algorithms are run on a conventional computer (a Turing machine) that has access to a true random number generator. It turns out that there are many algorithms that can run much faster if they have access to truly random numbers: these are called *randomized algorithms*. Until 2002 primality testing was known to be in *BPP*, because there was a randomized algorithm that did an arbitrarily good job determining if a number is prime or not. Then in 2002, Agrawal et al. developed an algorithm that can test if a number is prime or not in polynominal time.[47] This was a huge breakthrough. However, the algorithm is slower than the randomized algorithm, so in practice the randomized algorithm is typically used in preference to the 2002 algorithm.

Quantum mechanics gives quantum computers an unlimited supply of perfectly random numbers so *BQP* necessarily contains *BPP*: that is, every problem in *BPP* can be solved in polynomial time by a quantum computer. But we don't know if *BPP* is the same as *BQP* or contained in BQP. We write this mathematically as:

$$BPP \overset{?}{\subseteq} BQP \tag{1}$$

This means that we can write the complexity theory that we covered above succinctly as:

$$P \overset{?}{\subseteq} BPP \overset{?}{\subseteq} BQP \overset{?}{\subseteq} NP \tag{2}$$

---

[47] Agrawal, Kayal, and Saxena, "Primes Is in P" (2004).

And we have just gotten started! Today there are hundreds of complexity classes that have been formally defined – the online "Complexity Zoo" (www.complexityzoo.net) listed 545 such classes as of April 2021. (The website also has an easier-to-digest "petting zoo" that has just 17 complexity classes.) The good news is that not all of these classes little question marks over their relations: recall that it's straightforward to prove that $P \subset PSPACE$. But we won't prove it here! To see that proof, and many others, we recommend *Introduction to the Theory of Computation, 3rd Edition*.[48]

3.6 Computing Today

More than any other human technology, electronic computation has undergone phenomenal changes since its inception roughly 80 years ago. That improvement has come both from roughly a trillion-fold increase in the speed of computation and storage, as well as a speedup in the efficiency of algorithms that is surprisingly difficult to measure. But starting in the early 2000s, technology trends changed abruptly:

- Many of the tricks that semiconductor companies had used to speed up their computers since the 1960s started to sputter out. Companies like Intel responded by putting two, four, eight or more general-purpose computers on a single chip, what is now called *multi-core systems*. Companies like NVidia responded by putting hundreds and then thousands of restrictive, special-purpose cores on graphics cards, called *graphical processing units (GPUs)*. Programmers responded by adapting software to use this more difficult-to-program hardware.

- Companies like Amazon, Google, and Yahoo developed and deployed workable approaches for orchestrating thousands of individual computer systems to solve a individual complex problems. These approaches, alternatively called *cluster-computing*, *grid-computing*, and *warehouse-scale computing*, first appeared in the 1990s in the world of scientific computing, where engineers created systems with dozens and then hundreds of racks, each filled with very expensive, very reliable machines. The big breakthrough in the 2000s was the realization that companies could achieve better price-performance ratios by using commodity hardware. In today's warehouse-scale computing, each

---

[48]Sipser, *Introduction to The Theory of Computatio* (2012).

individual system isn't as fast or as reliable as the high-end systems used in scientific computing, but the individual computers are so much cheaper that many more computers can be purchased for the same cost, and fault-tolerant software can automatically reschedule work on a different computer if there is a hardware failure.

- Corporations that previously bought and ran their own computer systems transitioned to renting slices of computers at shared data centers. This approach, called *cloud computing*, gave organizations access to far more computing than was previously possible. The reason for this is that most organizations (and individuals) do not need a steady amount of computing power: they need it in bursts. Thus, just as it is more economically efficient for a home-owner wanting to dig a trench to rent rather than purchase an excavator (and perhaps an operator), in like manner, it is more efficient for a business that needs to solve a big problem to rent a few thousand virtual machines for a week, than to purchase a few dozen machines and run them for six months or a year.

The rate of technology change accelerates because one of the things that engineers can do with faster computers is create faster computers. For example, computer programs running on today's top-of-the-line integrated circuits not only help engineers design the next generation systems – today's computers can also *simulate* next years' systems to find out if the systems will work when they are finally constructed. Even though such simulations run significantly slower than will the future chips, they still help engineers find problems with the chips while they are still being designed, which saves money, shortens design cycles, and allows engineers to pursue more aggressive designs.

This feedback loop, what some people call a *virtuous circle*, is the reason that computers have become a trillion times faster, while aircraft and cars travel no faster today than in the 1960s: faster, more powerful vehicles don't make it possible to build faster, more powerful vehicles.

3.7 Conclusion

For most of its early history, computing has been a tool of governments to solve the kinds of problems governments have. Government and academic research in computing led to its adoption in other data-intensive activities. The trends of democratization of computing services through parallelization, cloud, and eventually the personal computer, brought these devices into our daily lives in unforeseen, wonderful ways.

The path and future of quantum computing could share characteristics with those of classical computing, but with important differences. Like classical computers, quantum computers need patronage from well-resourced and determined actors, and this often requires that government/military problems are on the front burner for applications. Classical computers experienced successive generations of speedups in hardware improvements from the relay, to the vacuum tube, to the transistor. Since the 1960s, classical computing has been transistor based. Quantum computing is still in the relay-vacuum tube stage and needs a breakthrough on the level of the transistor to scale up.

The introduction to complexity theory in this chapter lays the foundation for elucidating the kinds of applications that quantum computing will pursue most effectively. The press often focuses upon cryptanalysis as the problem that quantum computers will solve. However, complexity theory shows that much more interesting, yet more difficult-to-understand challenges, with far-reaching social implications, will be important domains for quantum computing.

The Birth of Quantum Computing

I N the 1940s England used its first electronic computers to crack enemy codes, while the US used its computers to perform computations for nuclear physics. Eight decades later, these same two applications are driving interest and investment in quantum computing. If the effort to build large-scale quantum computers is successful, these machines will surely be used to crack codes and model physics. But just as electronic computers eventually had many more applications than dreamed of in the 1940s, quantum computers will, in all likelihood, find work solving problems that are not even contemplated today.

This is the first of three chapters on quantum computing. We discuss this history in some depth in order to provide an intellectual foundation for understanding both how different quantum computers are from classical computers, and for helping readers to form an appreciation of just how early we are in the development of these machines. This appreciation will be relevant when we review policy issues in Chapter 8 and Chapter 9.

This chapter is based on both bibliographic research and interviews conducted with many quantum computing pioneers. Readers uninterested in this history can skip to Chapter 6, where we discuss the applications of quantum computing likely to be seen in the near future, the different kinds of quantum computers currently under development, the challenges facing the field, and the more distant future outlook for the technology.

4.1 Why Quantum Computers?

Quantum computers are strikingly different from classical computers, and billions of dollars have been spent developing them today without any payoffs other than papers in prestigious scientific journals. To date, this aggressive research program seeks to realize three specific applications for quantum computers: simulating physics and chemistry, factoring numbers, and searching for optimal solutions to specific kinds of mathematical computations ("optimization").

4.1.1 Richard Feynman and Quantum Computing

In 1981, the American physicist Richard Feynman (1918–1988) proposed that the kinds of mathematical problems that quantum physicists need to solve might be more efficiently worked on using a computer based on quantum mechanics than one based on classical physics.[1] Feynman was speaking at a conference exploring the physics of computation co-sponsored by MIT and IBM. Held at MIT Endicott House Conference Center, a converted mansion built in the style of a French manor house in the Boston suburbs, the conference brought together an eclectic collection of roughly 50 renowned physicists and computer scientists. Feynman was the conference's big draw, and his proposal makes this conference the proper birthplace of quantum computing.

Of course, all present-day computers are based on quantum mechanics: computers use the flow of electrons, and electrons are the quantization of electronic charge. But computer engineers (the professionals who design the hardware of computers) go out of their way to make electrons behave as if they are classical objects – as if they were little balls traveling along wires, like water through a pipe. Indeed, in the 1970s, as the feature size of semiconductor lithography got smaller and smaller, some scientists were concerned that the walls between those pipes were getting so thin that electrons might seep (or "tunnel") from one pipe to another, causing an error. Specifically, the fear was that *quantum tunneling*, a consequence of the Heisenberg uncertainty principle, might slow or even halt the relentless march of Moore's Law (see Section 3.5, "Moore's Law, Exponential Growth, and Complexity Theory" (p. 98)). So Feynman's idea that computer engineers might actually want to embrace the uncertainty, nondeterminism and inherent randomness that comes with quantum phenomena was a radical proposal indeed.

---

[1]Feynman, "Simulating Physics with Computers" (1982).

Quantum Confusion

Some popular accounts of quantum computing present key concepts inaccurately. Here we set the story straight.

Quantum computers are parallel machines, but they do not solve hard problems by trying all possible solutions at once. Quantum computers run in parallel, inasmuch as a machine with 50 qubits uses all 50 at once.[a] If we build a quantum system with 10 million qubits, all of those qubits will compute in parallel. While some quantum speedup comes from this parallelism, it is thought that more comes from the ability of quantum computers to compute with quantum wave equations.

Qubits are a superposition of two possibilities, but this does not mean that two qubits simultaneously have four values (00 , 01 , 10 and 11). Qubits do not simultaneously have two values any more than Schrödinger's cat is both alive and dead at the same time (see p. 523). A qubit has a single, definite quantum state when it is measured: that state is either a 0 or 1 , and the probability that it will be in one state or the other depends on the quantum calculation.

Quantum computers cannot store an exponential amount of information. Google has built a quantum computer with 53 qubits, but it cannot store 2^{53} bits (8192 TiB) of information. Google's quantum computer has *no storage at all* in the conventional sense. Each time the computer solves a problem, it selects a single 53-bit result from 2^{53} possible answers.

Quantum computers use superposition and entanglement, but they do not simultaneously consider every possible variation of complex puzzles. That would require cycling repeatedly forwards and backwards, performing additional computations with every cycle. Reusing space and time in this manner would be powerful, but this is not how quantum computers work, and it's probably not possible in our universe.[b]

---

[a]Likewise, even the 8-bit microprocessor in the original Apple II was a parallel machine, in that it could add and subtract 8 bits *in parallel* at a given time. To find a true serial machine, you need to go back to the very first digital computers and their so-called *bit-serial architectures*. These machines added 8-bit numbers a single bit at a time.

[b]For a discussion of time-travel computing, see Aaronson, "Guest Column: NP-Complete Problems and Physical Reality" (2005).

Before the dawn of quantum computing, computer engineers had always tried their best to hide the uncertainty and inherent nondeterminism of the quantum realm in every circuit that they designed. Computers built using tubes in the 1950s and transistors ever since do this by using large ensembles of electrons to represent each 0 and 1 – and by strenuously avoiding the roll of the dice that is inherent in all things involving quantum mechanics. Instead of building computers that are governed by probability, computer engineers have traditionally built machines that they hoped would be *deterministic*. That is, they hoped that the computer would always generate the same output given the same input. When their computers didn't, they called such behavior a bug. Nowadays, we enjoy the successes of computer engineers pursuing determinism. One's computer can process billions of bits a second and run for years without crashing.

Deterministic machines are great for running spreadsheets and typesetting books, but they are poorly suited for analyzing quantum systems, such as a the chemistry of a molecule. That is because the complexity of a quantum system scales exponentially with the number of particles that the system contains: it might take 16 times longer to analyze a molecule with eight atoms compared to a molecule with four. A molecule with 10 atoms might take 64 times longer to study.

Feynman's key insight was realizing that the exponential scaling inherent in modeling quantum systems with classical computers might be avoided by using a computer built from the ground up on the math of quantum mechanics – that is, a computer designed to preserve and embrace the nondeterminism of quantum states. But to do that, quantum computers would have to do something that conventional computers can't readily do: they would have to be able to run backwards.

4.2 Reversibility

The idea that quantum processes could represent digital information and be used for computing emerged slowly in the 1970s. One of the first building blocks, largely worked out at IBM and MIT, was the idea of reversible computing.

Reversibility is a property of both classical and quantum physics, and it has profound implications. In classical physics, reversibility means that astronomers can take the equations used to predict the motion of the Sun, Moon, planets, and stars in the future and *run*

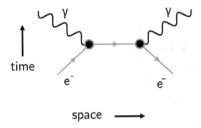

Figure 4.1. A Feynman diagram showing an electron–positron annihilation; rotate it 90° and you have an electron absorbing and then re-emitting a photon (γ). Note that time flows *up* in the diagram, along the vertical axis, while the three dimensions of space are represented as a projection along the horizontal axis.

the equations backwards to determine where those celestial bodies were located in the past. Indeed, taken from the vantage point of celestial mechanics, the direction of time is arbitrary.

In quantum physics, reversibility means that quantum processes can easily go forwards or backwards. In fact, at the quantum level, it is even possible to swap time with space.

4.2.1 The Arrow of Time

In 1948, Feynman, then a professor at Cornell University, came up with a visualization for describing how subatomic particles interact. The diagrams replaced the complex and hard-to-understand mathematics that physicists had previously used with pictures that can be understood even by a lay audience. They were so revolutionary and became so ubiquitous that today we call them *Feynman diagrams*.[2]

The Feynman diagram in Figure 4.1 depicts what happens when an electron (e^-) collides with a positron (e^+). Positrons are basically electrons that have a positive charge instead of a negative charge. Otherwise, electrons and positrons are identical. (When looking at the diagram, remember that time flows up from the bottom of the page to the top.)

---

[2]Feynman's diagrams were initially rejected by his peers, but gained popularity in the 1950s as Feynman successively refined his theory of how light and electrons interact. Feynman went on to share the 1965 Nobel Prize in Physics with Sin-Itiro Tomonaga and Julian Schwinger, "for their fundamental work in quantum electrodynamics, with deep-ploughing consequences for the physics of elementary particles." Flamboyant and commanding, today Feynman is also known for his ability to explain physics to lay audiences, for doing so with infectious enjoyment and captivating joviality, and for his work analyzing the NASA space shuttle *Challenger* disaster.

The positron is called an *antiparticle* because when it interacts with an electron, the two particles annihilate each other, leaving two gamma particles[3] traveling away from each other each at the speed of light. This is the classic "matter–antimatter" reaction popularized in the 1960s television series *Star Trek* – one of the many bits of science at the heart of the series' science fiction.[4]

Recall that time in the Feynman diagram flows from the bottom of the page to the top, while the width of the page depicts separation in space. One of the curious aspects of quantum physics, however, is that the choice of time's direction is arbitrary. Swap the direction of time, and Figure 4.1 equally well describes two photons colliding to produce an electron–positron pair.[5]

Given that time appears reversible at both the cosmic and the quantum level, why then does time to us appear to flow in one direction – that is, *why is there an arrow of time* that appears to point from the past to the future? This is an open question in both physics and philosophy.

One possible explanation is that time's arrow might be an illusion: perhaps time *does not* flow from the past to the future. Time's arrow might simply be a trick of consciousness. Perhaps time *is* consciousness, and all events in the past and future are already fixed in four-dimensional space. If true, this explains the pesky riddle of quantum entanglement – Einstein's spooky action at a distance – but it also closes the door on the possibility of free will. That is, the future might be fixed, but we simply aren't aware of how it will unfold. If the future is fixed, then everything that will happen has already happened, and we have already made all of our choices that we will ever make – we just don't know it yet. Although some people reject this explanation out-of-hand, anyone who has ever been surprised by the ending of a novel or a movie has experienced this effect first-hand.

4.2.2 The Second Law of Thermodynamics

Instead of resorting to metaphysical or religious explanations, physicists typically cite the Second Law of Thermodynamics as the expla-

---

[3]Gamma particles are highly energetic photons.

[4]*Star Trek* also featured the concept of teleportation – the Star Trek *transporter* – which we will revisit in Chapter 7, as well as one of the first popular depictions of computer forensics.

[5]Such reactions have never been observed, but there have been proposals for creating "gamma–gamma" colliders that would do just this.

> **Burnt Norton (Excerpt)**
>
> Time present and time past
> Are both perhaps present in time future,
> And time future contained in time past.
> If all time is eternally present
> All time is unredeemable.
> What might have been is an abstraction
> Remaining a perpetual possibility
> Only in a world of speculation.
>
> – T.S. Eliot (1936)

nation of time's arrow. The Second Law holds that the entropy of a closed system tends to increase with the passage of time. But this is a bit self-referential, since what we call the "Laws of Thermodynamics" aren't really laws at all – they are observations that physicists have made regarding how energy appears to move through the world around us.

The so-called Laws of Thermodynamics were worked out between 1850 and 1920 to explain the behavior of heat. They are "laws," not theories, because they describe *what* the scientists observed; they didn't try to explain the *why* behind the observations. And they aren't laws, because there is no penalty for violating them.[6]

The First Law of Thermodynamics holds that the energy of a closed system remains constant. The Second Law says when two objects touch, heat naturally flows from the warm object to the cold object and not the other way around. By the early twentieth century physicists had learned how to construct devices like heat pumps and refrigerators that use mechanical energy to move heat "uphill" – that is, to suck the heat out of cold objects to make them colder, dumping the energy someplace else, making that second place warmer. These devices don't actually violate the Second Law, however, when you take into account the entire system consisting of the object being cooled, the object being heated, the heat pump, and the energy source.

---

[6]The discipline of quantum thermodynamics derives the modern laws of thermodynamics from quantum mechanics, but since the "laws" of quantum mechanics are also simply mathematical equations that happen to fit observations made by physicists of the physical world, even these "laws" are not really laws.

The Third Law of Thermodynamics says that no matter how hard you work, you cannot cool an object to absolute zero Kelvin (−273.15 °C, or −459.67 °F). In fact, the colder a system gets, the more energy is required to cool it further.

There are many formulations for the Laws of Thermodynamics. Although most are mathematical, one is lyrical: *You can't win, you can't break even, and you can't get out of the game.*

Today the Second Law is widely understood in terms of *entropy.* A colloquial definition of entropy is that it is the amount of "disorder" that exists in a system – the more disorder, the more entropy. Another way of stating the Second Law is that the entropy of a closed system will tend to increase over time.

If you have ever made tea, you have experienced the Second Law. Take an empty teacup, drop in a tea bag, and fill the cup with boiling water. At first, the various organic molecules that make up the tea are all located inside the tea leaves.[7] The tea bag, its leaves, and the cup are all cold, the water is hot. This is a highly ordered system.

But as soon as the water and the tea mix, the organic molecules inside the tea leaves start to diffuse into the hot water, and within a few minutes the concentration of the molecules that we call "tea" dissolved in the water and still present in the tea leaves are roughly in equilibrium. Likewise, the temperature of the tea bag and the inside of the tea cup both rise, while the temperature of the water falls, until they too are roughly in equilibrium. If you wait long enough, the less agreeable molecules from the leaves will also migrate into the water, and the temperature of the water, the teacup, and the room will all come into equilibrium, and now you have ruined a perfectly good cup of tea in the service of science.

You may have also heard that there is a finite probability that all of the air molecules in a room will move into a corner, resulting in the asphyxiation of everyone in the room. In practice this never happens, because that finite probability is fantastically small. Likewise, there is a finite probability that the heat in the room will move back into the water, and that the bitter tea molecules will move back into the bag. But this is also very improbable – so improbable that you will never experience it, no matter how many cups of tea you forget on your kitchen counter.

---

[7]For a discussion of molecules that make up tea, see C.-T. Ho, Zheng, and Lib, "Tea Aroma Formation" (2015).

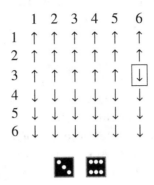

Figure 4.2. An exercise in entropy. The first roll rotates the arrow at row 3, column 6. Rotate enough arrows and the original pattern will be obscured.

You can also demonstrate the Second Law with 36 coins and a pair of dice. Place the coins in a 6-by-6 grid such that the top three rows show their heads pointed up and the bottom three rows show their heads pointed down. Once again, this is a highly ordered system – it's low entropy. Roll the dice and rotate the coin at the row specified by the first die and the column indicated by the second (see Figure 4.2). Repeat a hundred times, and you won't be able to see the original pattern. We have created a web-based version of this simulation that you can run at www.the-quantum-age.com/quantum-demos/.

Both tea diffusing into hot water and the coins rotating in accordance with dice rolls are randomized processes that are reversible in theory, but not in practice. This is probability at work. If you roll the dice twice and roll (3,6) followed by (3,6), you will end up with the original pattern. There is a 1 out of 36 (2.8 percent) chance that this will happen.[8] But if you role the dice *four* times, there are only 3888 sequences that will restore the original pattern, while there are 1 675 728 sequences of dice rolls that will not. The odds that a sequence of rolls will be such a restorative sequence grow *exponentially* worse with each additional pair of rolls. So while it is theoretically possible that you will one day see the initial checkboard pattern restored, the odds are vanishingly small. For example, the odds of restoring the board after six pairs of rolls is significantly

---

[8]Because we use the first die to represent the row and the second to specify the column, there are 36 distinct dice throws. For each of those 36 possible dice rolls, there is precisely one restorative sequence. Thus, there are a total of 36×36 dice rolls, of which 36 are restorative: $\frac{36}{36 \times 36} = \frac{1}{36} = 0.02\overline{7} = 2.7\%$.

Our Tools and Our Self-Conceptions

Einstein himself wrestled with the implications of quantum mechanics (see the sidebar "Man Plays Dice with Einstein's Words" on page 518), but for the average person these implications remain an abstraction in our day-to-day lives. As quantum technologies enter daily life, will we begin to see the world through the lens of quantum mechanics? After all, some ancients saw the universe as a geometric ballet, a reflection of the mathematics of the age. Clockwork and even steam technologies have served as metaphors to explain the celestial and our place in it. Consider this argument about our universe:

> The mechanism by which quantum mechanics injects an element of chance into the operation of the universe is called "decoherence." [...] Decoherence effectively creates new bits of information, bits which previously did not exist. In other words, quantum mechanics, via decoherence, is constantly injecting new bits of information into the world. Every detail that we see around us, every vein on a leaf, every whorl on a fingerprint, every star in the sky, can be traced back to some bit that quantum mechanics created. Quantum bits program the universe.[a]

How will we conceive of ourselves differently if the ideas in this book – the centrality of information and randomness – come to shape our worldview?

---

[a]Lloyd, "The Computational Universe" (2014).

less than the odds of winning any lottery on the planet. This is the Second Law at work, and it is all around us: time moves forward, eggs cannot be unscrambled, and people grow old. Feynman died of cancer, a disease caused by a random, uncorrected genetic mutation in a single cell. He was 69 years old.

4.2.3 Reversible Computation

There is a close relationship between the physics concept of entropy and the mathematical concept of information; in some formulations, entropy and information are actually the same thing.[9] There is also a close relationship between the operation of conventional computers – classical computers – and entropy. Specifically, when classical computers operate, entropy increases. To give you some intuition as to why this might be the case, we will examine what happens in a classical computer when numbers are sorted.

Now we will explore a bit more of the hypothetical computer language that we explored in the last chapter. Recall that there are two kinds of information stored in the computer's memory: variables and code. Each variable has a name and an initial value. The code is executed one line at a time. Code can store and retrieve numbers from locations in its memory (as specified by variable names).

Let's see what happens when our hypothetical computer executes this pseudocode program:

```
PROGRAM SORT_NUMBERS:
BEGIN VARIABLES
A: 3
B: 2
C: 1

BEGIN CODE
1: IF A > B THEN SWAP( A , B)
2: IF B > C THEN SWAP( B , C)
3: IF A > B THEN SWAP( A , B)
4: PRINT-VARIABLES
5: HALT
```

To run this program, we load it into the computer. That sets the initial values of the variables and then runs the program one line at a time. When the computer stops, the output looks like this:

```
SORT_NUMBERS OUTPUT:
A = 1
B = 2
C = 3
```

This program sorts the variables A, B, and C and prints the result.[10] These variables are each a physical place inside the computer's

---

[9]Frank, "The Physical Limits of Computing" (2002).

[10]The program implements a simple sort algorithm called bubble sort. Although generally bubble sort is viewed as an inefficient sorting algorithm, it's fine here.

Step	PC	A	B	C
start	0000 0000	0000 0011	0000 0010	0000 0001
1	0000 0001	0000 0010	0000 0011	0000 0001
2	0000 0010	0000 0010	0000 0001	0000 0011
3	0000 0011	0000 0001	0000 0010	0000 0011
4	0000 0100	0000 0001	0000 0010	0000 0011

Figure 4.3. The value of each variable as the program SORT_NUMBERS runs

memory that can store a number. Our computer is a classical digital computer, so A is actually a set of bits. (See the footnote on page 85 and page 86 for a discussion of bits.) In our computer, A has 8 bits, and we encode them as an *unsigned 8-bit integer* (see p. 88).

It turns out that there's another variable in this computer program that we haven't mentioned yet. This variable is called the *program counter* (PC): it keeps track of the current line that the computer is executing. The PC starts at the first line (0000 0001), and ends on the fifth line (0000 0101). Figure 4.3 shows the value of each of the registers at the completion of each line of the program.

Bits are not abstract things: there is a physicality to each bit inside a computer. In the case of our hypothetical computer, each bit is built from a little bucket that can hold electrons. Each 1 corresponds to a small electronic charge and each 0 represents the absence of charge. In the case of this specific hypothetical computer, each bucket can hold between 0 and 400 electrons (see Figure 4.4).[11]

The bucket controls a switch that the computer uses to determine if the number of electrons in the bucket represents a 1 or a 0 . If the bucket has no electrons, the switch is closed and the computer treats the bit as a 0 . If there are more than 200, the switch engages, and the computer treats the bit as a 1 . Every time the computer reads the bit, it then drains the bucket. If the computer reads a 1 , it reloads the bucket back to its full capacity of 400 electrons. This read combined with a write is called a *refresh* operation, and forcing each bit to be either a 0 or a 1 is called the *digital discipline*, which

---

[11]The buckets actually hold *excess electrons*, since the bucket itself is made out of atoms, and each of those atoms also have their own electrons. However, it is easier to ignore the electrons that are part of the register's walls and just think about the excess elections.

 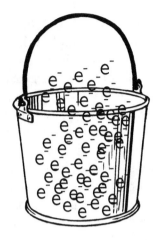

Figure 4.4. A bit in a computer's storage can be thought of as a bucket that can hold excess electrons. The bucket on the left holds no excess electrons and represents a 0. The bucket on the right holds 400 excess electrons and represents a 1.

we discuss in Section 3.3 (p. 84).[12]

From the First Law of Thermodynamics, we know that energy cannot be destroyed. When the computer starts up, the variables A, B, C, and PC are all 0000 0000 . The computer needs 4 bits (1600 electrons) to initialize A, B, and C. These electrons come from a massive reservoir of electrons called the computer's *ground*, which is drawn like this: ⏚.[13] Pulling those 400 electrons from the ground and dropping them into the buckets takes *work*. This work is performed using *energy* from the computer's power supply.[14]

As the computer program runs, electrons are being constantly sent from the memory back to the ground, and pulled back from ground into memory locations. For example, when the PC gets in-

---

[12]Readers with a background in electronics may realize that each bucket is actually a random access memory (DRAM) cell.

[13]On some computers, the computer's ground is actually connected to the third prong on of the electrical outlet – the ground prong – which connects to a green wire that eventually goes to the earth, hence the name *ground* on these computers is actually *the ground*! However, many computers these days don't have a wire connected to the earth. Instead, they have a *floating ground*, which is typically the negative terminal of a rechargeable battery.

[14]In a laptop or cell phone, the energy required to flip bits comes from a chemical reaction. In computers that are plugged into the wall, the energy might come from an electric dynamo powered by a wind turbine, or from photons sent to the Earth by the Sun.

cremented from `0000 0000` to `0000 0001`, those electrons come from the ground. When it gets incremented from `0000 0011` to `0000 0100`, the extra electrons go back to the ground.

All of this work generates heat, which is why a laptop gets warm when it is worked hard. The heat comes both from chemical reactions in the laptop battery to make the electronic energy that's needed to move the electrons, and from the movement of the electrons through the computer circuits, which also generates heat because the electrical wires have resistance. Overcoming that resistance also takes work.

Aside from the program counter, the SORT_NUMBERS program is pretty efficient in terms of electrons. All of the data movements are done with the SWAP function, which swaps the values in the two variables.[15] Nevertheless, when this program runs, information is destroyed. We know this because after the program runs, we've lost the original values of the variables A, B, and C, and there's no way to get them back.

In fact, there are other hidden sources of energy loss going on. The swap itself requires no energy, but IF statements in lines 1, 2, and 3 all generate a bit of information (whether to swap or not to swap) and then destroy that bit. And we are ignoring all of the bits that are set and then cleared when the computer executes PRINT-VARIABLES. Conservatively, that program is probably destroying billions of bits every time it runs.

4.2.4 The Landauer Limit

In 1961, Rolf Landauer (1927–1999) at IBM Research considered the operation of computers at the information-theoretic level. Landauer concluded that practical computation required that information be destroyed, resulting in the inevitable increase in entropy. Landauer showed[16] that even in an ideal computer, every bit of information that is lost must generate a tiny amount of heat – at least 3×10^{-21} J at room temperature. Today this is called the *Landauer limit*.

This amount of heat was insignificant compared to the other processes running inside IBM's computers of the early 1960s; no IBM

---

[15]Most computers have such swap instructions, although in practice what they do is far more complicated and electronically expensive than simply exchanging electrons between two memory locations.

[16]R. Landauer, "Irreversibility and Heat Generation in The Computing Process" (1961).

engineer had ever observed it. It's still insignificant today. A typical modern desktop computer consumes roughly 10×10^{-15} J to convert a 0 to a 1 – roughly a million times more than heat generated by the information loss.

Landauer became IBM's assistant director of research in 1965,[17] and became an IBM Fellow in 1969. In 1972, he recruited Charles H. Bennett (b. 1943) to join the research staff at Yorktown Heights. At the time, Bennett had been thinking about quantum information for nearly a decade (see the sidebar "The Birth of Quantum Cryptography" on page 137) and was working on a paper that challenged Landauer's fundamental finding – that is, he found a way around the Landauer limit. Bennett published that paper shortly after joining IBM.

Bennett's paper starts out by restating Landauer's conclusion:

> The usual digital computer program frequently performs operations that seem to throw away information about the computer's history, leaving the machine in a state whose immediate predecessor is ambiguous. Such operations include erasure or overwriting of data, and entry into a portion of the program addressed by several different transfer instructions. In other words, the typical computer is logically irreversible.[18]

But in the pages that follow, Bennett showed that Landauer had overlooked something: Landauer had assumed that computers necessarily had to destroy information when they operate. Bennett showed that this need not be the case: it is possible to compute entirely with *reversible* operations. Such a computer would be more complex than a computer built from conventional logic – computers like the ones that IBM was building in 1973 – but in theory could be just as powerful. That is, it would be a Turing machine, a generalized computer that can run any program and simulate any other computer. Bennett showed how to build a reversible Turing machine.

Bennett didn't actually *build* a reversible Turing machine, of course, any more than Alan Turing built a Turing machine when he published *On Computable Numbers*.[19] Bennett merely showed

---

[17]Physics Today, "Rolf Landauer" (2019).

[18]C. H. Bennett, "Logical Reversibility of Computation" (1973).

[19]Turing, "On Computable Numbers, with an Application to The Entscheidungsproblem" (1936).

that it is *theoretically possible* to build such a machine. Bennett also showed that such a machine would be significantly more complicated to design, harder to program, and would typically take twice as many steps as a non-reversible Turing machine to solve the same problem. But a reversible Turing machine would have a significant advantage over today's non-reversible systems: it would be liberated from Landauer's limit, and be able to compute with essentially no lower bound on energy loss.

As will be shown later in this chapter, reversible computation is also the key to solving problems on quantum computers.

4.3 Cellular Automata and Conway's Life

Bennett was not the only person in the 1970s interested in reversible computing. Another was Tommaso Toffoli, who developed his approach for reversible computation using a different approach to computing called *cellular automata*.

A graduate student at the University of Michigan, Toffoli had studied physics in Italy before moving to the US as part of the Fulbright Foreign Student Program. He eventually met up with Arthur Burks (1915–2008), a mathematician who had worked on the design of the EDVAC with John von Neumann (see the sidebar "John von Neumann" on page 138). After von Neumann's death, Burks completed and edited von Neumann's final book, which introduced the idea of *cellular automata*.[20] and explores many of their theoretical capabilities.

With his background in physics, Toffoli was interested in taking the research of von Neumann and Burks in a different direction. Specifically, he wanted to know if it was possible to build a *reversible* cellular automata. Toffoli recalled in an interview for this book that Burks and others thought that it wouldn't be possible to create such cellular automata, but Toffoli showed that it was, and published the work as his PhD thesis, with Burks as his thesis advisor.

4.3.1 Computing with CPUs, GPUs, and CA(s)

To understand the significance of Toffoli's question, and of what he discovered, we are going to look deeper into how computation works in a conventional computer.

The "brain" of the contemporary computer is a small device called the *central processing unit* (CPU). Inside the CPU there is

---

[20]von Neumann and Burks, *Theory of Self-Reproducing Automata* (1966).

The Birth of Quantum Cryptography

QIS pioneer Gilles Brassard (b. 1955) traces quantum cryptography's start to a friendship between Charles Bennett and Stephen Wiesner, who met while they were undergraduates at Brandeis University in the 1960s. Bennett went to Harvard to pursue his PhD, while Weisner went to Columbia University. Weisner came up with an idea he called "Conjugate Coding," which used a pair of entangled particles to do things like create electronic banknotes that would be impossible to counterfeit and create pairs of messages, of which only one could be read by the recipient. Wiesner submitted a paper on his thought experiment to *IEEE Transactions on Information Theory*, but the paper was rejected and Wiesner went on to other projects.[a]

The possibility of using entanglement for some kind of communication stuck with Bennett and he shared it from time to time with others. More than ten years later, Bennett and Brassard were at an IEEE conference in Puerto Rico, where Brassard was giving a talk that touched on quantum concepts. Bennett thought that Brassard might be interested in Weisner's idea of conjugate coding. Brassard was, and the two expanded the idea into the basic concept of "quantum cryptography," which they presented at the CRYPTO '82 conference. The following year, Bennett and Brassard presented their groundbreaking article, "Quantum Cryptography: Public Key Distribution and Coin Tossing"[b] (frequently called simply *BB84*). We will take up the story of quantum cryptography in the next chapter. And if you are interested in the original Conjugate Coding paper, you can read it too,[c] since the success of the BB84 convinced Wiesner to get his original paper published.

---

[a]G. Brassard, "Brief History of Quantum Cryptography: a Personal Perspective" (2005).

[b]C. H. Bennett and G. Brassard, "Quantum Cryptography: Public Key Distribution and Coin Tossing" (1984).

[c]Wiesner, "Conjugate Coding" (1983).

John von Neumann

Born in Budapest in 1903, John von Neumann was one of the most gifted scientists of the twentieth century. At the age of eight von Neumann was familiar with calculus; fluent in five languages, he published his first groundbreaking mathematical paper at the age of 19. Eager to escape Europe, he was offered one of the first professorships at the Institute of Advanced Study in Princeton, NJ, which he joined in 1933.

Von Neumann worked out the complex nonlinear equations describing the physics of shock waves; this had direct application to the design of explosives. Based on this work, he was invited to join the Manhattan Project in 1943, where he worked on the explosive "lens" for the implosion bomb. He was successful: the first implosion "gadget" detonated at the Trinity test site on July 16, 1945; the second detonated over the city of Nagasaki, Japan, on August 9, 1945, killing as many as 80 000 people.

At a chance meeting at the Aberdeen train station in August 1944, army lieutenant Herman Goldstine told von Neumann of a research project at the University of Pennsylvania to create a device that would be able to compute artillery tables far faster than the human "computers" that had been hired for the task.[a] Von Neumann joined the group, hoping that the Electronic Numerical Integrator and Computer (ENIAC) under construction, along with its successor Electronic Discrete Variable Automatic Computer (EDVAC), would be able to speed the Los Alamos bomb computations.

Goldstine typed up the group's design notes and gave them to von Neumann for editing during a train ride to New Mexico. When the report was distributed later that summer, *First Draft of a Report on The EDVAC*[b] carried von Neumann's name alone on its cover. This mistake is memorialized in the term *von Neumann architecture*, which describes the EDVAC's approach of storing both data and code in the computer's main memory. Today von Neumann architectures dominate the computer landscape. Quantum computers do not have von Neumann architectures, but they are controlled by conventional computers that do.

---

[a]See Grier, *When Computers Were Human* (2007) and LeAnn Erickson's 2011 documentary *Top Secret Rosies: The Female Computers of WWII.*
[b]von Neumann, *First Draft of a Report on The EDVAC* (1945).

a complex circuit where the actual computing – the addition, the subtraction, and so on – takes place. This circuit is literally the computer's processor, although on some computers it is called a *core*; until the early 2000s most home computers had a single core, whereas today most home computers have anywhere between two and twelve.[21] The rest of the computer exists to move data and code from the Internet into the computer's memory, and then from the computer's memory into CPU, and then to move the results back to the outside world.

Cellular automata take a different approach to computation. In these systems, computation takes place *in the memory itself*. Imagine a large rectangular grid of cells, like a massive checkerboard that extends to the horizon. Each square is a processor that has a small amount of memory and executes its small program in step with all of the other squares. Each square can also communicate with its neighbors. By itself, each square can't compute much, but the assemblage of all of the squares could be much faster than today's fastest computers, for the simple reason that more instructions are executing at any given moment. That is, whereas contemporary computers have between two and twelve cores, and whereas graphic processing units might have a few hundred or even a few thousand cores, a large system based on cellular automata principles might have millions or billions of cores.

The phrase *self-reproducing* in the title of von Neumann's last book asks not if it is possible to create a robotic factory that is programmed to produce robot factories, but if it is possible, using computation, to have an underlying mathematical pattern that can reproduce itself. Such a structure could be the core idea that empowered a robotic robot factory, but the underlying design pattern might show up in other systems as well.[22]

---

[21]Contemporary computers also typically have graphic processing units (GPUs), which can have dozens, hundreds, or even thousands of cores. These cores are less flexible than the cores in the CPU and are optimized for performing the kind of math necessary to render complex scenes. Each specialized GPU core is typically slower than a general purpose microprocessor core in the CPU, but the GPU has many more cores than the CPU, so the net result is that it runs much faster. Although, as their name implies, GPUs were originally created for graphical processing, another common use for GPUs today is performing the massive and repetitive mathematical algorithms required by contemporary artificial intelligence algorithms.

[22]Design patterns used in nature frequently show up in engineered systems, im-

That is, the self-reproducing automata that are the subject of von Neumann's book *could* be a factory of robots, placed in a complicated arrangement so that the factory of robots created new factories of robots. Alternatively, it might be a collection of math problems that, when solved, created a new set of *the same* math problems. This is fundamentally the advantage that von Neumann and Burks enjoyed by working with mathematical abstractions, rather than trying to actually build self-reproducing automata out of wires, relays, and engines: the abstract mathematical system allows the thinker to focus on the conceptually relevant part of the problem without worrying about the details. As theoreticians, they could consider their theoretical models and determine if the models would work (if they could possibly build the systems), or if the models wouldn't work (even if they spent their lifetimes trying to build the system perfectly). This interplay between theory and practice shows up again and again in the history of computing, and it is the reason why theoreticians believed that quantum computers would be so powerful even before the first quantum computer was ever constructed.

4.3.2 Life (The Game)

Probably the best known cellular automata is *Life*, invented by the British mathematician John Horton Conway FRS (1937–2020). Life is *not* reversible, but its influence is great to this day, so we use Life here to present the concept of cellular automata, which will then give us a tool for thinking about quantum computers.

Conway designed the rules of Life through trial-and-error; we present the rules in Figure 4.5. Conway's goal was to create a simple set of rules that nonetheless produced successive generations with unexpected complexity. Below we will look at a few simple examples that do not have such complexity, followed by two examples that remain fascinating to this day.

---

plying that the underlying requirements for both natural and engineered systems may share fundamental commonalities. For example, both bacterial and computer programs called *quines* are self-reproducing automata that are structured in two parts: the first part is the genetic material or information that describes the machinery necessary to reproduce, and the second part is the machinery itself, which reads the information and reproduces both the information and the machinery. A factory of computers that built computers would probably be based on similar principles. See also Bratley and Millo, "Computer Recreations: Self-Reproducing Programs" (1972).

The rules of Life:
1. Gameplay is on a square grid of cells, such that each cell has eight neighbors.
2. Each cell can either be empty or alive.
3. There is a global clock. Each time the clock ticks, every empty cell that is surrounded by exactly three live cells transitions from empty to alive. (A "birth.")
4. Alive cells that have two or three live neighbors remain alive.
5. Alive cells with less than two alive neighbors become empty. (They die of "loneliness.") Alive cells with four or more alive neighbors become empty. (They die of "overpopulation.")

Figure 4.5. Rules for John Conway's "Life"

A grid with no live cells remains eternally empty:

A single live cell also becomes empty and remains that way forever:

The three possible arrangements of two live cells also die out:

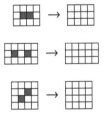

With three live cells there are three possibilities. A triangle of three live cells becomes a 2-by-2 square, which is eternally stable:

Three cells arranged in a diagonal will take two generations to die out. More exciting are three cells arranged in a horizontal row: they became a vertical row, which then became a horizontal row again. This repeating pattern is called a *blinker*:

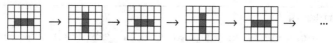

Start with five live cells and things get complicated fast. For example, there is the *glider*, which moves one cell to the right and one cell down every four generations, as demonstrated by this progression:

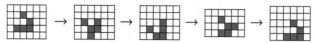

A slightly different collection of five cells called R-pentomino produces a staggering amount of complexity. The initial pattern runs without a repeat for 1103 generations and ends up producing eight 2x2 blocks, six gliders, four six-celled "beehives," four blinkers, and a collection of other objects. It must be watched on a computer screen to see this in all its glory. Below are the results at 150 generations using the web-based LifeViewer.[23] Look carefully and you can see that three of the pattern's gliders have already been launched and are sailing off to infinity:

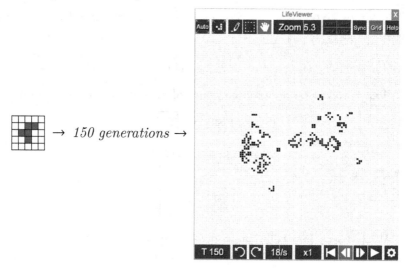

→ *150 generations* →

Conway invented Life in 1969 and sent a typewritten letter about it to Martin Gardner (1952–2000), editor of the popular "Mathematical Games" column in *Scientific American*. Gardner featured the game a few months later,[24] igniting an interest in both Life and cel-

---

[23]See www.conwaylife.com/wiki/R-pentomino

[24]Gardner, "The Fantastic Combinations of John Conway's New Solitaire Game 'Life'" (1970).

lular automata that continues to this day. Indeed, when Conway died in 2020 (one of the early notable deaths in the COVID-19 pandemic), the obituary in the *New York Times* quoted Gardner stating that at the peak of its popularity, "one quarter of the world's computers were playing [Life]."[25] The obituary also quoted musician Brian Eno, who said "Conway's LIFE changed mine ... Conway himself thought it rather trivial, but for a nonmathematician like me, it was a shock to the intuition, a shattering revelation – to watch glorious complexity emerging from staid simplicity."

Life Is Turing Complete

Although this fact wasn't discovered for many years after its invention, the rules of Life have sufficient complexity that they are *Turing complete*. That is, with clever programming, the rules of Life and a starting configuration of sufficient complexity can implement the central processing unit of a computer that can read, execute, and modify its own program. This basic idea was created by Alan Turing, another English mathematician, in the 1930s.[26] Turing's great discovery was that a mechanical calculating device can compute *any computable function in all of mathematics* if it 1) can read instructions from a tape; 2) write new instructions back to the tape; 3) move the tape forwards or backwards; and 4) has logic for executing the instructions. This means that you could use a large grid running Conway's Life to compute the mathematical constant π (pi) to a million places if you wanted to. You could even use a grid running Life to simulate a top-of-the-line Intel microprocessor, which means that you could use it to run the Windows or Macintosh operating system, provided that you had a grid that was large enough.[27] (We discuss computing and what it means to be Turing complete in Section 3.4 (p. 91).)

Like all Turing Machines, the Life Turing Machine (LTM) has control logic, memory cells, and the ability to read and write to a massive "tape." One of the repeated patterns used by the Life Turing Machine is the *glider gun*, first developed by famed MIT hacker Bill Gosper (b. 1943), which repeatedly "shoots" gliders across the grid.

---

[25]Roberts, "John Horton Conway, a 'Magical Genius' in Math, Dies at 82" (2020).

[26]Turing, "On Computable Numbers, with an Application to The Entscheidungsproblem" (1936).

[27]Rendell, "A Universal Turing Machine in Conway's Game of Life" (2011).

The Life Turing Machine (LTM) uses the gliders to communicate between its various parts.

The LTM requires 11 040 Life generations for one Turing Machine cycle. Whether or not that is "slow" depends on how fast the underlying cellular automata runs: in a browser at 18 generations a second, it's slow from the point of view of a human watching the screen; on some kind of theoretical stringy fabric that can crunch 600 trillion generations per second (600 THz), it would be considerably faster than any computer in existence today. Likewise, if a cell in the Life array is the size of the array we show above, a LTM large enough to run a web browser would probably be larger than our planet. But if each cell in the Life array were on the order of 10^{-35} m – that is, a distance on the scale of the smallest quantum effects (see Appendix A) – then the entire computer would likely fit into the space of a single hydrogen atom.

Turning a massive, parallelized, conceptually clean cellular automata into an ornately complex contraption built from glider guns and mathematical tape may seem itself more like a mathematical diversion than a practical exercise in computing. The point of the exercise is to demonstrate that the underlying computational medium of Life's cellular automata is universal: it can therefore compute anything that is computable. Building a computer with glider guns and tapes is no more strange than building one with relays, tubes, or semiconductor transistors.

Where could one go with these observations? Recall Toffoli's interest in recasting physics as computation. Conway's Life is one of an infinite number of possible cellular automata systems, each with its own set of rules. A cellular automata could have rules that just consider each cell's north, south, east and west neighbors, for example. Cells could have a third state, *young*, which would prevent them from counting towards a birth. Cells could eventually die from old age. The game could be played on a hexagonal grid, or a three-dimensional grid, or even a five-dimensional grid. The key thing that makes it a cellular automata is that every cell follows a set of rules – typically the same rules – and that each runs more-or-less independently. Beyond that, everything is up for grabs.

Conway's Life demonstrates that even simple underlying rules can produce complex and unforeseen outcomes. Could our own reality be described by the rules of a cellular automata? What if the fundamental stuff of the universe, deep down, actually *is* a cellular automata?

4.4 Digital Physics

The idea that reality itself might be nothing more than a program running on some cosmic computer was *not yet* a common idea among academics and science fiction authors in the 1970s. Nevertheless, it was increasingly clear that there was something fundamental about computation and information – not just at the societal level, but in the underlying fabric of biology and physics.

For example, there was the matter of life itself. In 1953 Watson and Crick published the structure of deoxyribonucleic acid (DNA),[28] the molecular basis of heredity, and started to unravel the entire process by which information encoded in DNA is synthesized into proteins. By 1970, scientists were increasingly comfortable with the idea that most (if not all) biological processes were based on the movement of *information* carried by molecules.[29]

Likewise, by the 1970s the philosophical implications of quantum mechanics – for example, whether Schrödinger's cat could be both alive and dead at the same time (see p. 523) – were increasingly being discussed and accepted outside the rarefied world of theoretical physics. In 1974, Stephen Hawking showed that quantum uncertainty causes black holes to radiate small amounts of energy – now called *Hawking radiation* – setting off what American theoretical physicist Leonard Susskind called "The Black Hole War"[30] over the question of whether or not information was destroyed by black holes or conserved in Hawking radiation.[31] So the idea that reality itself might be fundamentally based on information – that reality might *be* in-

---

[28]Watson and Crick, "Molecular Structure of Nucleic Acids: a Structure for Deoxyribose Nucleic Acid" (1953).

[29]The role of information in shaping the form of physical reality easily dates back to ancient Greece, where Heraclitus posed the question of whether or not the ship that Theseus had used to sail from Crete to Athens was the same ship after centuries afloat in Athenian harbor, despite the fact that all of its oars and timbers having been incrementally replaced over the years.

[30]Susskind, *The Black Hole War: My Battle with Stephen Hawking to Make The World Safe for Quantum Mechanics* (2008).

[31]Meanwhile in the popular press, the bestselling books Capra, *Tao of Physics: an Exploration of The Parallels between Modern Physics and Eastern Mysticism* (1975) and Zukav, *The Dancing Wu Li Masters* (1979) both drew similarities between quantum mechanics and eastern mysticism.

formation – wasn't necessarily so far-fetched, at least to those who thought about it.

Even if the underlying fabric of the universe is not actually a cellular automata, being able to describe it as such might give scientists a powerful alternative formulation for quantum physics. But in order to do that, the cellular automata certainly wasn't going to be the kind described by the rules of Conway's Life. That is because the Game of Life is not reversible, but physics is.

4.4.1 Edward Fredkin and Project MAC

Project MAC was established at MIT in 1963 to develop interactive computer systems and explore applications for their use.[32] Roberto Mario Fano (1917–2016) was the founding director of Project MAC, followed by the legendary J. C. R. Licklider (1915–1990), an American psychologist and computer scientist, who ran Project MAC from 1968 until 1971. Edward Fredkin (b. 1934) was Project MAC's third director, from 1971 until 1974, when Fredkin moved to California to spend a year learning quantum mechanics from Richard Feynman, with Fredkin teaching Feynman about computers in return. Fredkin's tenure as director was unlike the others, in that it was the only time that Project MAC (or its successors) had been run by a wealthy, ex-military, college drop-out who had made his fortune when his AI startup went public.[33].

Fredkin was born in southern California in 1934 into a family that once owned a chain of radio stores but lost them at the start of the

---

[32]"MAC" was an unstable acronym, variously standing for "Multiple Access Computer" (the project pioneered timesharing, allowing a computer to be accessed by more than one person at once), "Machine-Aided Cognition" (one of the project's original goals), "Man And Computer" (and later "Men Against Computers," because the project's members were overwhelmingly male and the computers somewhat buggy), and even "Minsky Against Corby" (recognizing the long-running feud between MIT professors Marvin Minsky and Fernando José Corbató – a stress that ultimately led Minsky's Artificial Intelligence Lab to break with Project MAC and go its own way). Following Fredkin's tenure, Project MAC was renamed the Laboratory for Computer Science and run by Michael Dertouzos from 1974 to 2001, and then by Victor W. Zue from 2001 to 2003, at which point the Laboratory for Computer Science and the AI Lab merged back together to form the MIT Computer Science and Artificial Intelligence Laboratory (CSAIL) MIT Institute Archives, "Laboratory for Computer Science (LCS)" (2011).

[33]Garfinkel (aut.) and Hal Abelson (ed.), *Architects of The Information Society* (1999).

Quantum Physics and Free Will

We know that many experts and organizational systems can embrace probabilities in a contingent world, but does that embrace have limits? Moving from the level of legal systems to the individual, quantum technologies could erode our assumptions about human morality.

Our assumptions about human morality are based in nondeterminism – one implication of which is that we have free will, that our choices are ours, along with the moral responsibility of them. Could more familiarity with quantum mechanics begin to alter our assumptions about determinism and ultimately, assumptions of free will?

Novelist Ted Chiang writes an exhilarating story that explores the moral responsibilities of a many-world universe in *Exhalation*.[a] In the story, Chiang imagines a version of the many worlds theory, one where the universe splits and is duplicated every time quantum decoherence occurs. In Chiang's world, people can consult an oracle that reveals how they acted in other worlds split from one's own by quantum decoherence. One character regrets an act, consults the oracle, and finds that other versions did not engage in the bad act. He thus concludes that his bad act in this world was an anomaly, one that does not stain his character too deeply, because in other worlds, he took a different set of actions. The philosophy of personal responsibility is woven together amongst these series of different worlds according to this character. Others however are crushed by the events in alternate worlds and regret the actions taken in their own world. If people begin to see their lives as deterministic, as one version of themselves in a reality of many versions, might they start to believe that they are not really responsible for their acts in this world?

---

[a]Anxiety is the Dizziness of Freedom in Chiang, *Exhalation* (2019).

Great Depression.[34] That is, they had known money, but now they were poor, and even though Fredkin hadn't experienced wealth himself, the family's loss nevertheless affected him deeply. Fredkin grew up experimenting with electricity and chemicals, got poor grades in high school, but got accepted to the California Institute of Technology on the strength of his entrance examinations. CalTech did not give Fredkin any financial aid, so Fredkin worked multiple jobs. It still was not enough money, and his grades were still terrible, so in his second year he dropped out of CalTech and volunteered to be an Air Force officer – it was better than the alternative of being drafted to serve in the Korean War.[35]

The Air Force first trained Fredkin to fly jets, then to be an intercept controller. "It's like air traffic control, except we're trying to get them to the same place at the same time," he later explained.[36] After that, the Air Force sent Fredkin to MIT's Lincoln Laboratory to learn about computers so he could oversee the testing of the new-fangled computerized air defense systems that were then under construction. It turned out that Fredkin was quite good with computers: when the Air Force had trouble with a computer at MIT's Haystack Observatory that was tracking rocket launches, Fredkin was sent to figure out what was going wrong. (He found an overflow error in the computer's programming.) Fredkin also created one of the first computer assembler languages, and then taught a course at Lincoln on how to use it.[37]

Fredkin got along well with computers, but not with the Air Force. He left military service and took a civilian job at Lincoln Lab working with the same computers. But Fredkin had bigger plans. On his own, he placed an order for one of the world's first commercial computers, a Royal McBee Librascope General Purpose 30, which was a tube-based machine first manufactured in 1956 that had a retail price of $47 000 (equivalent to $465 000 in 2021). Fredkin recalls that he only had $500 to spare, but the computer had a long delivery time, so he figured that he would find the money before he needed to pay up. His plan was to offer programming courses at area companies so that they could then provide contract programmers to

[34]Wright, "Did The Universe Just Happen?" (1988).
[35]Fredkin, interviewed by Garfinkel in September 2020.
[36]E. F. Fredkin, "Oral History of Ed Fredkin" (2006).
[37]Walden, "Early Years of Basic Computer and Software Engineering" (2011), p. 52.

the government, use the tuition money to pay for the computer, and make a profit in the process.

Fredkin made a list of his prospects; at the top of the alphabetized list was Bolt Beranek & Newman (BBN), a 10-year old MIT spin-off specializing in contract research. At BBN Fredkin met Licklider, who soon convinced Fredkin to drop his plan to be an itinerant teacher-with-a-computer and instead join BBN's research staff. Licklider then convinced BBN to assume Fredkin's purchase commitment for the LGP-30, at the reduced price of $30 000. BBN had no obvious need for the machine; Licklider pushed. "If BBN is going to be an important company in the future, it must be in computers," Licklider told Leo Beranek, one of the company's founders. Beranek agreed to the purchase, even though BBN had "never spent anything like that on a single research apparatus."[38,39]

BBN soon acquired a second computer, the PDP-1, which it leased from another MIT spin-off called Digital Equipment Corporation (alternatively shorted to Digital or DEC over the company's 41-year life).[40] Not a full-size computer like the ones sold by IBM, DEC called the PDP-1 a *mini-computer*. This was right around the time that Project MAC was getting started at MIT, and Fredkin was convinced that the PDP-1 could be logically partitioned into four *even smaller pieces* so that the single machine could serve multiple people at the same time, an approach called *time sharing*. At Fredkin's suggestion BBN brought in two MIT faculty as consultants: Marvin Minsky (1927–2016) and John McCarthy (1927–2011) – two of the computer scientists who had coined the phrase "artificial intelligence" just a few years earlier.[41] Working together, Fredkin

---

[38]Beranek, "Founding a Culture of Engineering Creativity" (2011).

[39]BBN *did* become an important company in the future. The company designed and produced the Interface Message Processors (IMPs) that routed packets on the ARPANET and early Internet. It also created and spun off Telenet, Inc., the company that built and sold service on the world's first public packet-switched network. BBN was variously publicly traded and private, and was ultimately acquired by Raytheon in 2009 for $350 million. A major player in quantum technologies, with scores of academic publications along with applied research into photonics, superconducting qubits, graphene, control systems, and cryogenic systems, BBN now holds over 20 patents in quantum technologies.

[40]DEC eventually created 53 PDP-1 computers; BBN got the first. Another was given to the MIT for students to use; in 1962 Steve Russell and others used it to create the video game SpaceWar!

[41]McCarthy et al., "A Proposal for The Dartmouth Summer Research Project on Artificial Intelligence" (1955).

and McCarthy successfully implemented time sharing on the PDP-1. Fredkin also experimented with cellular automata on the PDP-1's graphics screen.[42]

Still focused on getting rich, Fredkin left BBN in 1962 and founded Information International Incorporated, an early AI startup. Minsky and McCarthy joined the board as founders. Triple-I, as it was known, did early work with the LISP programming language and in robotics, but the company's ultimate success came after Fredkin designed and the company started selling the first high-resolution film scanner for motion picture film. Fredkin took the company public six years later, becoming rich in the process. (Triple-I was eventually acquired by Agfa-Gevaert in 2001.[43]) With his newfound wealth, Fredkin would ultimately purchase a mansion, an island in the British Virgin Islands, and a television station.

Licklider left BBN in 1962 to head the Information Processing Techniques Office (IPTO) at the US Department of Defense Advanced Research Projects Agency (ARPA, later renamed DARPA), where he put in place research projects that directly led to the creation of the Internet. He worked at IBM from 1964 to 1967, and rejoined the MIT faculty in 1968 as Director of Project MAC.

Fredkin rejoined MIT the same year as Licklider and also went to work for Project MAC, although the two events were not connected. Fredkin was recruited by Minsky, with whom he had formed an enduring friendship, to be the AI Lab's co-director. The idea was for Fredkin to help steady the lab, using his combination of technical skills and business acumen. In 1972 Fredkin became Project MAC's director, and was promoted to full professor (perhaps in an attempt to erase the embarrassment of having the lab run by a college dropout who didn't have a PhD).

Running Project MAC did not suit Fredkin. He soon hired his own replacement, then moved out to California for a year-long sabbatical, spending the 1974–1975 school year back at Caltech, this time as a Fairchild Distinguished Scholar at the invitation of Richard Feynman.[44] Upon returning to Boston, Fredkin resumed his profes-

---

[42]Wolfram, *A New Kind of Science* (2002), p. 876.

[43]Wright, "Did The Universe Just Happen?" (1988).

[44]Minsky introduced Fredkin to Feynman back in 1962 – three years before Feynman won the Nobel Prize, when Feynman was considerably less famous. "Feynman showed us a mathematical problem he had been working on. He had a notebook and the notebook had all these pages of mathematics," Fredkin recalled in

sorship at the Project MAC, which had been renamed the Laboratory for Computer Science, and continued working on the project he had started in California with Feynman: reversible computing.

4.5 Reversible Computing and Supercomputing

The basic idea of a reversible computer is that it is a computing machine that can go forwards or backwards in time for any sequence of computations. We discussed the idea of reversible computers earlier in this chapter while exploring Bennett's idea of a reversible Turing machine (p. 135), but it's not clear if Fredkin or Feynman were aware of Bennett's work at IBM.

4.5.1 A Most Successful Term Paper

Instead of building his reversible computer using the theoretical mathematical constructs of a Turing machine, Fredkin's reversible computer reflected his own practical orientation. His first approach was a model of a computing machine based entirely on billiard balls careening around a friction-less obstacle course and having perfectly elastic collisions. He called this the *billiard ball computer* and ultimately published the idea in 1982.[45] You can't actually build such a computer, of course, because we don't have frictionless billiard balls that undergo perfectly elastic collisions. That's why the computer

---

our interview. "He said, 'Look – this mathematical problem is something we need to solve. I tried to solve it, a graduate student also did it.' He showed us – he had a notebook with about 50 pages of dense mathematics in it, handwritten, and he kept circling great big expressions and giving them names. And he said, 'Look, I've done all the math here, and I get a final expression. Murray Gell-Mann has also done it, and a graduate student has done it, and all we know is that the three of us got three different results that are not compatible. So our conclusion is that no one can do this much mathematics without doing errors. Can you guys do something about it?'" (Murray Gell-Mann (1929–2019), was awarded the 1969 Nobel Prize in Physics "for his contributions and discoveries concerning the classification of elementary particles and their interactions.") When Minsky said that symbolic algebra was a problem that the lab was working on, Feynman added that he refused to type on a computer, so the symbolic algebra system *also* needed to be able to read his handwriting and convert it to computer notation. On the flight back from Los Angeles, Minsky said that he would have a graduate student work on the algebra, and Fredkin would work on the handwriting recognizer. In retrospect, the Minsky–Feynman–Fredkin meeting didn't result in any breakthroughs in handwriting recognition or symbolic math computation, but it did set the groundwork for the invention of quantum computing two decades later.

[45]E. F. Fredkin and Tommaso Toffoli, "Conservative Logic" (1982).

is just a theoretical model: it's a way for thinking about building a reversible computer without actually having to build one.

The actual reversible computer that Fredkin proposed building would be built out of semiconductors. To do that, Fredkin needed a new set of basic circuits that themselves were reversible, and that could be used to build a reversible computer. Today we call Fredkin's basic circuit the *Fredkin gate*.

The Fredkin gate has three inputs (C, I_1, and I_2), and three outputs (C, O_1, and O_2). The fact that the number of inputs matches the number of outputs is not an accident: it is required by the basic rules of reversibility. That is, every input to the gate must have a unique output: this makes it possible to run the gate backwards for any output and learn its original input. Eight possible inputs with eight corresponding outputs is the smallest number of combinations that produces a device that is both reversible and universal.

In addition to being reversible, The Fredkin gate is *universal*, in that any digital circuit can be built from a combination of Fredkin gates. (In today's computers, the NAND gate is sometimes used as a universal building-block, because any electronic circuit can be built using a combination of NAND gates. See Section 3.3.2 (p. 90).) Because it is a binary logic gate, each of the inputs and outputs can be either a 0 or a 1. If C is 0, the output bits are each the same as the corresponding input bits. If C is 1, then the output bits are swapped. The Fredkin gate is thus also called a *controlled swap*, or CSWAP. It is shown in Figure 4.6.

Tommaso Toffoli completed his dissertation in 1976 and submitted a journal article proving that reversible automata could be constructed; the article was published the following year.[46] Toffoli recalls interviewing with Charles Bennett at IBM and with Fredkin at MIT and decided to become a Research Scientist in Fredkin's group, which he joined in 1977.

The following spring Fredkin taught an eclectic graduate course at MIT called Digital Physics (Figure 4.4). The course consisted of Fredkin sharing his intuition about the nature of reality with graduate students, and then trying to get students to develop formal mathematical proofs of these conjectures as their final projects. One of those projects was Bill Silver's term paper, "Conservative Logic,"

---

[46]Tommaso Toffoli (1977).

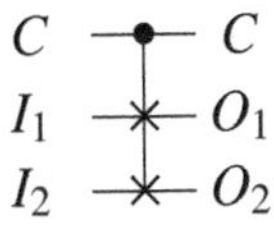

C I_1 I_2	C O_1 O_2
0 0 0	0 0 0
0 0 1	0 0 1
0 1 0	0 1 0
0 1 1	0 1 1
1 0 0	1 0 0
1 0 1	1 1 0
1 1 0	1 0 1
1 1 1	1 1 1

$C \;\longrightarrow\; C$
$I_1 \;\longrightarrow\; O_1$
$I_2 \;\longrightarrow\; O_2$

Figure 4.6. The Fredkin gate (CSWAP)

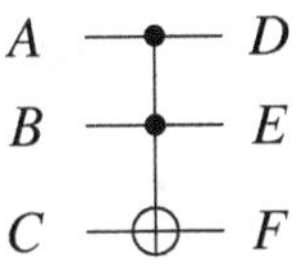

A B C	D E F
0 0 0	0 0 0
0 0 1	0 0 1
0 1 0	0 1 0
0 1 1	0 1 1
1 0 0	1 0 0
1 0 1	1 0 1
1 1 0	1 1 1
1 1 1	1 1 0

$A \;\longrightarrow\; D$
$B \;\longrightarrow\; E$
$C \;\longrightarrow\; F$

Figure 4.7. The Toffoli gate (CCNOT)

in which Silver worked out detailed proofs regarding the properties of Fredkin's gate.[47]

In 1980, Toffoli came up with an improvement to the Fredkin gate that is somewhat better suited for designing complex circuits. Today it's called the Toffoli gate. Whereas the Fredkin gate is called a controlled swap (CSWAP), the Toffoli gate is called a *controlled controlled NOT* (CCNOT).[48] This gate is shown in Figure 4.7.

Like the Fredkin gate, the Toffoli gate is also universal, meaning that it can be used to create any kind of digital electronics currently in use (or imaginable, for that matter). Both gates can also be generalized to more than three inputs. In practice Toffoli gates are used more often than Fredkin gates when discussing quantum circuits, perhaps because they offer more flexibility.

4.5.2 Reversible Computing Today

Heat was not a major concern for most computers in the 1980s, but it is today. Nevertheless, mainstream computer companies are not building their conventional systems with reversible logic. Here are some reasons why they aren't:

---

[47]Silver left MIT in 1981 to join his classmate Marilyn Matz and MIT Lecturer Dr. Robert J. Shillman in a startup venture called Cognex Corporation, which sought to develop and commercialize computer vision systems. Cognex went public in 1989 and is currently listed on the NASDAQ as CGNX with a market cap of $14B.

[48]In a controlled NOT gate, a control bit determines whether a data bit is inverted or not. In a controlled controlled not (CCNOT), *both* control bits need to be 1 in order for the data bit to be inverted.

DIGITAL PHYSICS
Edward Fredkin
January 17, 1978

6.895 Digital Physics
(New)

Preq.: Permission of Instructor
Year: G(2)
3-0-9

An inquiry into the relationships between physics
and computation. 6.895 is appropriate for both computer science and physics students. Models of computation based on systems that obey simple physical
laws and digital models of basic physical phenomena.
Tutorial on conventional digital logic. Information,
communication, memory and computation. A formal
model of computer circuitry, conservative logic, will
be used to model computers at various levels of complexity from simple logic gates to processors, memory, conventional computers and Turing machines.
Questions about reversibility and about the conservation of information during computation. Minimum
energy requirements for a *unit* of computation. Generally reversible iterative processes. Tutorial on some
areas of the quantum mechanics. Digital time and
space. Universal cellular automata. Digital model of
the zero-dimensional Schrodinger equation. Proof of
the conservation of probability in the digital model.
Three dimensional digital Schrodinger equations. Digital Newtonian mechanics. Digital determinism. The
laws, physical constants and experimental tests of digital physics. Atomism. Questions of the ultimate nature of reality. Metaphysics and cosmogony.
E. Fredkin

Figure 4.8. Announcement for Fredkin's Digital Physics course.

- The computer industry has nearly a hundred years' experience working with computer designs that are not reversible, while there has been comparatively little work done with reversible computing. The switching cost of moving from our current technology stack to a new one would be substantial, even if this other stack offers theoretical advantages. Similar switching costs are observed in other industries, such as the nuclear industry's failure to shift to a thorium-based fuel cycle, or the failure of the US to shift to the metric system.

- Although computers do convert electrical energy into heat when those 1 s are sent to ground, a significantly larger source of wasted energy is from semiconductor effects such as *resistance* (the fact that semiconductors do not perfectly pass electricity) and *leakage* (the movement of charge from one electronic device to another in a manner not aligned with the electronic circuit). What's more, leakage gets worse as transistors get smaller, placing a limit of just how small silicon electronics can get. Another limiting factor is the wires that carry signals between semiconductor devices: they have both resistance and capacitance, which again limits how energy-efficient, and how fast, signals can be carried between devices.

- Reversible computing requires more than reversible gates: it requires replacing large chunks of the technology stack. For example, there is a need to develop efficient reversible algorithms, presumably written in new computer languages that support reversible computing.

- Reversible computers require more transistors than traditional computers because they need to retain all of the information necessary to reverse the computation.

- Given that the computing industry hasn't hit the limits of non-reversible technology, there has been no reason to pursue reversible computing. Instead, the industry has exploited other approaches – most notably parallel computing – to achieve the significant speedups we have experienced over the past four decades. Whereas in the 1990s it was common for desktops and laptops to have a single CPU, today systems typically have between four, eight or even more.

- An even bigger speedup has taken place on the other side of the Internet, in the "cloud" that delivers web pages to a desktop computer or information to applications running on a smart phone. Cloud computing has made it possible for each query to use hundreds or thousands of computers for an instant, getting a tremendous speedup.[49]

While reversible computing doesn't currently make sense for electronic computers, it is an area of active research. Meanwhile, reversibility is a basic requirement of computing on a quantum computer. The reason has to do with entanglement and superposition: the quantum part of a quantum computation stops when the wave function collapses, which happens the moment a non-reversible action takes place and a measurement is performed. So a quantum computer that implements any sort of logic has to use reversible logic *by necessity*.

Today it is common for quantum computer engineers to express the complexity of their algorithms in terms of the number of Toffoli gates that their algorithm and problem require, just as electronic computer engineers describe the complexity of their systems in terms of the number of electronic NAND gates or transistors. For example, in 2019 Google released a paper describing an approach for factoring the large integers used in cryptography *in hours*, stating that such a machine would require twenty million state-of-the-art (e.g. "noisy") qubits, and "$0.3n^3 + 0.0005n^3 \lg n$ Toffolis."[50] With a standard encryption key size, $n = 2048$, this comes to roughly 2.6 billion Toffoli gates.

While that may seem like a lot of gates, in November 2020 the Apple M1 system-on-chip contained 16 billion transistors.[51] Although the two kinds of gates are fundamentally different, the comparison shows that it is within the realm of today's technology to build a device with billions of active components. We will return to Google's paper in the next chapter.

---

[49]For example, in 2010 a single search at Google used more than a hundred computers, but each for just two-tenths of a second. See Dean, "Building Software Systems At Google and Lessons Learned" (2010).

[50]Gidney and Ekerå, "How to Factor 2048 Bit RSA Integers in 8 Hours Using 20 Million Noisy Qubits" (2019).

[51]Apple Computer, "Apple Unleashes M1" (2020).

4.5.3 Defense Money

What made all of this research possible was a spigot of money from the US Department of Defense flowing into MIT's various computing projects during the 1960s and 1970s. This is not a new story, of course. The first computers built in Germany, England, and the US were all built to help with the war effort. It was the awarding of the SAGE missile defense system to IBM that cemented the company's position as the dominant computer manufacturer in the world. In 1961 IBM built its first transistorized supercomputer, the IBM 7030 "Stretch," for the US National Security Agency, apparently to assist in some way in the business of code-cracking. By the 1970s investments in supercomputing were helping to make sophisticated stealth aircraft a reality and to make the mathematical modeling of nuclear explosions so accurate that the US was able to stop physically testing nuclear weapons.[52]

Even before the simulations and models became crazy accurate, conducting physics experiments inside a computer had many advantages that made them a strong complement to experiments conducted in the lab or in the deserts of Area 51. Three such advantages are speed, scalability, and repeatability:

- **Speed** is the most obvious advantage: in the world of a computer, setting up a new experiment typically means editing a few files and reserving time on the computer system. This makes it easy for scientists to try a wide range of different ideas.

- **Scalability** means that scientists can run more experiments in a period of time simply by buying more computers. Scalability is not so easy in the lab, where running multiple experiments at the same time means having more lab space, as well as having more flesh-and-blood researchers to conduct the experiments.

- **Repeatability** is an often-overlooked advantage of conducting experiments in simulation. With complex experiments in a physical lab it is often difficult to repeat the experiment and get nearly the same result. This is because the outside world is always intruding. A truck may drive by, causing the ground to

---

[52]The US signed the Comprehensive Nuclear Test Ban Treaty on September 27, 1996, in part because the computer modeling had become so powerful as to make testing itself obsolete.

vibrate; a solar flare may eject a shower of high-energy atoms, ions, and electrons into space, causing a light show in the northern sky and interfering with sensitive electronic instruments down here on Earth. All of this must be taken into account when conducting physical experiments. With computerized experiments, the only real risk is bugs in the software.

Realizing these goals requires machines that are easy to program, reliable, secure, and accessible – hence the government's interest in funding basic research into software design, operating systems, security, and networking. The world we live in today – the hardware and software that was used to write the book you are reading – are direct beneficiaries from these government funding decisions.

A key to enabling the creativity and productivity of this basic research was the way that the funding agencies gave the researchers flexibility to set their own agenda. At MIT, the Laboratory for Computer Science and the Artificial Intelligence Laboratory were funded in no small part by a series of master agreements with DARPA, such as Office of Naval Research contract N00014-75-C-0661, which moved millions of research dollars from Washington to Cambridge. The money was delivered as a block grant, with individual faculty members needing to simply write project proposals describing what each planned to do with their share of the pie. As long as the faculty projects advanced the overall goal of building computers that were faster, better at solving problems, or easier to program, funding was all but guaranteed.

In November 1978, Fredkin and Toffoli included in MIT's proposal to DARPA a 20-page project description titled "Design principles for achieving high-performance submicron digital technologies."[53] The proposal expanded the ideas of conservative logic, showing how it would be possible to use reversible gates to cheat the power loss associated with conventional digital electronics. It then proposed approaches for using even less power, such as using superconducting switches with Josephson Tunneling Logic (also called *Josephson junctions*). The only mention of cellular automata was a reference to Toffoli's 1977 journal article, and while the proposal mentions Landauer's work, it doesn't mention Bennett's. But it does

---

[53]Twenty-four years later, the proposal was finally published (E. F. Fredkin and Tommaso Toffoli, "Design Principles for Achieving High-Performance Submicron Digital Technologies" (2001)).

cite Fredkin's unpublished lecture notes from 1975–1978, and Bill Silver's MIT term paper. If nothing else, the proposal shows scientific progress is not linear, and the mere fact that scientific work has been published is no guarantee that others working in the exact same field will see it (or at least take notice of it) in a timely manner.

Fredkin and Toffoli's proposal was funded (likely a foregone conclusion), marking the beginning of the group's support by DARPA.

4.6 The Conference on The Physics of Computation (1981)

In the 1930s H. Wendell Endicott (1880–1954), a successful industrialist and philanthropist,[54] built a French-style manor house on a hill crest of his 25-acre suburban estate overlooking the Charles River in Dedham, Massachusetts. Endicott's will stated that the house should be donated "to an educational, scientific or religious organization." The property was offered to MIT when Endicott died, and the Institute turned it into a luxurious conference center.

When academics start developing a new field, it's common to hold some kind of meeting for early innovators to meet and exchange ideas. Always thinking big, in 1980 Fredkin decided to hold a conference at Endicott House and invite the biggest names he could get in physics and computing to discuss his up-and-coming ideas. Fredkin knew that he would need to have a big name to get the other big names to come, so he called up Richard Feynman, who agreed to give a keynote speech. Fredkin invited IBM Research to co-sponsor the conference. Rolf Landauer readily agreed, and both he and Charles Bennett agreed to attend.

Fredkin, Landauer, and Toffoli were the official organizers. Then came the invitations! Fredkin had earlier met Konrad Zuse, the German inventor who had built one of the world's first digital computers during World War II (see Chapter 3), so Zuse got an invite. The prominent physicists Freeman Dyson and John Wheeler were invited. Also invited were a number of up-and-coming researchers, including Paul Benioff (b. 1930), who went on to create the first mathematical model of a quantum computer; Hans Moravec (b. 1948), best known now for his work in robotics and artificial intelligence, and his writings as a futurist; and Danny Hillis (b. 1956), who went on to create the supercomputing company Thinking Machines, after which he became a Fellow at Walt Disney Imagineering. In total

---

[54]MIT Endicott House, "Our History" (2020).

1 Freeman Dyson	13 Frederick Kantor	25 Robert Suaya	37 George Michaels
2 Gregory Chaitin	14 David Leinweber	26 Stand Kugell	38 Richard Feynman
3 James Crutchfield	15 Konrad Zuse	27 Bill Gosper	39 Laurie Lingham
4 Norman Packard	16 Bernard Zeigler	28 Lutz Priese	40 P. S. Thiagarajan
5 Panos Ligomenides	17 Carl Adam Petri	29 Madhu Gupta	41 Marin Hassner
6 Jerome Rothstein	18 Anatol Holt	30 Paul Benioff	42 Gerald Vichnaic
7 Carl Hewitt	19 Roland Vollmar	31 Hans Moravec	43 Leonid Levin
8 Norman Hardy	20 Hans Bremerman	32 Ian Richards	44 Lev Levitin
9 Edward Fredkin	21 Donald Greenspan	33 Marian Pour-El	45 Peter Gacs
10 Tom Toffoli	22 Markus Buettiker	34 Danny Hillis	46 Dan Greenberger
11 Rolf Landauer	23 Otto Floberth	35 Arthur Burks	
12 John Wheeler	24 Robert Lewis	36 John Cocke	

Photo courtesy Charles Bennett.

Figure 4.9. The Physics of Computation Conference, MIT Endicott House, May 6–8, 1981

roughly 60 researchers attended. Financial support for the conference was provided by the MIT Laboratory for Computer Science, the Army Research Office, IBM, the National Science Foundation, and the XEROX Corporation.[55] Norman Margolus, a PhD student in Fredkin's group, recorded and took notes of every lecture. These notes were then turned into articles and eventually published.

Before the conference, Feynman told Fredkin that he refused to focus his keynote on computers and physics, because computers and physics had nothing to do with each other. Physics is all about probability and randomness, Feynman said, whereas the whole goal of computing for the previous 50 years had been building machines that were reliable and predictable – the very opposite. Fredkin told Feynman that he could talk about anything he wanted, just come.

---

[55]E. Fredkin, Rolf Landauer, and Tom Toffoli, "Physics of Computation" (1982).

Fredkin recalls that when Feynman got up, the physicist started telling the story of how Fredkin had invited him to talk about computation and physics, and that he had refused to do so. "And I've changed my mind, and I'm going to talk about what he originally wanted," Feynman reportedly said in his matter-of-fact way.

Feynman's talk at the Endicott conference marks the birth of quantum computing, an idea that was unknowingly conceived by Feynman and Fredkin during Fredkin's year-long sabbatical at CalTech. It was a crazy idea. At roughly the same time that computer engineers were worrying that quantum mechanical effects in the form of quantum tunneling and uncertainty might pose real limits to computation by making machines act nondeterministically, Feynman proposed embracing the nondeterminism of quantum mechanics to build computers that could solve a problem that was simply too complicated to solve any other way – and that problem was quantum physics itself.

Feynman started his talk with a straightforward question: "What kind of computer are we going to use to simulate physics?"[56] After briefly suggesting that such a computer should have elements that are *locally connected* (like a cellular automata or a Thinking Machines' Connection Machine), he showed that the probabilistic nature of quantum physics means that quantum physics simulations necessarily have exponential complexity. The only way around this, Feynman said, was by using computing elements based on quantum mechanics itself, because the quantum wave equations would then match the systems that they were simulating. (Feynman says a lot of other things in his talk as well, but that's the gist of it.)

The rest of the conference was a fun mix of physics and computer science. Toffoli delivered a talk suggesting that physics might receive fresh insights from computing if computing is modeled with reversible computation.[57] Paul Benioff discussed and further developed his model of quantum mechanical Turing machines.[58] Fredkin and Toffoli significantly extended Bill Silver's MIT term paper and presented their ideas on Conservative Logic.[59] Danny Hillis presented his ideas on how to build massive computers using mesh networks

---

[56]Feynman, "Simulating Physics with Computers" (1982).

[57]Tommaso Toffoli, "Physics and Computation" (1982).

[58]Benioff, "Quantum Mechanical Hamiltonian Models of Discrete Processes That Erase Their Own Histories: Application to Turing Machines" (1982a).

[59]E. F. Fredkin and Tommaso Toffoli, "Conservative Logic" (1982).

with only local connectivity and routing – the basis of the Connection Machine that he was building.[60] Landauer discussed the impact of Heisenberg's Uncertainty Principle on the minimal energy requirements of a computer.[61] Marvin Minsky speculated that if the vacuum of the Universe is composed of discrete "cells, each knowing only what its nearest neighbors do," then "classical mechanics will break down ... and strange phenomena will emerge" such as the phenomena described by both relativity and quantum mechanics,[62] possibly pointing the way towards a theory of quantum gravity. Other contributions included those by Donald Greenspan,[63] and John Wheeler,[64] all of which appeared in two successive issues of the *International Journal of Theoretical Physics*. It was not a top journal, but it was the best peer-reviewed journal that would take the collection.

4.7 Russia and Quantum Computing

Invention is rarely a straight line, and insight rarely comes in a single flash. It is common for good ideas to be invented and re-invented.

As we have seen, Toffoli and Fredkin were developing reversible logic at roughly the same time that Paul Benioff developed the idea of a quantum Turing machine.[65] These academics soon found each other, thanks to the milieu of papers, conferences, phone calls and email that American academics enjoyed in the 1970s and 1980s.

What about on the other side of the Iron Curtain?

Historians of quantum computing frequently point out that in Russia, R. P. Poplavskii wrote a 1975 Russian-language article, "Thermodynamical Models of Information Processing"[66] in which it was observed that classical computers would be insufficient for simulating quantum systems that do not have a simple solution: "The quantum-mechanical computation of one molecule of methane requires 10^{42} grid points. Assuming that at each point we have to perform only

---

[60]W. D. Hillis, "New Computer Architectures and Their Relationship to Physics or Why Computer Science Is No Good" (1982).

[61]Rolf Landauer, "Physics and Computation" (1982).

[62]Minsky, "Cellular Vacuum" (1982).

[63]Greenspan, "Deterministic Computer Physics" (1982).

[64]Wheeler, "The Computer and The Universe" (1982).

[65]Benioff, "The Computer As a Physical System: A Microscopic Quantum Mechanical Hamiltonian Model of Computers As Represented by Turing Machines" (1980); Benioff, "Quantum Mechanical Models of Turing Machines That Dissipate No Energy" (1982b).

[66]Poplavskii, "Thermodynamical Models of Information Processing" (1975).

10 elementary operations, and that the computation is performed at the extremely low temperature $T = 3 \times 10^{-3}$ K, we would still have to use all the energy produced on Earth during the last century."[67] In 1980, Yuri Manin wrote *Vychislimoe i nevychislimoe* (*Computable and Uncomputable*), which further explored such ideas. The language barrier, combined with the very real travel barrier imposed by the Soviet Union, prevented these works from being influential in the West.

Today we can read excerpts of Manin's 1980 article in English, thanks to his leaving Russia and publishing an English-language 2007 edition of his essays. "We need a mathematical theory of quantum automata," Manin wrote. "Such a theory would provide us with mathematical models of deterministic processes with quite unusual properties. One reason for this is that the quantum state space has far greater capacity then the classical one: for a classical system with N states, its quantum version allowing superposition (entanglement) accommodates e^N states."[68]

Some journalists and historians of science cite these articles by Poplavskii and Manin as evidence for the idea that quantum computing arose on both sides of the Iron Curtain. However, these articles do not appear to have spawned conferences or investment in Russia, as their counterparts did in the United States (see discussion of nation-state investment in quantum information science in Section 9.2, "Industrial Policy" (p. 380)).

We believe that these publications are similar to Feynman's 1959 talk, in which Feynman posits that at the atomic scale computation can be performed not with circuits, "but some system involving the quantized energy levels, or the interactions of quantized spins."[69] Such a statement is a long way from Feynman's detailed proposals for quantum computing that would come two decades later, and there is no intellectual approach for drawing a line from Feynman's 1959 talk to modern-day quantum computing (or to modern-day nanotechnology, for that matter), because that line points back to Fredkin and Toffoli, and then to Burks and von Neumann. While Poplavskii and Manin were certainly walking down intellectually intriguing paths,

---

[67]As quoted in Manin, "Classical Computing, Quantum Computing, and Shor's Factoring Algorithm" (1999).

[68]Manin, *Mathematics As Metaphor: Selected Essays of Yuri I. Manin* (2007).

[69]Feynman, "There's Plenty of Room at The Bottom: An Invitation to Enter a New Field of Physics" (1959).

the historical record implies that their paths were never explored beyond the first few steps.

4.8 Aftermath: The Quantum Computing Baby

Feynman returned to California, where he delivered several more lectures on the promise of quantum computing. He published an article about the idea in a special publication marking the 40th anniversary of the Los Alamos laboratory;[70] a revised version appeared in Optics News.[71] Another version of the article appeared in Foundation of Physics the following year.[72,73]

4.8.1 Growing Academic Interest

Three years after the MIT conference, the British physicist David Deutsch wrote an article discussing the relationship between computing, physics, and the possibility of quantum computing for the *Proceedings of the Royal Society of London*, one of the world's oldest and most prestigious scientific journals. "Computing machines resembling the universal quantum computer could, in principle, be built and would have many remarkably properties not reproducible by any Turing machine,"[74] Deutsch hypothesized. The statement is literally true, because quantum computers as he proposed them would have access to both a source of perfect randomness and the ability to create entangled states. Such a machine *would* be able to model quantum physics and quantum chemistry to any arbitrary precision (discussed in Chapter 5), and create unbreakable cryptographic codes (discussed in Chapter 7). This article helped to legitimize the idea of quantum computing and present it to a broader scientific and technical community that had not previously encountered it. "To view the Church–Turing hypothesis as a physical principle does not merely

---

[70]Feynman, "Tiny Computers Obeying Quantum Mechanical Laws" (1985b).

[71]Feynman, "Quantum Mechanical Computers" (1985a).

[72]Feynman (1986).

[73]Feynman's son, Carl Feynman, was an MIT classmate of Danny Hillis. Feynman learned of Thinking Machines when the company was being formed and offered to spend the summer helping out. He was hired as a consultant shortly after the company was founded, becoming its first employee. Feynman soon found that the Connection Machine's mesh architecture was surprisingly well-suited to performing the complex computations required for simulating quantum mechanics and other kinds of physical systems, paving the way for the company's early sales. See W. D. Hillis, "Richard Feynman and The Connection Machine" (1989).

[74]Deutsch, "Quantum Theory, The Church–Turing Principle and The Universal Quantum Computer" (1985).

make computer science into a branch of physics. It also makes part of experimental physics into a branch of computer science."

Reading Deutsch's article 35 years after its publication, a confusing aspect is the fact that he differentiates a "quantum computer" from something he calls a "Turing-type machine." The article conveys that a Turing-type machine is limited in that it can only execute steps sequentially, while Deutsch suggests that a quantum computer will be able to solve some problems faster because it will be able to consider many states at once, in part because it is based on quantum computing, and "quantum theory is a theory of parallel interfering universes." What is confusing about this today is that the Church–Turing hypothesis is not concerned with the *speed* with which a computation can be performed – it is only concerned with whether a computation can be performed *at all*.[75] In 1984 it was not immediately clear whether quantum computers would face the same limitations of Turing machines, or if they might implement a stronger, more powerful form of computation. Today computer scientists have shown that quantum computers may be more efficient at solving certain kinds of problems, but they cannot solve problems that are fundamentally different than Turing machines – or if they can, we haven't figured out how to express such power.[76] Surprisingly, even this perceived efficiency of quantum computers is a belief – it has not been mathematically proven, for reasons described in the following chapter.

In 1985 Asher Peres at Technion, the Israel Institute of Technology, published an article further exploring how a quantum computer might do something extremely simple: adding together 1-bit numbers. In working through his example, Peres showed that a quantum mechanical computer would necessarily require some kind of error

[75] For example, a sequential Turing machine with a clock speed of a billion cycles per second is likely faster at computing problems than a parallel Turing machine with a thousand processors all running with a clock speed of a 10 cycles per second, but both machines are universal. By *universal*, we mean that either of these machines, given enough memory and enough time, could compute what any other Turing machine can compute.

[76] Quantum cryptography is fundamentally different from quantum computing, in that today we know mathematically that systems that use quantum cryptography can do something that it is simply impossible to do with conventional cryptography, and that is exchange messages in a way that they cannot be intercepted without detection by an attacker. However, Quantum cryptography is not strictly solving a problem, and it doesn't use quantum computing, so it doesn't disprove the sentence referenced in the paragraph above.

correction. Ideally, Peres wrote, with such a system "it should be *impossible to keep a record* of the error,"[77] because errors would ideally cancel out each other. He ended the article by noting that quantum computers need not be digital computers: "Ultimately, a quantum computer making full use of a continuous logic may turn out to be more akin to an old-fashioned analog computer, rather than to a modern digital computer. This would be an ironic twist of fate." (The D-Wave quantum computer resembles an analog computer; we discuss traditional analog computers in Chapter 3.)

In October 1992, the Dallas IEEE Computer Society and Texas Instruments sponsored the Workshop on Physics and Computation. "This workshop was long overdue since the first major conference on the Physics of Computation was held at MIT over a decade ago," wrote Doug Matzke, the workshop's chair. Landauer was the keynote sponsor; Fredkin "gave a stimulating and entertaining talk" at the banquet.[78] A follow-up conference was scheduled for two years later, in 1994.

In June 1994, Peter Shor, then a researcher at AT&T Bell Labs, published a technical report at the Center for Discrete Mathematics & Theoretical Computer Science (DIMACS), at the time a joint research project between Bell Labs and Rutgers University. An "extended abstract" based on the technical report was presented at the Foundations of Computer Science (FOCS) 1994 conference, which took place between November 20–22 in Santa Fe, New Mexico. Shor's paper showed that if a certain kind of quantum circuit could be built on an as-yet non-existent quantum computer, then laws of quantum mechanics could be combined with number theory in such a way as to solve a particular math problem very efficiently. Solving *that* particular math problem would make it possible to efficiently factor large numbers.[79] And factoring large numbers would have a huge impact on the world, because the world's most sophisticated encryption systems at the time (and still today) depended upon the fact

---

[77] Peres, "Reversible Logic and Quantum Computers" (1985).

[78] Matzke, "Message From The Chairman" (1993).

[79] Shor uploaded "an expanded version" of his FOCS paper to arXiv on August 30, 1995, and updated that version in January 1996. The papers can be found at arxiv.org/abs/quant-ph/9508027. This version of the paper was published as Shor, "Polynomial-Time Algorithms for Prime Factorization and Discrete Logarithms on a Quantum Computer" (1997).

that we are unable as a species, on Earth, today, to rapidly factor large numbers.

It is hard to overstate the significance of Shor's algorithm for the development of quantum computing. Before Shor's announcement and subsequent publication, quantum computers were nonexistent theoretical constructions that were largely a curiosity of the physics and theoretical computer science communities. Shor's algorithm showed that there would be serious, real-world implications for quantum computers that would directly impact national security. It was the starting gun of the quantum computing race. Charles Clark at the US National Institute of Standards and Technology organized the NIST Workshop on Quantum Computing and Communication, held in August 1994 at the agency's campus in Gaithersburg, Maryland.[80] Based on a discussion at the workshop, NIST had a working quantum circuit with two qubits based on trapped ions operational in July 1995.[81] (David J. Wineland, one of the paper's authors, would later share the 2012 Nobel Prize with Serge Haroche "for groundbreaking experimental methods that enable measuring and manipulation of individual quantum systems.")

Also in the summer of 1995, the MITRE Corporation's "JASON" summer study, funded by DARPA, focused on quantum computing. The report identified factoring and simulating quantum physics, but presented diagrams for how to create a quantum adder and multiplier, and discussed the importance of quantum error correction. The report had three main recommendations.

- "Establish a research program to investigate possibilities for quantum computing beyond Shor's algorithms."

- "Seed research in various communities for quantitative minimization of algorithmic complexity and optimum circuit."

- "Supplement ongoing experimental research related to the isolation and control of discrete quantum systems suitable for quantum logic."

---

[80]Gaithersburg, MD: National Institute of Standards and Technology, "NIST Jump-Starts Quantum Information" (2018).
[81]C. Monroe et al., "Demonstration of a Fundamental Quantum Logic Gate" (1995).

Also in 1995, Benjamin Schumacher coined the word *qubit* in his article "Quantum Coding."[82] In the article, Schumacher compares the information-theoretic differences between traditional bits of information and "Shannon entropy" and quantum bits, which had previously been called two-state quantum systems, and which Schumacher termed *qubit*. But whereas Shannon's seminal 1948 article[83] contemplated the information capacity of a noisy channel, Schumacher considered the information capacity of a noiseless quantum communications channel. He then considers the impact of entanglement between quantum states. In the article's acknowledgments, Schumacher notes: "The term 'qubit' was coined in jest during one of the author's many intriguing and valuable conversations with W. K. Wootters, and became the initial impetus for this work. The author is also grateful to C. H. Bennett and R. Jozsa for their helpful suggestions and numerous words of encouragement."

4.8.2 The First Quantum Computers

Three years after NIST created the first quantum circuit, two separate teams of researchers proposed, developed and published similar approaches for using nuclear magnetic resonance (NMR) in liquids as the medium for quantum computation.[84] "Although NMR computers will be limited by current technology to exhaustive searches over only 15 to 20 bits, searches over as much as 50 bits are in principle possible, and more advanced algorithms could greatly extend the range of applicability of such machines," observed Cory et al.

The challenge with NMR-based quantum computers is that the NMR spectrum increases in both complexity and density with each additional qubit. At some point the spectrum becomes too complex, and too noisy, to make sense of the computation's result. But these computing systems demonstrated that the theoretical ideas first proposed by Feynman and later refined by Shor actually worked: in 1998 the first algorithm was run on an NMR-based quantum computing system (see Section 5.3 (p. 210)), and in 2001 Shor's algorithm was run for the first time on an actual quantum computer, an NMR system with 7 qubits, successfully factoring the number 15 to get

---

[82]Schumacher, "Quantum Coding" (1995).

[83]Shannon, "A Mathematical Theory of Communication" (1948).

[84]Gershenfeld and Chuang, "Bulk Spin-Resonance Quantum Computation" (1997); Cory, Fahmy, and Havel, "Ensemble Quantum Computing by NMR Spectroscopy" (1997).

its prime factors, 3 and 5.[85] We will further discuss Shor's break-through and the race for quantum factoring in the next chapter (see Section 5.2, p. 188).

4.8.3 Coda

In the past 25 years, the world has seen quantum computers go from theoretical constructs to working machines that can solve real problems. But progress on quantum computers has been much slower than progress during the first 25 years of classical electronic computers.

The London Mathematical Society published Alan Turing's model for computation in 1936. By March 1940 Turing had built the first code-breaking Bombe at Bletchley Park. Together with the Colossus machines, Bletchley Park was able to decrypt thousands of messages a day, and had a significant impact on the war effort. In fact, the impact was so significant that the existence of these machines was kept secret for decades. Meanwhile, by the end of World War II there were stored program computers in various states of design, operation, and construction in Germany (where the effort was largely ignored by the Nazi military), the United Kingdom, and the United States. Early electronic computers used a variety of different technologies for computing and storage, including relays, tubes, and spinning magnetic drums, but the industry was profitable from the very start. By 1965 the industry had firmly settled upon transistorized logic. IBM manufactured the first hard drive in 1956, and in 1970 Intel publicly released the first commercial DRAM (dynamic random access memory) chip. Governments and corporations bought these computers to solve problems that required organizing information and performing computations.

Quantum computing, in contrast, was first proposed in the 1970s. It wasn't until 1994 that there was a clearly articulated reason for creating such a machine: not to simulate physics, but to crack codes. Unlike the first electromechanical and electronic computers, the first quantum computers could not crack any messages of any significance whatsoever: the most impressive mathematical feat that one of the machines accomplished was to factor the number 15 into the prime numbers 3 and 5. Unlike the work at Bletchley Park, the work on quantum computing has taken place in public, with multinational

---

[85]Vandersypen et al., "Experimental Realization of Shor's Quantum Factoring Algorithm Using Nuclear Magnetic Resonance" (2001).

Quantum Computers: Not Just Fancy Analog Devices

At the dawn of the computer age there was considerable interest in so-called *analog computers* for solving a variety of scientific problems. Some of these machines were mechanical, with rods, gears and curves milled into metal,[a] while others were electronic. Indeed, much of the recent success in artificial intelligence is based on a computing model that is essentially analog (and was first created with analog computers), and analog computers are making a comeback in some areas.[b]

But quantum computers are not simply a new take on analog computers:

- The physical things that represent information inside an analog computer are one-dimensional vectors, such as position (in mechanical analog computers) or voltage (in electronic analog computers). Quantum computers use two-dimensional vectors (the complex numbers used to compute quantum wave functions).

- Analog computers don't rely on superposition or entanglement, with the result that all of the information stored within an analog computer is not potentially interacting with all of the other information stored inside an analog computer. Put another way, the individual parts of a large analog computer appear to experience local causality and statistical independence; the lack of these makes quantum computing possible.

- As such, information can be copied out of an analog computer without destroying the information it contains. It is thus possible to covertly eavesdrop on an analog computer. Quantum computers and networks, in contrast, can detect eavesdropping because it destroys their computations.

- Analog computers can't efficiently run quantum algorithms such as Shor's algorithm or Grover's algorithm.

---

[a]Clymer, "The Mechanical Analog Computers of Hannibal Ford and William Newell" (1993).
[b]Tsividis, "Not Your Father's Analog Computer" (2017).

teams engaging in a friendly competition within the pages of scientific journals. Today, 23 years after the first successful quantum computation, there is still no agreement on what media should be used for quantum computation, and whether it is better to run machines in vats of liquid helium cooled close to absolute zero, or if they can be run at room temperature. Whereas technologies for storing digital information preceded Turing's paper by more than a century,[86] approaches for storing quantum information are still on the drawing board.

Unquestionably, computing with superposition and wave equations that we describe in Appendix B – what we call quantum computing – is much harder than computing with relays, tubes, and transistors – classical computing – that we describe in Chapter 3. Having recounted the history of quantum computing from 1961 through 1998, we explain in the next chapter why governments and corporations continue to pursue quantum computing. We discuss the kinds of devices being made, their intended uses, the competitive landscape, and the outlook for the technology.

---

[86]Joseph Marie Jacquard (1752–1834) patented his punch-card operated loom in 1804.

Quantum Computing Applications

"A good way of pumping funding into the building of an actual quantum computer would be to find an efficient quantum factoring algorithm!"[1]

THE risk of wide-scale cryptanalysis pervades narratives about quantum computing. We argue in this chapter that Feynman's vision for quantum computing will ultimately prevail, despite the discovery of Peter Shor's factoring algorithm that generated excitement about a use of quantum computers that people could understand – and dread.

To explain this outcome, we canvass the three primary applications that have been developed for quantum computing: Feynman's vision of simulating quantum mechanical systems, factoring, and search. The next chapter discusses today's quantum computing landscape.

For Feynman, a quantum computer was the only way that he could imagine to efficiently simulate the physics of quantum mechanical systems. Such systems are called *quantum simulators*.[2] Quantum simulation remains the likely first practical use of quantum comput-

---

[1] Berthiaume and Gilles Brassard, "Oracle Quantum Computing" (1994), written hours before Peter Shor discovered such an algorithm.

[2] The term *quantum simulators* is confusing, because it is also applied to programs running on conventional computers that simulate quantum physics. For this reason, some authors use the terms *Feynman simulators* or even *Schrödinger–Feynman simulators*.

ers. Oddly, this application is *not* responsible for most of the public interest in quantum computers, which has instead been fueled by the desire to make super-machines that can crack the world's strongest encryption algorithms. Since then, without dramatic demonstrations of other capabilities, and with the underlying complexity of achievements that have been made, many news articles cast quantum computing in a single, privacy-ending narrative.

We believe that prominence of cryptanalysis in public interest and government funding over the past two decades is because a working quantum computer that could run Shor's algorithm on today's code would give governments that owned it an incredible advantage to use over their adversaries: the ability to crack messages that had been collected and archives going back decades. But while this advantage may be responsible for early funding of quantum computing, we believe that the cryptanalytic capabilities of initial quantum computers will be limited and outshone by the ability of these machines to realize Feynman's vision. And Feynman's vision, unlike cryptanalysis, confers first-mover advantage, since a working quantum physics simulator can be used to build better quantum physics simulators. That is, quantum physics simulations are likely to create a virtuous circle, allowing the rate of technology change to increase over time.

The last section of this chapter turns to search, and explains the kinds of speedups quantum computers are likely to provide. Understanding those likely speedups further advances our prediction that the future of quantum computing will be Feynman's.

5.1 Simulating Physical Chemistry

In this section we explore how one might actually go about simulating physics with quantum computers. Despite the similarity of titles, this section is not an extended discourse on Feynman's articles. Instead, it is a discussion of how chemists actually simulate the physics of chemical reactions with classical computers today, and how they might do so with quantum computers tomorrow.

Classical computers – like the computers used to write and typeset this book – are designed to execute predetermined sequences of instructions without error and as reliably as possible. Computer engineers have made these machines steadily faster over the past 80 years, which makes it possible to edit this book with graphical editors and typeset its hundreds of pages in less than a minute. Both of those activities are fundamentally a sequence of operations applied to a sequence of bits, starting with an input stream of 0 s and 1 s, and possibly a character typed on a computer keyboard) and deterministically creating a single output stream (the PDF file that is displayed on the computer's screen).

Modeling molecular interactions is fundamentally different from word processing and typesetting. When your computer is running a word processing program and you press the [H] key, there is typically only one thing that is supposed to happen: an "H" appears at the cursor on the screen. But many different things can happen when two molecules interact: they might stick together, they might bounce, or an atom might transfer from one molecule to the other. The probability of each of these outcomes is determined by quantum physics.

To explore how two molecules interact, the basic approach is to build a model of all the atomic nuclei and the electrons in the two-molecule system and then compute how the wave function for the system evolves over time. Such simulations quickly become unworkable, so scientists will consider a subset of the atoms and electrons, with the hope that others will stay more-or-less static. This hope was formalized in 1927 and today is known as the Born–Oppenheimer approximation, named after Max Born and J. Robert Oppenheimer who jointly proposed it. Other approximations exist, such as assuming that the nuclei are fixed in space and are point charges, rather than wave functions themselves. High school chemistry, which typically presents the electrons as little balls of charge spinning around the nuclei, is a further simplification.

Many of the chemistry discoveries of the twentieth century were possible because the Born–Oppenheimer approximation is largely correct, but it is not perfect. For example, it may not apply in exotic materials, such as graphene.[3] More generally, it may not apply to certain kinds of surface chemistry. "There is growing evidence that the usual approach to modelling chemical events at surfaces is incomplete – an important concern in studies of the many catalytic processes that involve surface reactions."[4]

Most chemistry can be understood working with the time-independent Schrodinger equation, in which the chemist simply looks for the most likely configuration of the atoms and electrons. Systems

---

[3]Pisana et al., "Breakdown of The Adiabatic Born – Oppenheimer Approximation in Graphene" (2007).
[4]Sitz, "Approximate Challenges" (2005).

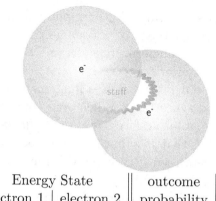

| Energy State | | outcome |
electron 1	electron 2	probability
low	low	2%
low	high	5%
high	low	90%
high	high	3%

Figure 5.1. The possible energy states of two electrons in a hypothetical quantum system.

that cannot be studied this way can be modeled using what's called Monte Carlo methods (or a Monte Carlo simulation), in which the chemist creates a probabilistic model and then runs the simulation multiple times, examining the range of possible outcomes.

We present a simplified Monte Carlo simulation in Figure 5.1. To keep things simple, we have assumed that there are only two electrons of interest, and that each will end up in either a *low* or *high* energy state. Facing this system, a scientist can use modeling software to determine the probability of each of these outcomes. Here our hypothetical scientist has used a conventional computer to run this experiment many times, tabulate the results, and report them in the rightmost column as an *outcome probability*.

Our scientist would take a fundamentally different approach to solve this problem on a quantum computer. Instead of modeling the probabilities, the scientist designs a *quantum circuit* that directly represents (or simulates) the chemistry in question. With most quantum computers today, the scientist would then turn on the quantum computer, placing each of its quantum bits (called *qubits*) into a superposition state. The quantum circuit plays through the quantum computer, changing how the qubits interact with each other over time. This "playing" of the quantum circuit is performed by a second computer – a classical computer – that controls the quantum computer. When the circuit is finished playing, the second computer measures each qubit, collapsing the superposition wave function and revealing its quantum state. At this point each qubit is *either* a 0 or a 1 .

In this example, each qubit might directly represent an energy state of an electron that was previously modeled. So if our scientist designed a quantum circuit and ran it on our hypothetical quantum computer, the result might look like this:

Trial	qubit 1	qubit 2
#1	1	0

It looks like the quantum computer has found the right answer instantly!

Actually, no. Because if the scientist ran the experiment a second time, the answer might be different:

Trial	qubit 1	qubit 2
#2	1	1

In an actual quantum computer, the experiment would run multiple times:

Trial	qubit 1	qubit 2
#3	1	0
#4	1	0
#5	0	0
#6	1	0
#7	1	0
#8	1	0
#9	1	0
#10	1	0

After these trials, the results are tabulated to get a distribution of possible answers. The statistics that are similar to those produced by the classical computer, but a little different:

qubit 1	qubit 2	Trial #s	Count	Probability
0	0	#5	1	10%
0	1	–	0	0%
1	0	#1, #3, #4, #6, #7, #8, #9, #10	8	80%
1	1	#2	1	10%

Notice that the quantum computer does not generally produce the same results as the classical computer. This may be because we did not run sufficiently many trials to get results with the same statistical distribution as the results produced by the classical computer. It might also be because the model run on the classical computer is incomplete. More likely, both models are incomplete, but incomplete in different ways. (Even if they were identical models, it's unlikely that identical statistics would emerge with just 10 runs.)

It is important to remember that in this simulation, as in real quantum systems, *there is no right answer.* Instead, there is a range of possible answers, with some more probable and some less probable.

In practice, efficient quantum computing algorithms are designed so that "correct" or desired answers tend to generate constructive interference on the quantum computing circuits, while answers that are not desired tend to cancel each other out with destructive interference. This is possible because what quantum computers actually do is to evolve carefully constructed probability waves in space and time. These waves "collapse" when the final measurement is made by the scientist (or, more specifically, by the classical computer that is controlling the quantum computer). For a discussion of quantum mechanics and probability, please see Appendix B.

The advantage of a quantum computer becomes clear as the scale increases. Exploring the interaction of 32 electrons, each of which could be in two states, requires exploring a maximum of 4 Gi[5] combinations. A classical computer would need to explore *all* of those combinations one-by-one. Exponential growth is really something: simply printing out those 4 Gi combinations at 6 lines per inch would consume 11 297 linear miles of paper. Today for certain problems, quantum computing scientists have discovered algorithms that run more efficiently on quantum computers than the equivalent classical

---

[5]4 Gi means 4 Gigi, which is the SI prefix that denotes powers-of-two rather than powers-of-ten counting. 4 Gi is $4 \times 1024 \times 1024 \times 1024 = 2^{32} = 4\,294\,967\,296$, or roughly 4.2 billion.

algorithms that exist to solve the problems on conventional computers. Generally speaking, the more qubits a quantum computer has, the more complex a system it can simulate.

Approaches for programming quantum computers are still in their infancy. Because the machines are small – with dozens of qubits, rather than millions – programmers need to concern themselves with individual qubits and gates. In some notable cases quantum computers are being constructed to solve specific problems.[6] This is reminiscent of the way that the first computers were built and programmed in the 1940s, before the invention of stored programs and computer languages: in England the Colossus computers were built to crack the Germans' Lorentz code, while in the US the ENIAC was created to print artillery tables. Programming quantum computers will get easier as scientists shift from single-purpose to general machines and as the machines themselves get larger.

In addition to the number of qubits, the second number that determines the usefulness of a modern quantum computer is the stability of its qubits. Stability is determined by many things, including the technology on which the qubits are based, the purity of the materials from which the qubits are manufactured, the degree of isolation between the qubits and the rest of the universe, and possibly other factors. Qubits that are exceedingly stable could be used to compute complex, lengthy quantum programs. Such qubits do not currently exist. In fact, an entire research field explores ways to shorten quantum algorithms so that they are compatible with short-lived qubits.

Quantum engineers use the word noise to describe the thing that makes qubits less stable. *Noise* is a technical term that engineers use to describe random signals. The reason we use this term is that random signals fed into a speaker literally sound like a burst of noise, like the crackle between stations on an AM radio, or the sound of crashing waves. Noise in the circuit does not help the quantum computer achieve the proper distributions of randomness and uncertainty described by quantum mechanics. Instead, noise collapses the wave functions and scrambles the quantum computations, similar to the way that jamming the relay contacts in the Harvard's Mark II computer caused it to compute the wrong numbers on September 9, 1947.[7] Early computers only became useful after computer engineers

---

[6]Zhong et al., "Quantum Computational Advantage Using Photons" (2020).

[7]A moth was found pinned between the contacts of Relay #70 Panel F. Grace Hopper, a developer and builder of the Mark II, taped the insect into her laboratory

Quantum Error Correction

The quantum computing applications that we discuss in this chapter all assume the existence of a working, reliable quantum computer with sufficient qubits, able to run quantum circuits with sufficient size and complexity for a sufficiently long period of time.

Although an absolutely reliable quantum computer is a useful theoretical construct for thinking about quantum computing algorithms, actual quantum computers will probably need to use some form of *quantum error correction*, in which multiple noisy qubits are used to simulate a smaller number of qubits that have less noise.

Although quantum error correction is powerful, today's techniques do not appear to be up to the task of sustaining a single quantum computation for time periods that would be long enough to pose a threat to modern cryptographic systems.

learned how to design circuits that reduced noise to the point of irrelevance. They did this using an engineering technique called *digital discipline* that is still used today (see p. 84), but that approach won't work with quantum computers.

Instead, companies like Google, IBM, and Rigetti have created machines that have noisy qubits. As a result, most quantum programs today are small and designed to run quickly. Looking towards the future, many noisy qubits can be combined to simulate cleaner qubits using an error-correcting technique called surface codes,[8] but today's machines do not have enough sufficient noisy qubits for this to be practical. Another approach is to use a quantum computing media that is largely immune to noise; that's the approach being taken by Microsoft with its so-called *topological qubits*, although other approaches using photonic qubits or ion traps might produce similar noise-free results. But for today, noise significantly limits the complexity of computations that can be done on quantum computers, even if we could build machines with hundreds or thousands of noisy qubits.

---

notebook with the notation "first actual case of bug being found."

[8]Fowler et al., "Surface Codes: Towards Practical Large-Scale Quantum Computation" (2012).

Even so, some companies are eager to get a head start, and are having their scientists and engineers learn to program these machines today. As a result, IBM is able to generate revenue with its "quantum experience" by giving free access over the Internet to machines with only a few qubits, and renting time to institutions who want access to IBM's larger machines. Likewise, Amazon Web Services has started making small quantum computers built by other companies available through its "Bracket" cloud service. However, the power of these machines is dwarfed by Amazon's conventional computing infrastructure.

Finally, there is an important point that we need to make: there is no mathematical *proof* that a quantum computer will be able to simulate physics faster than a classical computer. The lack of such a proof reflects humanity's fundamental ignorance on one of the great mathematical problems of time, *NP completeness* (see Section 3.5.4, "NP-Complete and NP-Hard" (p. 110)). What we do know is that *today's* quantum simulation algorithms get exponentially slower as the size of the problem being simulated increases in size, and the simulation algorithms that we have designed for quantum computers do not. But this may reflect the limits of our knowledge, rather than the limits of classical computers. It might be that work on quantum computing leads to a breakthrough in mathematics that allows us to create dramatically faster algorithms to run on today's classical computers. Or it may be that work on quantum computing allows us to prove that quantum computers really are fundamentally more powerful than classical computers, which would help us to solve the great mathematical question of NP completeness. What we know today is that quantum computers can take advantage of quantum physics to run so-called *BQP algorithms*, and that today's BQP algorithms run more efficiently than the fastest algorithms that we know of to run on classical computers. (See Section 3.5.4 (p. 110) and Section 3.5.6 (p. 116) for a more in-depth discussion of these topics.)

5.1.1 *Nitrogen Fixation, without Simulation*

To put efforts to develop a quantum computer into context, this section explores how such a machine might help develop more efficient approaches for "fixing" nitrogen.

Nitrogen, in the form of organic nitrates, is both vital for biological life and in surprisingly short supply. The productivity of pre-industrial agriculture was often limited by the lack of nitrogen, rather than limitations of water or sunlight. Industrial agriculture has solved this problem through the industrial production of nitrogen-based fertilizers.

The need for added nitrogen is surprising given the fact that plants are surrounded by nitrogen in the form of air. Nearly 80 percent of dry air is nitrogen. The problem is that nitrogen in the air is N_2, also written $N \equiv N$, with a triple chemical bond between the two nitrogen atoms. This triple bond has the charge of six electrons, making it difficult to break. As a result, the nitrogen in air is inaccessible to most plants.

Nitrogen fixation is the process of taking N_2 and turning it into a more usable form, typically ammonia (NH_3). The overall chemical reaction is not very complex:

$$Energy + N_2 + 3\,H_2 \longrightarrow 2\,NH_3 \qquad (1)$$

Most of the natural nitrogen fixation on Earth happens in the roots of alfalfa and other legumes, where nitrogen-fixing bacteria live in a symbiotic relationship with the plant host.[9] Instead of hydrogen gas, biological nitrogen fixation uses ATP (adenosine triphosphate) produced by photosynthesis, some spare electrons, and some hydrogen ions (present in acid) that just happen to be floating around. The products are ammonia (containing the fixed nitrogen), hydrogen gas, ADP (adenosine diphosphate), and inorganic potassium (written as Pi below):

$$N_2 + 16\,ATP + 8\,e^- + 8\,H^+ \longrightarrow 2\,NH_3 + H_2 + 16\,ADP + 16\,Pi \qquad (2)$$

The plant then uses photosynthesis and sunlight to turn the ADP back into ATP.

In 1909, the German chemist Fritz Haber discovered an inorganic approach to nitrogen fixation using high pressure and the chemical element osmium, which somehow helps the electrons to rearrange. Chemists say that osmium *catalyzes* the reaction. Haber was awarded the Nobel Prize in Chemistry in 1918, "for the synthesis of ammonia from its elements."[10]

---

[9] There is also a small amount of nitrogen fixation that results from lightning.
[10] Haber is also known as the "father of chemical warfare" for his work weaponizing the production and delivery of chlorine gas as part of Germany's efforts during World War I, and for his institute's development of Zyklon A. Despite this service

Haber sold his discovery to the German chemical firm BASF, which assigned Carl Bosch the job of making the process commercially viable. Osmium has 76 electrons that are exquisitely arranged, which presumably is the reason for its catalytic prowess, but it is also one of the rarest chemicals on the planet, so Bosch and his colleague looked for a cheaper catalyst. They discovered that uranium also worked, but settled on a catalyst made by treating iron with potassium. (Iron is in the same column of the periodic table as Osmium because they have the same arrangement of "outer" electrons, with the result that they have some similar chemical properties.) Today modern industrial catalysts for nitrogen fixation include mixtures of aluminum oxide (Al_2O_3), potassium oxide (K_2O), zirconium dioxide (ZrO_2), and silicon oxide (SiO_2). For this work, Carl Bosch received the 1931 Nobel Prize in Chemistry, which he shared with Friedrich Bergius, another BASF employee.

Chemically, the modern Haber–Bosch process looks something like this:

$$Energy + N_2 + 3\,H_2 \xrightarrow{\;Fe,\,Fe_3O_4,\,Al_2O_3\;} NH_3 + H_2 \tag{3}$$

The energy comes from temperatures in the range from 750°F to 3000°F, with pressures as great as 350 times atmospheric pressure at sea-level, and the hydrogen comes from natural gas. Today the world is so hungry for nitrogen that the Haber–Bosch process is responsible for 3 percent of the world's carbon emissions and consumes roughly 3 percent of the world's natural gas. Not surprisingly, scientists are constantly looking for ways to improve nitrogen fixation. Areas of current research including finding better catalysts[11] and how biolog-

---

to his country and the fact that he had converted from Judaism to Christianity, Haber was considered a Jew by the Nazi regime, and fled to England after the Nazis rose to power. "[S]cientists there shunned him for his work with chemical weapons. He traveled Europe, fruitlessly searching for a place to call home, then suffered heart failure in a hotel in Switzerland in 1934. He passed away shortly thereafter at the age of 65, but not before repenting for devoting his mind and his talents to wage war with poison gasses." See King, "Fritz Haber's Experiments in Life and Death" (2012). Zyklon A ultimately led to the development of Zyklon B, the gas that was used in the Nazi extermination camps.

[11] Ashida et al., "Molybdenum-Catalysed Ammonia Production with Samarium Diiodide and Alcohols or Water" (2019).

ical systems work.[12,13,14] After all, alfalfa is able to fix nitrogen at room temperature with just air, water, sunlight, and some clever microbes.

5.1.2 Modeling Chemical Reactions

One way for industry to develop improved nitrogen fixation catalysts would be to better understand what is happening at the atomic level when nitrogen gas becomes ammonia inside those microbes. Chemists think of this process in terms of some chemical bonds being broken while new chemical bonds are created. Much of modern chemistry is devoted to describing and predicting the behavior of such chemical bonds.

Except there is really no such thing as a chemical bond! While students in high school chemistry class learn to visualize bonds as little black lines connecting letters (e.g., $N \equiv N$), "bonds" and indeed our entire model of chemical reactions are really just approximations for Schrödinger wave equations that evolve over time and describe the probability that a collection of mass, charge and spin will interact with our measuring devices. It is just far too hard to write down such wave equations, let alone solve them. Meanwhile, the mental models of chemical bonds and other approximations developed over the past 150 years all work pretty well, especially with ongoing refinements, and so chemists continue to use these approximations.[15]

More accurate models that do a better job incorporating the underlying quantum physics would let chemists create more accurate predictions of how these things we call atoms rearrange during the course of a chemical reaction. Highly accurate models would let chemists design and try out catalyst candidates in a computer, with-

---

[12]Molteni, "With Designer Bacteria, Crops Could One Day Fertilize Themselves" (2017).

[13]*Biological Nitrogen Fixation: Research Challenges – A Review of Research Grants Funded by The US Agency for International Development* (1994).

[14]Manglaviti, "Exploring Greener Approaches to Nitrogen Fixation" (2018).

[15]A current textbook about the chemical bond reminds its readers that there are no electrons spinning around the atoms, only a "charge wave surrounding the nucleus." (I. D. Brown, *The Chemical Bond in Inorganic Chemistry: The Bond Valence Model, 2nd ed.* (2016), Chapter 2.) (See Figure 5.2 in this book, p. 186.) Nevertheless, the author continues, "chemists have largely rejected this simple wave picture of the atom in favor of a hybrid view in which the charge is composed of a collection of electrons that are not waves but small particles, [with the] density of the charge wave merely represent[ing] the probability that an electron will be found at a given location."

out having to go to the trouble of actually synthesizing them in a lab. This is the world of *computational chemistry*, also called quantum chemistry, or even computational quantum chemistry, which uses the math of quantum mechanics to answer questions about the chemical nature of the world around us.

Wave equations describe probabilities, so predicting the behavior of atoms at the quantum level requires programs that explore probability distributions. One way to do this is with a Monte Carlo simulation (see the sidebar "The Monte Carlo Method" on page 189). Simulations take exponentially longer to run as the number of electrons in the system increases – a good rule of thumb is that each additional electron doubles the simulation's running time.

In the Haber–Bosch nitrogen fixation equation presented above, there are 14 electrons among the two nitrogen atoms and 6 hydrogen electrons for a total of 20 electrons. But do not forget that all-important catalyst: that is where the chemical dance of the electrons is happening. Iron has 26 electrons per atom, while Fe_3O_4 has 110, and Al_2O_3 has 50. There must be some extraordinarily complex chemistry happening at the interface of the gaseous nitrogen and the solid catalyst.

To understand that complex chemistry, a computational chemist creates a simulation of the electrons and nuclei. Into the simulation the chemist programs physical constants that have been measured over the decades as well as mathematical functions that represent the laws of quantum mechanics. The more electrons and nuclei, the more complex the simulation.

The math of quantum physics is based on probability, so all of those probabilistic interactions – many coin flips – become inputs to the simulation. For example, some of the random draws might have less electron charge in a particular location between the two nitrogen nuclei and more charge between the nitrogen and iron nuclei that are interacting with some oxygen. This might sometimes push the two nitrogen nuclei slightly further apart – their electrostatic charges repel, after all – which might sometimes cause the charge probability to rearrange a little more, and then all of a sudden ... *wham!* ... the two nitrogen nuclei can now pick up some free floating protons, and the physics simulation has converted simulated nitrogen into simulated ammonia!

Running this simulation with a classical computer requires many random draws, many crunchings of quantum mathematics, and a lot

Figure 5.2. McMaster University Professor Emeritus I. David Brown observes: "An electron is the smallest quantum of charge that can have an independent existence, but the free electrons that are attracted to a nucleus in order to form a neutral atom cease to exist the moment they are captured by the nucleus. They are absorbed into the charge wave and, like Lewis Carroll's (1865) Cheshire Cat that disappears leaving only its smile behind, the electron disappears bequeathing only its conserved properties: charge, mass and spin, to the charge wave surrounding the nucleus." I. D. Brown, *The Chemical Bond in Inorganic Chemistry: The Bond Valence Model, 2nd ed.* (2016), chapter 2.

of matrix mathematics. Remember, classical computers are *deterministic by design*. To explore what happens when four random variables encounter each other, the computer takes random draws on each four variables and crunches the math. One cannot simply explore what happens when the most-probable value of each variable happens, because there might be some important outcome when three of the variables are in a low-probability configuration.

If it takes 10 seconds to simulate a single random variable, it will take on the order of $10 \times 10 \times 10 \times 10 = 10^4 = 1000$ seconds to simulate four random variables. With 10 random variables (and without any optimization), it will take 10^{10} seconds or $115\,740$ days – roughly 317 years.

These days, a computation that takes 317 years is not a big deal, provided that the computation consists of many individual problems that can be run in parallel. Good news: quantum simulations are such a problem! As we write this book in 2021, cloud providers will rent a computer with 96 cores for roughly \$5/hour. One can rent 100 of those computers for \$500/hour and solve the 317-year problem in 12 days for \$6000. Alternatively, one can rent 1000 of those computers and solve the problem in 29 hours – for the same price of \$6000. (This demonstrates why cloud computing is so attractive for these so-called *embarrassingly parallel* workloads.)

Today's massive cloud computing data centers provide only *linear* speedup for these hard problems: if 1000 computers will solve the problem in 29 hours, then $10\,000$ computers will solve the problem in 2.9 hours. And there's the rub: absent a more elegant algorithm, each additional electron in our hypothetical simulation increases the problem's difficulty *exponentially*. With 20 electron variables, the problem takes on the order of 10^{20} seconds or $3\,168\,808\,781\,402$ years – 3168 billion years! – which is more time than anyone has.[16] Even with a million 96-core computers (a speedup of 96 million), our hypothetical computation would take $33\,008$ years, which is still too long. Classical computers are simply not well-suited to simulating probabilistic quantum physics.

Some people believe that quantum computers may be able to efficiently solve problems involving quantum modeling of chemical reactions. Even the "quantum simulators" discussed here, special-

---

[16]Current estimates are that the universe is somewhere between 15 and 20 billion years old.

purpose machines constructed to solve a specific problem, should be dramatically faster than all of the world's computers working forever ... provided that we can scale the quantum simulators to be large enough. As such, quantum chemistry simulation is likely to be the first application for quantum computers in which they are used for something other than doing research and writing papers about quantum computers.

Critics, meanwhile, argue that today's software packages (both commercial and open-source) are based on well-understood, validated approximations that have worked for decades, and that limitations of these systems might be solved merely with more conventional computing power. For example, a September 2020 article by Elfving et al. works real-world physical chemistry problems and concludes that a practical quantum computer that could solve these problems in hours, rather than years, would require millions of physical qubits. The authors' nuanced conclusion is that while quantum computing may one day produce systems that can make meaningful contributions to physical chemistry, a far more promising near-term solution would be to rewrite today's chemistry simulation packages to take advantage of graphical processing units.[17]

5.2 Quantum Factoring (Shor's Algorithm)

As we explained in Section 4.8, "Aftermath: The Quantum Computing Baby" (p. 164), Peter Shor's discovery of an algorithm that can rapidly break numbers down into their prime factors sparked the world's interest in quantum computing. In this section we will describe why Shor's algorithm was so important, how it became a driver of quantum computing, and why it is no longer a driver – at least, not in the public, commercial world. (See Section 3.5.6 (p. 116) for a discussion of what we mean by "rapidly.")

To understand why Shor's algorithm is such a big deal, we start with a discussion of public key cryptography. In Section 5.2.3 (p. 199) we discuss how a quantum computer makes factoring faster. We will then explore whether Shor's algorithm running on a quantum computer would truly be faster than anything that could ever run on a classical computer, or whether we just need better math.

---

[17]Elfving et al., "How Will Quantum Computers Provide an Industrially Relevant Computational Advantage in Quantum Chemistry?" (2020).

The Monte Carlo Method

Modeling nuclear reactions was one of the first uses of electronic computers in the 1940s. Stanislaw Ulam at Los Alamos was trying to create a mathematical model for the movement of neutrons through material. He couldn't create an exact model, so he ran hundreds of individual mathematical experiments, each modeling the probabilistic interactions between a neutron and the material and finding a slightly different path. Ulam called this the *Monte Carlo method*, named after the casino where his uncle frequently gambled.[a]

Ulam shared his idea with fellow scientist John von Neumann, who directed the team at University of Pennsylvania to program the ENIAC to carry out the computations.

One requirement of algorithms like Monte Carlo is that the random numbers must be truly random. Generating such numbers requires physical randomness, something that the early computers didn't have. Instead, the systems of the day used algorithms to generate sequences of numbers that *appeared* random, but which were actually determined from the starting mathematical "seed." von Neumann later quipped: "Anyone who considers arithmetical methods of producing random digits is, of course, in a state of sin."[b]

It is necessary to use algorithms such as the *Monte Carlo method* when modeling quantum interactions, because it is not possible to solve the Schrödinger wave equation for even mildly complex systems.[c]

Ulam's success was evidenced by the fusion bomb test in November 1952 and decades of employment for physicists at weapons laboratories around the world. By the 1990s modeling had gotten so good that it was no longer necessary to even test the bombs, and the United States signed (but did not ratify) the Comprehensive Nuclear-Test-Ban Treaty.

---

[a]Metropolis, "The Beginning of The Monte Carlo Method" (1987).

[b]von Neumann, "Various Techniques Used in Connection with Random Digits" (1951).

[c]Random sampling can also be used to find approximate integrals to complex mathematical functions: instead of attempting to find an exact solution, the approach is to evaluate the function at a number of randomly chosen locations and interpolate. This is similar to statistical sampling, except that what's being sampled is a *mathematical universe*, rather than a universe of people or objects.

5.2.1 An Introduction to Cryptography

In modern usage, we use the word "cryptography" to describe the body of knowledge involved in creating and solving secret codes. Here the word "code" means a system for representing information, while "secret" implies that something about the code allows people who know the secret to decode its meaning, while people who do not know the secret cannot.

Secret Key Cryptography

One of the oldest known codes it the "Caesar cipher," which was reportedly used by Julius Caesar for messages to his generals. Messages are encrypted character-by-character by shifting each letter forward in the alphabet by three positions, so T becomes Q, H becomes E, E becomes B, the letter C wraps around to Z, and so on. To decrypt messages simply shift in the other direction. QEB ZXBPXO ZFMEBO FP KLQ SBOV PBZROB, that is, THE CAESAR CIPHER IS NOT VERY SECURE.

The Caesar cipher is called a *secret key algorithm* because the secrecy of the message depends upon the secrecy of the key, and the same key is used to encrypt and decrypt each message. It's not a very good secret key algorithm, because once you know the secret – shift by three – you can decrypt any encrypted message. We call this number three the *key* because it is the key to decrypting the message! You can think of the Caesar cipher as a lock that fits over the hasp used to secure a wooden box, and the number *three* as a key that opens the lock.

We can make the algorithm marginally more complicated by allowing the shift to be any number between 1 and 25: that creates 25 possible encryption keys, so an attacker needs to figure out which one is in play. It's still not very hard to crack the code.

There are lots of ways to make this simple substitution cipher stronger, that is, to make it harder for someone to decrypt or "crack" a message without knowing the secret piece of information used to encrypt the message in advance. This is directly analogous to making the lock on the box stronger. For example, instead of shifting every letter by the same amount, you can make the encrypted alphabet a random permutation of the decrypted alphabet. Now you have a word puzzle called a cryptogram. These can be easy or hard to solve depending on the length of the message, whether or not the message uses common words, and the number of times each letter is present in the message.

Humans solve these puzzles by looking for patterns in the encrypted message, called a *ciphertext*. We can eliminate such patterns by encrypting each letter with a different key. Now there are no patterns! This kind of encryption algorithm is sometimes called a Vernam cipher (named after its inventor, Gilbert Vernam) or more commonly a *one-time pad* (because spies of yore had encryption keys written on pads of paper, with instructions to use each key once and then destroy it). One-time pads are hard to use in practice, because the key needs to be both truly random and as long as the original message. We discuss them more in Section 7.4 (p. 276).

Public Key Cryptography

For all of human history until the 1970s, cryptography existed as a kind of mathematical deadbolt, in which each encrypted message was first locked and then later unlocked by the same key. There were thus four principal challenges in creating and deploying a working encryption system: 1) Assuring that the sender and the intended recipient of an encrypted message had the same key; 2) Assuring that no one else had a copy of the correct key; 3) Assuring that the correct key could not be guessed or otherwise discovered by chance; 4) Assuring that the message could not be decrypted without knowledge of the key. (See Figure 5.5.)

All of this changed in the 1970s with the discovery of public key cryptography, a term used to describe encryption systems in which a message is encrypted with one key and decrypted with a second.

Originally called *non-secret* encryption, it is now generally believed that public key cryptography was discovered in 1973 by James Ellis, Clifford Cocks, and Malcolm Williamson[18] at the Government Communications Headquarters (GCHQ), the United Kingdom's signals intelligence and information assurance agency (roughly the UK's equivalent of the US National Security Agency (NSA)). The UK intelligence agency reportedly shared the discovery with the NSA,[19] but neither sought to exploit the invention. The basic idea was then rediscovered at Stanford by Professor Whitfield Diffie and Professor Martin Hellman, whose paper "New Directions in Cryptography" in-

---

[18]Ellis, Cocks, and Williamson, "Public-Key Cryptography" (1975).
[19]Levy, "The Open Secret" (1999).

spired Ronald Rivest, Adi Shamir, and Leonard Adleman at MIT to create a working public key system.[20,21]

The basic concept of public key cryptography is a mathematical lock that is locked with one key and unlocked with a second. The key that locks (encrypts) is called the *public key*, while the key that unlocks (decrypts) is the *private key*. The two keys are mathematically linked and need to be made at the same time.[22]

A locked suggestion box is a good mental model for how public key cryptography works: to encrypt something, write it on a piece of paper and drop it into the locked box. Now the only way to get that message back is by unlocking the box and retrieving the message. In this example, the slot in the box represents the public key, and the key that unlocks the padlock represents the private key (Figure 5.3).

The great advantage of public key cryptography is that it dramatically simplifies the problem of key management. With public key cryptography, each person in an organization simply makes their own public/private keypair and then provides their public key to the organization's central registry, which then prints a phone book containing each employee's name and public key, then sends each employee their own copy. Now any employee can send an encrypted message to any other employee by simply looking up the intended recipient's

---

[20]Ronald L. Rivest, Adi Shamir, and Len Adleman, "A Method for Obtaining Digital Signatures and Public-Key Cryptosystems" (1978).

[21]The RSA crypto system was published first in Martin Gardner's column in *Scientific American* (Gardner, "Mathematical Games: A New Kind of Cipher That Would Take Millions of Years to Break" (1977)), in which the RSA-129 number that we will discuss on p. 261 was first published. In that article, the MIT professors famously offered US$100 to anyone who could factor the 129-digit number or otherwise decrypt the message that they had encrypted with it. The professors also offered a copy of their technical paper to anyone who sent a self-addressed stamped envelope to their offices at MIT. Rivest discusses this in his Turing award lecture (Ronald L. Rivest, "The Early Days of RSA: History and Lessons" (2011)), following Adleman's lecture (Leonard Adleman, "Pre-RSA Days: History and Lessons" (2011)), and followed by Shamir's (Adi Shamir, "Cryptography: State of The Science" (2011)).

[22]There is a more refined version of public key technology called *identity-based encryption* (IBE) that allows the keys to be made at separate times by a trusted third party. IBE was proposed by Adi Shamir in 1984 (Adi Shamir, "Identity-Based Cryptosystems and Signature Schemes" (1984)). Two working IBE systems were developed in 2001, one by Dan Boneh and Matthew K. Franklin (Boneh and Franklin, "Identity-Based Encryption From The Weil Pairing" (2001)), the other by Clifford Cocks of GCHQ fame (Cocks, "An Identity Based Encryption Scheme Based on Quadratic Residues" (2001)).

Figure 5.3. A locked suggestion box is a good metaphor for public key cryptography. To protect your message, just drop it through the slot. To retrieve your message, you must unlock the padlock and open the lid. Photograph by Hashir Milhan (CC BY 2.0) of a suggestion box in Sri Lanka.

key in the directory, using that key to encrypt a message, and then sending the message using the corporate email system. Nobody will be able to decrypt the message – not even the system administrators who run the corporate email system or the employee who printed the phone book.

Public key cryptography can also be used to create a kind of *digital signature*. In this case, the encrypting key is retained and the decrypting key is published. To sign a document, just encrypt it with your private key, then publish the result as a *signature*. Anyone who has access to your public key (from the directory) can decrypt your signature and get back to the original document. If you practiced good cryptographic hygiene and no one has obtained your private key, now called the *signing key*, then we now have good proof that you alone could have signed the document.

It is still possible for employees to send and receive messages within an organization without using public key cryptography, but the procedures are more involved. One possibility is for the central authority to create a different secret key for every pair of employees that needs to communicate, then to send each pair of employees all of the keys that they need in a sealed envelope. This approach has the feature that individuals can only exchange encrypted email with other individuals with whom they are authorized to exchange messages. Another feature is that the central key-making authority can in theory decrypt any message exchanged by a pair of employees if it retains that pair's key, although the authority can choose to destroy its copy if it wishes to allow the pair to communicate without the possibility of eavesdropping. This is the sort of system that military organizations traditionally set up, and it is presumably what GCHQ and the NSA were using in the 1970s, which is why they saw no need to develop the non-secret encryption that Cocks and Ellis had invented: GCHQ and NSA already had a system that was well-developed and deployed to meet their organizational requirements, and the benefits of digital signatures were not immediately obvious.

For the academics at Stanford and MIT, however, the discovery of public key cryptography opened the door on a new area of intellectual pursuit that combined the fields of number theory and computation. It was an academic green field, full of wonder, possibility, and low-hanging fruit. For example, in 1978, an MIT undergraduate named Loren Kohnfelder realized that digital signatures made it unnecessary for an organization to publish a directory of every employee's public key. Instead, the organization could have a single private/public keypair for the organization itself, and use the private key to sign each employee's public key. The employees could then distribute to each other their own public keys, signed by the organization's public key, to other employees as needed. As long as each employee had a copy of the organization's public key, they could verify each other's keys, and the organization would not need to send out a directory with every employee's public key. Today we call these signed public keys *digital certificates* and the central signing authority a *certificate authority*. With his 1978 undergraduate thesis, Kohnfelder had invented public key infrastructure (PKI).[23]

The following year, Ralph Merkle's PhD thesis[24] introduced the idea of cryptographic hash functions. A hash function is a mathematical function that takes an input of any size and produces an output of a fixed size. The basic concept was invented by IBM engineer Hans Peter Luhn in the 1950s.[25] Merkle's innovation was to have hash functions that produced an output that was both large

---

[23]Kohnfelder, "Towards a Practical Public-Key Cryptosystem" (1978).
[24]Merkle, *Secrecy, Authentication and Public Key Systems* (1979).
[25]Stevens, "Hans Peter Luhn and The Birth of The Hashing Algorithm" (2018).

– more than a 100 bits – and unpredictable, so that it would be computationally infeasible to find an input that produced a specific hash. Given such a function, you don't need to sign an entire document, you just need to sign a hash of the document. Today we call such things *cryptographic hash functions* and there are many, the most prominent being the US Government's Secure Hash Algorithm version 3 (SHA-3).

In the end, the discovery catalyzed interest and innovation in cryptography. Academics and entrepreneurs were attracted to the field; they launched companies and ultimately set in motion the commercialization of the Internet, which was only possible because public key cryptography allowed consumers to send their credit card numbers securely over the Internet to buy things.

A Demonstration of RSA Public Key Cryptography

The most widely used public key encryption system today is RSA, named after its inventors Rivest, Shamir, and Adleman. The system is based on math that is beyond this book but it is easy to find if you have interest, and easy to understand if you understand basic number theory. For the purpose of this demonstration we will just assume that you have a set of magic dice that always roll prime numbers and a box that given these two prime numbers p and q outputs two sets of numbers: your public, encrypting key $\boxed{e,n}$ and your private, decrypting key $\boxed{d,n}$.

We roll the prime number dice and get two prime numbers:

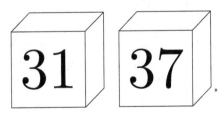

We drop these into our key generator and get two keys:

public key		private key	
e	7	d	463
n	1147	n	1147

To encrypt a plaintext message P (which is a number) to produce an encrypted message C (which is another number), we use this mathematical formula:

$$C = P^e \ (\text{mod } n) \tag{4}$$

This means multiply the number P by itself e times and then take the integer remainder after dividing the resultant by n. For example, the number 53 (which represents the letter "S") encrypts as 914:

$$C = 53^7 \ (\text{mod } 1147) = 1\,174\,711\,139\,837 \ (\text{mod } 1147) = 641 \tag{5}$$

To decrypt the number 914, we follow roughly the same procedure using the values for d and n:

$$P = C^d \ (\text{mod } n) = 641^{463} \ (\text{mod } 1147) = 53 \tag{6}$$

We haven't expanded 641^{463} above; the number is 1300 digits long. RSA implementations use a variety of mathematical tricks to avoid naively computing these numbers – for example, you can apply the modulo after *each* multiplication to prevent the intermediate number from getting too large – but it's easy enough to do the math directly using the Python programming language if you want to check our work.

The RSA algorithm is secure as long as you can't compute the number d knowing e and n (and provided that you follow some implementation guidance that was developed after the algorithm was first published[26]). It turns out that it's easy to compute d, however, if you can factor n. Not a lot was known about the difficulty of factoring numbers in 1977, although the best factoring numbers took exponentially more time as the length of the number being factored increases. That's still the case today. This may be something inherent in the nature of factoring, or it may reflect a limitation in our knowledge. After more than 40 years of intensely studying the question, mathematicians, computer scientists, and cryptographers still don't know.

5.2.2 Forty Years of Public Key Cryptography

Despite the fact that humanity is still unsure about the fundamental hardness of factoring, we have learned a lot about cryptography over the past 40 years. Here we focus on three significant improvements: speed, algorithmic improvements, and key length.

---

[26]For an example of such guidance, see Housley, "Use of The RSAES-OAEP Key Transport Algorithm in Cryptographic Message Syntax (CMS)" (2003).

Cryptographic Speed

The computers of the 1970s were too slow for public key cryptography to be practical: a single RSA encryption or decryption on a computer could take as long as 30 seconds. By the 1980s computers were fast enough that it took just a few seconds, and some companies developed and marketed *cryptographic co-processors* that could accelerate the math required to make RSA run fast as well as store the RSA private keys in tamper-proof hardware. By the 1990s general purpose microprocessors were fast enough that special purpose hardware was no longer needed, and these days most microprocessors include special instructions and dedicated silicon that can be used to accelerate both secret and public key cryptography.

As a result, cryptography has gone from being a technology that was only used occasionally, when it was absolutely needed, to a protection that is always enabled. For example, the early web used encryption just to send passwords and credit card numbers, sending everything else over the Internet in plaintext. These days encryption is the default, and web browsers warn when any page is downloaded without encryption.[27]

Algorithmic Improvements

Working together, cryptographers and security engineers have also made stunning improvements to cryptographic systems, making them both faster and more secure.

Although the underlying math of RSA is sound, cryptographers developed many subtle nuances to use it in practical applications. For example, if we simply encrypt letters one code at a time, as we did in the example above, an adversary has a straightforward method to attack the ciphertext. The adversary can encrypt all possible combinations of messages using the public key until a match emerges with the ciphertext. The attacker can do this because the attacker always has access to the target's public key – that's the core reason we are using public key cryptography. This approach of trying every possible combination is called a *brute-force attack* or a *key-search* attack. For this reason, whatever message that's encrypted is always combined with a random string of bits, called a pad. With a long pad it's

---

[27]Our understanding of Internet security has also expanded, so now we know that a single advertisement, image, or font downloaded without encryption over the Internet can be leveraged by an attacker to compromise your computer's interactions with a remote website.

Elliptic Curve Public Key Cryptography

In the 1980s, cryptographers Neal Koblitz and Victor S. Miller independently suggested that mathematical constructs called "elliptic curves over finite fields" might provide the sort of functionality operations required to build a working public key cryptography system.[a] They were right, and elliptic curve cryptography (ECC) was developed and standardized in the 1990s, culminating with the adoption of the Elliptic Curve Digital Signature Algorithm (ECDSA) in 1999. Following that, the US National Security Agency aggressively promoted ECC over RSA. Since relatively short ECC keys were just as secure as RSA keys that were much longer, ECC systems led to faster computations that required less power.

At first, the primary disadvantage of ECC was the need to license patents from Certicom, the Canadian company founded in 1985 to commercialize ECC technology. Whereas RSA was protected by a single US patent that expired in 2000,[b] Certicom aggressively patented many different aspects of both the ECC math and efficient ECC implementations.

More recently, security experts have raised some concerns regarding the technology – specifically that the number theory of elliptic curves is less well-studied than the number theory that underlies the RSA algorithm. In 2015, Neal Koblitz and Alfred Menezes noted that the NSA was moving away from elliptic curve cryptography.[c]

Like RSA, the math that underlies ECC is also vulnerable to quantum computers. And since the ECC keys are significantly shorter than RSA keys, quantum computers will be able to crack the ECC keys in use today long before they are able to crack today's RSA keys. Assuming that there are no fundamental scientific limits to scaling up the quantum computer, "it's just a matter of money," observed Koblitz and Menezes.

---

[a]Neal Koblitz, "Elliptic Curve Cryptosystems" (1987); Miller, "Use of Elliptic Curves in Cryptography" (1986).

[b]L. M. Adleman, R. L. Rivest, and A. Shamir, "Cryptographic Communications System and Method" (1983).

[c]N. Koblitz and Menezes, "A Riddle Wrapped in an Enigma" (2016).

impossible for the attacker to try every combination; padding also assures that the same message will always encrypt differently, which makes cryptanalysis harder. RSA without a pad is called *Textbook RSA*: it's good enough for textbooks, but it doesn't actually protect your message.

Engineers developed clever encryption protocols that limit the number of public key operations that need to be computed. This is done by combining public key cryptography with traditional secret key cryptography. For example, an hour of HD video (roughly 10 GB of data, with compression) can be encrypted with a single public key operation. This is done by first encrypting the video with a randomly generated secret key, and then encrypting the secret key with a public key algorithm. This approach is sometimes called a *hybrid system*; it is the approach that is used by both the Trusted Layer Security (TLS) protocol and the Secure Shell (SSH) protocols used to send information over the Internet.

5.2.3 *Cracking Public Key with Shor's Algorithm*

Here is one measure of public key technology's success: today the vast majority of information sent over the Internet is encrypted with TLS, the hybrid system described above that uses public key technology to exchange a session key, and then uses the session key to encrypt the information itself. If you are viewing web pages, you are probably using TLS.

TLS is sometimes called a *pluggable protocol*, meaning that it can be used with many different encryption algorithms – it's as simple as plugging-in a new algorithm implementation. When you type a web address into your browser, your browser opens a connection to the remote website and the remote website sends to your browser the website's public key certificate, which is used to establish the website's identity. The two computers then negotiate which set of algorithms to use based on which algorithmic plug-ins the web server and the web browser have in common. Today there are tools built into most web browsers to examine website certificates and the TLS connections, but these tools can be confusing because the same website can appear to provide different certificates at different times. This is typically because a single "website" might actually be a collection of several hundred computers, all configured with different certificates.

Because the public key certificate is sent over the Internet when a web page is downloaded, anyone who can eavesdrop upon and capture the Internet communications now has all of the information that they need to decrypt the communications, provided that they have sufficient computing power to derive the website's matching private key from its public key – that is, to "crack" the public key. In the case of RSA, this is the very factoring problem posed by decrypting the *Scientific American* message that was encrypted with RSA-129. In the case of elliptic curve algorithms, other mathematical approaches are used to crack the public key.

Before the invention of Shor's algorithm, the fastest factoring algorithms required exponentially more time to execute as the number of bits in the public key increased. Shor's algorithm uses an approach for factoring that has only *polynominal complexity*: longer keys still take longer to factor, just not exponentially longer. The catch is that Shor's algorithm requires a working quantum computer with enough stable qubits to run a quantum algorithm that helps to factor the number in question: with perfect qubits, factoring the numbers used in modern cryptographic system would require thousands of qubits. But if the qubits have even the smallest amount of noise, then it will be necessary to use quantum error correction, increasing the number of qubits needed to roughly a hundred million (see p. 206).[28] Of course, the first computer to use transistors was built in 1953 at Manchester University: it had just 92 point-contact transistors that had been constructed by hand. Today's Apple M1 microprocessor has 16 billion transistors, built with a feature size of just 5 nanometers.

Shor's algorithm contains a classical part and a quantum part. The classical part contains some of the same number theory that powers RSA encryption, which isn't terribly surprising since both are based on prime numbers, factoring, and Euler's Theorem. To use RSA, the *code-maker* randomly chooses two prime numbers, p and q. These numbers are multiplied to compute N and also used to create the public key and private key. With Shor's algorithm, the attacker just has the public key, which contains N. The attacker also has access to a quantum computer that can perform two quantum functions: the quantum Fourier transform and quantum modular exponentiation. With these functions, the attacker can factor N, learning p and q, and re-generate the code-maker's private-key.

---

[28] Mohseni et al., "Commercialize Quantum Technologies in Five Years" (2017).

With this private key, the attacker can decrypt any message that was encrypted with the code-maker's public key.

At a high level, one might consider Shor's algorithm as a carefully designed collection of dual-slit experiments, where the slits are arranged according to the public key, N, in such a way that the interference pattern displayed on the screen reveals information about the factors p and q. One might think of the quantum computer as taking an X-ray of the number N. If the bits of N are arranged in just the right way, if they are connected to just the right quantum circuit, and if the X-rays are sent from just the right directions, then the diffraction pattern (see Appendix B) will reveal properties of p and q.

Alas, explaining either the classical *or* the quantum aspects of Shor's algorithm requires more math and physics than we require for readers of this book, so we refer interested readers with sufficient skills to other publications, including the second version of Shor's 1997 paper[29] which can be downloaded from arXiv,[30] as well as the Wikipedia article on Shor's algorithm.[31]

If you had a quantum computer with sufficiently many stable qubits to run Shor's algorithm, and *if* you had recorded the complete encrypted communication between a web server and a web browser at anytime from the dawn of the commercial Internet through today, *then* decrypting that communication would be straightforward.

For example, consider an unscrupulous internet service provider (ISP) that wants to eavesdrop on a user's email. Before 2008, the ISP merely needed to capture the user's packets and reassemble them into web pages – a fairly trivial task.[32] But since 2008 Google has allowed users to access the server using encryption,[33] and in 2010 Google made encryption the default. Once the user started using encryption, the nosy ISP would be out of luck: the web pages

---

[29] Shor, "Polynomial-Time Algorithms for Prime Factorization and Discrete Logarithms on a Quantum Computer" (1997).

[30] arxiv.org/abs/quant-ph/9508027v2

[31] With some amusement, we note that in June 2021 the quantum algorithm section of the Wikipedia article contained this note: "This section **may be too technical for most readers to understand**. Please *help improve it* and *make it understandable to non-experts*, without removing the technical details." We encourage any of our readers with sufficient skill to accept this challenge.

[32] Ohm, "The Rise and Fall of Invasive ISP Surveillance" (2009b); Bellovin, "Wiretapping The Net" (2000).

[33] Rideout, "Making Security Easier" (2008).

would be encrypted using RSA cryptography. However, if the ISP had recorded these packets and later rented time on a sufficiently large quantum computer, all the ISP would need to do is to extract Gmail's public key certificate, factor N, apply the RSA key generation algorithm to compute the private key, use the private key to decrypt something that the *master secret* was used to encrypt the web pages, and then use the master secret to decrypt the individual pages. This is not hard to do – there exists software that readily performs all of the reassembly and decryption – provided that you have a copy of the server's private key.

If you had captured the packets and *didn't* have a quantum computer, there are still other ways to get that private key. You might be able to get it by hacking into Google's server and stealing it. Alternatively, you might be able to bribe someone at Google, or even obtain a court order against Google to force the company to produce its private key or use it to decrypt the captured transmission.

In 2011, Google made a change to its computers to remove the risk that a stolen private key could be used to compromise the privacy of its service users: Google implemented *forward secrecy* by default.[34] Also known as *perfect forward secrecy*, the term is applied to security protocols that use session keys that are not revealed even if long-term secrets used to create or protect those session keys are compromised. In the case of web protocol, forward secrecy is typically assured by using digital signatures to certify an ephemeral cryptographic key created using the Diffie–Hellman key agreement protocol, which is an interactive public key encryption algorithm that allows two parties to agree on a shared secret.[35]

Google's 2011 move to forward secrecy is a boon for privacy: it means that after the conclusion of communications between a user's web browser and the Gmail server, not even Google can use its own private key to decrypt communications that might have been covertly recorded. This is because Google's Gmail server destroys its copy of the ephemeral encryption key that was used to encrypt the session when the session concludes.

---

[34]Langley, "Protecting Data for The Long Term with Forward Secrecy" (2011).

[35]Diffie–Hellman is an *interactive* algorithm because performing the protocol requires the two parties to exchange information with each other and act upon the exchanged information. In this way it is different from RSA, which is a *non-interactive protocol*, because it is possible for one party to encrypt or decrypt information using RSA without the active participation of the other party.

It turns out that the forward secrecy algorithm used by Google, the Diffie–Hellman key agreement protocol, is also vulnerable to an attacker that has a quantum computer. This is because the security of the Diffie–Hellman algorithm depends on the difficulty of computing something known as a *discrete logarithm*, and the quantum part of Shor's algorithm can do that as well. So those packets recorded by the ISP in our scenario are still vulnerable to some future attacker with a large-enough quantum computer.

5.2.4 Evaluating The Quantum Computer Threat to Public Key Cryptography

Factoring is clearly a problem that quantum computers will be able to solve faster than classical computers if they become sufficiently large. Will quantum computers ever actually be large enough to pose a threat to public key cryptography? *We don't know the answer to this question today.*

In 2001, a 7-qubit bespoke quantum computer constructed by Isaac Chuang's group at IBM Alamaden Research Center successfully factored the number 15 into its factors 3 and 5.[36] The number 15 is represented in binary by four bits: 1111 . The number 15 is also, not coincidentally, the smallest number that is not prime, not even, and not a perfect square. So realistically, it's the smallest number that the IBM team could have meaningfully factored.[37]

The quantum "computer" that IBM used doesn't look anything like our modern conception of a computer: it was a tube containing a chemical that IBM had synthesized especially for the experiment, a chemical called a "perfluorobutadienyl iron complex with the inner two carbons," and with chemical formula (Figure 5.4):

$$F_2C=C(Fe(C_5H_5)(CO)(CO))CF=CF_2$$

The quantum circuit was played through the tube as a series of radio frequency pulses, and the qubits were measured using nuclear magnetic resonance (NMR), a procedure in which a material is placed in a strong magnetic field and probed with radio waves at

---

[36]Vandersypen et al., "Experimental Realization of Shor's Quantum Factoring Algorithm Using Nuclear Magnetic Resonance" (2001).

[37]Even numbers are easy to factor: just divide them by two. Numbers that are perfect squares are also easy to factor: just take their square root, which can be quickly computed using Newton's method. The number 15 is the smallest non-even number that is the product of two different primes: three and five.

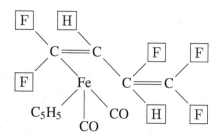

Figure 5.4. The perfluorobutadienyl iron complex with the inner two carbons that IBM scientists synthesized in 2001 for the purpose of factoring the number 15. The seven qubits are represented by the five fluorine (F) and two hydrogen (H) atoms shown surrounded by a box .Vandersypen et al., "Experimental Realization of Shor's Quantum Factoring Algorithm Using Nuclear Magnetic Resonance" (2001).

different frequencies. We discuss NMR-based quantum computers in Section 4.8.2 (p. 168).[38]

Since IBM's demonstration, other researchers have factored other numbers on quantum computers. None of these approaches have managed to factor a number out of reach of a conventional computer. Most of the numbers factored can be factored with pen and paper. For example, in 2012 a team led by Nanyang Xu at the University of Science and Technology of China, Hefei, successfully factored the number 143 using "a liquid-crystal NMR quantum processor with dipole–dipole couplings."[39] The factors were 11 and 13, of course. What's exciting is that the researchers used a different factoring approach called *adiabatic quantum computation* (AQC), using only four qubits. In 2014, Nikesh Dattani at Kyoto University and Nathaniel

---

[38]It may seem implausible that a tube containing a solution of a specially synthesized compound inside a scientific instrument is actually *computing*, at least in the way that we typically think of the term. But the IBM experiment demonstrated that the computational media responded in a way that was consistent with factoring the number 15, producing the numbers 3 and 5.

It turns out that computing is more fundamental than electronics, and there are many different media that can be used for computation. For example, in the 1970s Danny Hillis created a computer from Tinkertoy rods and wheels that could play Tic-Tac-Toe. "It could have been built by any six-year old with 500 boxes of tinker toys and a PDP-10," Hillis wrote at the time (D. Hillis and Silverman, "Original Tinkertoy Computer" (1978)). Another improbable computing medium is the seemingly haphazard but highly structured collection of lipids, proteins, nucleic acids, small amine molecules, amino acids, and neuropeptides that make up the human neurological system.

[39]Xu et al., "Quantum Factorization of 143 on a Dipolar-Coupling Nuclear Magnetic Resonance System" (2012).

Bryans at University of Calgary posted a follow-up article to the arXiv open-access archive purportedly showing that the published results of the Chinese researchers could also be used to factor the numbers 3599, 11 663, and 56 153.[40][41] The work on AQC factoring is exciting because it suggests that research in quantum computing may eventually lead researchers to make fundamental discoveries about factoring or even the nature of computation, with results that could then be applied to both quantum and classical computers. Although there have been no such discoveries to date, the field of quantum factoring is still quite young compared with other branches of number theory.

As of January 2019, the current record for factoring published in the peer-reviewed literature is held by Chinese scientists, who factored the 7-digit (20-bit) number 1 005 973 using 89 qubits on a D-Wave quantum annealing machine. The team noted that by using a factoring algorithm based on quadratic unconstrained binary optimization (QUBO), the team was able to constrain the factoring problem to the type of qubits that D-Wave provides. "Factoring 1 005 973 using Shor's algorithm would require about 41 universal qubits, which current universal quantum computers cannot reach with acceptable accuracy," the authors noted wryly.[42] This development was exciting because it demonstrated a new use for the D-Wave annealer, discussed further in Chapter 6, which is limited to certain kinds of applications. The scientists reasoned that because D-Wave scaled its annealer from just 128 bits to 2000 in just seven years, perhaps a machine capable of factoring the kinds of numbers used to secure today's commercial Internet might soon be constructed.

We disagree: such a capacity would require a D-Wave computer with significantly more qubits than seems likely for the foreseeable future. (As of June 2021, D-Wave's largest system, the Advantage, has just 5000 qubits.[43]) To crack the RSA systems that are used

---

[40]Dattani and Bryans, "Quantum Factorization of 56153 with Only 4 Qubits" (2014).

[41]The Dattani/Bryans work was covered by the news site Phys.org (Zyga, "New Largest Number Factored on a Quantum Device Is 56,153" (2014)), but the work did not appear in the peer-reviewed literature.

[42]Peng et al., "Factoring Larger Integers with Fewer Qubits via Quantum Annealing with Optimized Parameters" (2019).

[43]D-Wave Systems Inc., "D-Wave Announces General Availability of First Quantum Computer Built for Business" (2020).

to protect today's commercial Internet would require the ability to factor 2048- or 4096-bit numbers.[44]

Even with this work on factoring – perhaps because of it – there is still wide agreement in the scientific community that a practical application of quantum computing to factoring is far off. It is unclear whether the winning system will be a universal quantum computer with stable qubits that can also factor, or a special purpose device designed to perform factoring quickly. The advantage of the first machine is generality. The advantage of the second is that it could likely be developed years before a general-purpose quantum computer, and it could probably be developed for less money, and possibly in secret.

Yet another threat could come through a quantum-classical approach where the factoring problem is solved in parts with a quantum computer, and a classical computer is used to combine and process these parts to come to a full solution.[45] The partial analysis approach might afford today's small quantum computers a role in cryptanalysis.

Google scientists have projected that factoring a conventional RSA public key in use on the commercial Internet today "would take 100 million qubits, even if individual quantum operations failed just once in every 10 000 operations."[46] A National Academies group assessed in 2019 that "to create a quantum computer that can run Shor's algorithm to find the private key in a 1024-bit RSA encrypted message requires building a machine that is more than five orders of magnitude larger and has error rates that are about two orders of magnitude better than current machines, as well as developing the software development environment to support this machine." The authors of the report stated that it is "highly unexpected" that a quantum computer that can break a 2048-bit RSA key will be built before 2030.[47]

---

[44]For comparison, as of February 28, 2020, the largest RSA challenge number to be publicly factored is RSA-250, a 250-digit, 829-bit number (Boudot et al., "Factorization of RSA-250" (2020)). The total amount of computer time required to perform the computation "was roughly 2700 core-years, using Intel Xenon Gold 6t130 CPUs as a reference (2.1Ghz)," the authors reported (Goodin, "New Crypto-Cracking Record Reached, with Less Help Than Usual From Moore's Law" (2019)).

[45]Ekerå and Håstad, "Quantum Algorithms for Computing Short Discrete Logarithms and Factoring RSA Integers" (2017).

[46]Mohseni et al., "Commercialize Quantum Technologies in Five Years" (2017).

[47]Grumbling and Horowitz, *Quantum Computing: Progress and Prospects* (2019).

DNA-Based Computing and Storage

DNA (deoxyribonucleic acid) is the polymerized molecule inside cells that carries inheritance information used to synthesize proteins. It has been called "the building block of life."

Before the event of quantum computers, some researchers thought DNA's ability to encode and to reproduce information might also make DNA a useful substrate for computing. One proponent was Leonard Adleman (the "A" of RSA), who is frequently credited with inventing the field.

Adleman encoded a small graph into a DNA molecule and then used biomolecular reagents "to solve an instance of the directed Hamiltonian path problem."[a] This was highly significant, as the Hamiltonian Path problem is NP-complete. If DNA computing could solve it efficiently, and if the system can be scaled up, DNA can be used to solve any other NP problem. In particular, a DNA computer would be able to factor efficiently.[b]

Work on DNA computing has continued, with researchers developing a variety of DNA-based algorithms.[c] A recent review of "DNA-based Cryptanalysis"[d] found that the field remains promising. But it has been eclipsed by quantum computing.

There have been significant breakthroughs in using DNA to encode information directly. In June 2019, a Boston-based startup called Catalog announced that it had encoded all 16 GB of Wikipedia into a set of DNA strands the size of a pencil eraser.[e] DNA is also stable over long periods of time; DNA is now routinely recovered from humans that lived thousands of years ago. Since DNA is the basis of life, the ability to transcribe DNA is likely to be re-invented by any future biologically based civilization on Earth, should our current technological society fail. DNA thus makes an excellent backup medium not just for organizations, but also for the intellectual heritage of our civilization.

---

[a]Leonard Adleman, "Molecular Computation of Solutions to Combinatorial Problems" (1994).

[b]Factoring is not NP-complete, but it is contained within NP.

[c]W. Chang, Guo, and M. S. Ho, "Fast Parallel Molecular Algorithms for DNA-Based Computation: Factoring Integers" (2005).

[d]Sadkhan and Yaseen, "DNA-Based Cryptanalysis: Challenges, and Future Trends" (2019).

[e]Shankland, "Startup Packs All 16GB of Wikipedia Onto DNA Strands to Demonstrate New Storage Tech" (2019); Catalog Technologies, Inc., "Catalog" (n.d.).

5.2.5 Post-Quantum Cryptography

Fully realized, large-scale, and sufficiently error-free, quantum computers will mean that public key encryption systems based on the RSA, Diffie–Hellman, and Elliptic Curve systems are no longer secure. But this will not mean the end of public-key cryptography.

Since the discovery of public key cryptography in the 1970s, dozens of public key encryption algorithms have been devised. Of these, many do not depend on the difficulty of factoring or computing a discrete logarithm, and as such these algorithms would not be crushed by Shor's algorithm and a suitably large quantum computer. In fact there are so many choices and they are all so significantly different that it is not immediately clear which is the best.

To help the world choose, in 2016 NIST embarked on the Post-Quantum Cryptography (PQC) Standardization effort. At the time, NIST stated that the competition for a PQC asymmetric algorithm would likely be more complex than its successful competitions to pick the Advanced Encryption Standard (AES) and the Secure Hash Algorithm 3 (SHA-3). "One reason is that the requirements for public-key encryption and digital signatures are more complicated. Another reason is that the current scientific understanding of the power of quantum computers is far from comprehensive. Finally, some of the candidate post-quantum cryptosystems may have completely different design attributes and mathematical foundations, so that a direct comparison of candidates would be difficult or impossible."[48]

NIST started with a field of 82 algorithm candidates, which was reduced to 26 algorithms in early 2019. In July 2020 NIST announced the "Round 3 candidates" for the competition, with four public-key and key-establishment algorithms under consideration as "finalists:" Classic McEliece,[49] CRYSTALS-KYBER,[50] NTRU,[51] and SABER.[52] Another three algorithms are under consideration for digital signature algorithms: CRYSTALS-DILITHIUM,[53] FALCON,[54] and Rainbow.[55] Each algorithm is being presented in a web-based

---

[48]National Institute of Standards and Technology, "Post-Quantum Cryptography" (2017).

[49]classic.mceliece.org

[50]pq-crystals.org

[51]ntru.org

[52]www.esat.kuleuven.be/cosic/pqcrypto/saber/

[53]pq-crystals.org

[54]falcon-sign.info

[55]www.pqcrainbow.org

seminar open to the public, with the previous presentations and videos archived on the NIST website. It is unclear when the process will be finished, but it is likely that the scientific community will have standardized a new family of asymmetric algorithms long before the availability of quantum computers with sufficient power to crack the algorithms in use today.

In the meantime, all of the algorithms that NIST is evaluating are published, several with accompanying intellectual property statements stating that the authors do not hold patents on the algorithms, have not filed for patents, and have no intention to file for patents. This means that the algorithms are available for experimentation now! And indeed, July 2016, Google announced that it had deployed its experimental CECPQ1 key agreement protocol in "Chrome Canary," the experimental, nightly build version of its popular Chrome web browser.

"Quantum computers exist today but, for the moment, they are small and experimental, containing only a handful of quantum bits," Google's software engineer wrote in the company's Security Blog.[56] "However, a hypothetical, future quantum computer would be able to retrospectively decrypt any internet communication that was recorded today, and many types of information need to remain confidential for decades. Thus even the possibility of a future quantum computer is something that we should be thinking about today."

Google's protocol uses the conventional and PQC algorithms in parallel, so that both must be successfully attacked together, during the same session, in order for the contents of a protected session to be compromised.

One of the reasons that Google decided to experiment with PQC is that the PQC data structures are significantly larger and slower to compute than the data structures used today. Thus, it makes sense to experiment with this technology now, on a limited scale.

In 2019 Google and the webhosting company Cloudflare continued the experiment, jointly deploying an improved algorithm called CECPQ2. "With Cloudflare's highly distributed network of access points and Google's Chrome browser, both companies are in a very good position to perform this experiment."[57]

---

[56]Braithwaite, "Experimenting with Post-Quantum Cryptography" (2016).

[57]Kwiatkowski, "Towards Post-Quantum Cryptography in TLS" (2019).

If you are interested in learning more about the PQC algorithms, Kwiatkowski's illustrated blog post does a great job explaining them, although it would be useful to have first taken a course in college-level algebra.

5.3 Quantum Search (Grover's Algorithm)

Two years after Shor showed that a large enough quantum computer would be able to factor the numbers used to secure the Internet, Lov Grover (also at Bell Labs) made a startling discovery: a properly constructed quantum computer could speed up all sorts of computations that have a certain mathematical property. The speedup was not as significant as Shor's: instead of turning a problem that is computationally intractable into one that can be solved in just a few hours, Grover's algorithm gives a square-root speedup: if solving a problem takes on order of N steps without Grover, typically abbreviated $O(N)$, it now takes on the order of the square root of N steps – that is, $O(\sqrt{N})$. On the other hand, whereas Shor's algorithm can only be applied to the relatively obscure domain of number theory, Grover's algorithm can be broadly applied to a wide range of practical problems. Grover's algorithm is the second major quantum computing algorithm.

Later in this section we will discuss how Grover's algorithm can be used to crack a version of one of the world's most popular encryption algorithms. We'll show why this was such a big deal at the time, and then discuss why it's not really a big deal any more. After that, we'll discuss other applications for Grover's algorithm. To get started, though, we need to further explore the world of cryptography and code cracking.

5.3.1 Symmetric Ciphers: DES and AES

In 1977 the US Government adopted a standard algorithm for encrypting data that it unceremoniously named the Data Encryption Standard. Before the adoption of the DES, the few companies that sold data security equipment to the public generally made up their own encryption algorithms and asserted that they were secure. This created a difficult commercial environment, because most customers (including most government customers) were not equipped to evaluate the vendors' claims. The DES solved this problem: after it was adopted, vendors could simply follow Federal Information Processing Standard 46: no longer did they need to claim that the algorithm they had cooked up in their labs was mathematically secure. This is the function of standards, and with the DES the standardization process worked beautifully. Both inside and outside the US government, the algorithm was rapidly adopted and deployed.

The adoption of the DES was not without controversy, however. In choosing the DES, the National Bureau of Standards did not use an existing military encryption algorithm. Instead, NBS (the precursor to today's National Institute of Standards and Technology) invited submissions from industry and academia. The first submission round was unsuccessful. For the second round, IBM submitted an algorithm it had developed called Lucifer, based on a novel construction created by the German-born mathematician Horst Feistel (1915–1990).[58]

Ideally, symmetric block cipher algorithms like DES and Lucifer have the property that the only way to decrypt an encrypted message is by knowing (or guessing) the correct key. Clearly, one way to attack such a cipher is to try all possible keys – the brute-force approach. In practice there are other kinds of attacks; such attacks make it possible to correctly guess the decryption key without explicitly trying all of them.

The original Lucifer algorithm had a 128-bit key length (see the sidebar "Key Length" on page 213), but after analysis by the National Security Agency, the algorithm's internals were changed somewhat and the key shortened to 56 bits. (It was widely assumed at the time that the US Government had intentionally weakened Lucifer because US intelligence agencies didn't want an encryption algorithm adopted as a national standard that was too difficult to be cracked. In fact, we now know that the final DES algorithm with its 56-bit keys was *stronger* than the 128-bit algorithm: unlike Lucifer, DES was resistant to a cryptanalysis technique called "differential cryptanalysis" that was not widely known in the 1970s and would not be

---

[58]Feistel's family fled Germany in 1934. He enrolled at MIT in Physics and graduated in 1937, then proceeded to earn a master's degree at Harvard. At the outbreak of World War II Feistel immediately came under suspicion because of his German citizenship, but his talents were well recognized by others in the US government: Feistel was granted US citizenship on January 31, 1944, and awarded a top secret security clearance the following day. He worked at the US Air Force Cambridge Research Center, MIT's Lincoln Laboratory, and MITRE, before moving to IBM.

discovered by academic cryptographers until the 1990s.[59])

When DES was adopted in 1977 it was not feasible for an attacker to try all $2^{56} = 72\,057\,594\,037\,927\,936$ possible keys to crack a message, but this proved to be possible by the 1990s. To make DES stronger, some organizations adopted a variant called *triple-DES* in which DES was used three times over, each time with a different key, to encrypt a message. This produced an effective key size of 168 bits, but it was also three times slower than a single encryption. There were also lingering doubts as to whether or not the DES had vulnerabilities that had been intentionally hidden by its creators which might make even triple-DES suspect.

In the late 1990s, NIST ran a second public competition to select a new national encryption standard. This time the vetting process was public as well. After two years, NIST adopted the Advanced Encryption Standard (AES), a symmetric block encryption algorithm developed in the 1990s that is better than DES in every possible way.

AES has three primary modes of operation: AES-128, AES-192, and AES-256, with 128-bit, 192-bit, and 256-bit keys respectively. In practice, only AES-128 and AES-256 are widely used: AES-128 is the fastest, for applications that require the fastest possible algorithm, and AES-256 for the applications where speed is not the most important factor. Because the strength of the algorithm doubles with each additional bit, AES-256 is at least 2^{128} times stronger than the 128-bit version.

In fact, the number 2^{128} is so impossibly large that it is not possible to crack a message encrypted with AES-128 using brute-force search on a classical computer: there is simply not enough time. For example, if you had five billion computers that could each try 90 billion AES-128 keys per second, it would take 24 billion years – roughly the age of the Universe – to try all possible AES-128 keys. Without a functioning quantum computer running Grover's algorithm, the only way that an AES-128 message will be cracked will be if a significant underlying mathematical vulnerability is found in the AES algorithm itself. Today such a discovery does not seem likely.

However, it may be possible to crack such messages using Grover's algorithm running on a sufficiently large quantum computer. We dis-

---

[59]Coppersmith, "The Data Encryption Standard (DES) and Its Strength against Attacks" (1994).

Key Length

The most visible change in cryptography over the past 40 years is the way that cryptographic keys have steadily increased.

Key length is traditionally expressed in bits. A key length of two means that there are four possible secret keys: 00 , 01 , 10 , and 11 . With a key length of three, there are eight possible secret keys: 000 , 001 , 010 , 011 , 100 , 101 , 110 , and 111 . With 4 bits there are 16 possible keys, and with 8 bits there are 256. Concisely, if there are n bits, there are 2^n possible secret keys – the number of keys grows *exponentially* as the number of bits increases. With a strong secret key algorithm, it is necessary to try every possible key in order to crack the message: there are no algorithmic short-cuts.

Whereas adversaries will attack a message encrypted with a secret-key algorithm by trying to decrypt the message, attacks against public-key algorithms typically involve attacking the public key itself. In the case of RSA, such attacks involve factoring the product of the two prime numbers p and q. Such factoring is harder with longer public keys. As a result, engineers have used longer and longer public keys as computers have gotten better at factoring.

In the early days of the commercial Internet, web browsers supported an intentionally weak 512-bit RSA algorithm and a stronger 1024-bit algorithm. The idea was that the weakened algorithm was to be used outside the US and for non-commercial applications, and the 1024-bit version was to be used within the US for commercial applications. Today there are no significant export restrictions on cryptographic software and 2048-bit RSA (617 decimal digits) is widely used, although 4096-bit RSA (1234 decimal digits) systems are increasingly being deployed. For comparison, the original RSA-129 number is 426 bits (129 decimal digits), and the number 1147 used in the example on page 195 is 11 bits (4 decimal digits).

cuss this below in Section 5.3.3 (p. 218).

5.3.2 Brute-Force Key Search Attacks

As we mentioned above, messages encrypted with symmetric encryption algorithms can be forcibly decrypted, or "cracked," by trying all possible keys in sequence. In Table 5.1 we show how this works in practice. We have an 8-character message that has been encrypted with a key that was specially created for this text. The first few attempts fail, but eventually we find one that succeeds. In an actual brute force search, the computer stops when it finds a decryption succeeds, but in the table we keep going until we've tried all 72 quadrillion possibilities.

There are two technical challenges to conducting a key search attack: the time it takes to try all possible keys, and the difficulty of

Figure 5.5. A safe with a combination lock on its door is a good metaphor for secret key cryptography and symmetric ciphers. To protect your message, just enter the combination lock on the panel, open the safe, put in your message, and close the door. To retrieve your message, enter the same combination on the panel, open the door, and retrieve your message. Photograph by Dave L. Jones (EEVBlog), Wikimedia Commons Account Binarysequence (CC BY-SA 4.0).

Table 5.1. Decrypting a message encrypted with the Data Encryption Standard by trying all possible keys. Each DES key is 56 bits long; there are roughly 72 quadrillion keys. Characters that are not printable are displayed with a bullet (•). Notice that when the correct key is found, all of the decrypted characters are printable. In this case the key was found roughly halfway through because it starts with the bit sequence 1000. The same approach can be used with AES, except that there are $2^{128} = 340\,282\,366\,920\,938\,463\,463\,374\,607\,431\,768\,211\,456$ possible keys in its weakest implementation.

Trial	Binary Key (56-bits)		Decrypted Output	Text
0	0000	... 0000	BE 47 A1 7A 2E 81 0E 8C	¾G¡z.•••
1	0000	... 0001	62 59 0B B1 CB 67 8F 3A	bY•±Ëg•:
2	0000	... 0010	B3 9B 0D 12 1F C5 A9 7C	³••••Å©|
3	0000	... 0011	84 19 9D C6 B0 F5 AD 75	•••Æ°õ•u
4	0000	... 0100	D4 E6 90 8D 8F 77 EA 07	Ōæ•••wê•
		...		
38 326 038 678 974 151	1000	... 0111	42 65 72 6B 65 6C 65 79	Berkeley
		...		
72 057 594 037 927 935	1111	... 1111	FB 90 3D D5 99 A3 27 3D	û•=Õ•£'=

recognizing a correct decryption.[60] The time is determined by how many keys per second your code-cracking machine can attempt, and how many code-cracking machines you happen to have. For example, at Bletchley Park during World War II, the Bombe (see p. 80), designed to crack the three-rotor version of the Germans' Enigma code, could cycle through all 17 576 possible rotor combinations in 20 minutes. With two of these machines, the British could try half the combinations on one machine and half on the other, and crack a message in 10 minutes. Or they could attack two messages with the two machines, and use the full 20 minutes to crack each. Of course, 20 minutes to crack a message was the *worst case*; on average a message would be cracked after half of the rotor positions had been tried. It was also necessary to detect when the correct rotor position was found. The Germans made this easier by their tendency to begin their encrypted messages with the same sequence of characters.

---

[60]Many treatises on cryptography and code-breaking ignore the challenge of detecting when text is correctly decrypted. In practice, this challenge is readily overcome, provided that the attacker knows something about the format of the decrypted messages. This is called a *known plaintext attack*. In some cases the attacker can arrange for a message of its choosing to be encrypted by the system under attack; this is called a *chosen plaintext attack*.

When the US Data Encryption Standard was adopted by the National Bureau of Standards (NBS) in 1977, Hellman wrote a letter to NBS arguing that the reduction of the DES keysize from 64 bits to 56 bits suggested that it was done "to intentionally reduce the cost of exhaustive key search by a factor of 256."[61] In a follow-up article, Diffie and Hellman hypothesized that it should be possible to create a special-purpose DES-cracking microchip that could try a million keys each second. With a million such chips, it would be possible to try all 2^{56} keys in a day. They estimated the cost of constructing such a machine at $20 million in 1977 dollars; assuming a five-year life of the machine and a daily operating cost of $10 000, the average cost of cracking a DES-encrypted message in 1977 would be just $5000, including the cost of developing the machine.[62] With expected improvements in microelectronics, the Stanford professors estimated that the cost of their hypothetical DES-cracking machine would be just $200 000 by 1987. In fact, it actually took 20 years. In 1998 the Electronic Frontier Foundation (EFF) announced that it had spent $250 000 and constructed the fabled DES Cracker. The EFF machine tried 90 billion 56-bit DES keys every second, and cracked its first challenge message after only 56 hours of work.[63] The project is widely credited with putting the last nail into the coffin of weak symmetric encryption schemes.

When cracking symmetric encryption systems with a brute force attack, each additional bit of key length doubles the difficulty of the attack, because each additional bit doubles the number of keys that need to be searched. With 4 bits, there are 16 keys to search; with 8 bits there are 256, and so on. For a while, the US Government's proposed replacement for DES was the so-called "Clipper" chip, which supported an 80-bit key, making it 2^{24} or roughly 16 million times harder to crack – except that each Clipper chip was gimmicked so that the government *didn't need* to perform such an attack to decrypt a message encrypted with Clipper. That's because the Clipper implemented the government's "Escrowed Encryption Standard" (FIPS-185), which meant that every Clipper had its own secret decryption key that could be used to decrypt any message that

---

[61]Blanchette, *Burdens of Proof: Cryptographic Culture and Evidence Law in The Age of Electronic Documents* (2012).

[62]Diffie and Hellman, "Special Feature Exhaustive Cryptanalysis of The NBS Data Encryption Standard" (1977).

[63]Electronic Frontier Foundation, *Cracking DES* (1998).

the chip encrypted, and the government kept copies of these keys so that messages could be decrypted for legal process or in the event of a national security emergency. To prevent companies from creating software-only Clipper chips that didn't implement key escrow, the government declared that the encryption algorithm used by the chip had to be kept secret in the interest of national security.

As might be expected, Clipper chip was a commercial failure.

When the National Institute of Standards and Technology initiated its efforts to create a replacement algorithm for the Data Encryption Standard in the late 1990s, it committed itself to an open, unclassified project. NIST invited submissions for the new algorithm, held two academic conferences to discuss the submissions, and ultimately adopted an algorithm invented outside the United States by a pair of Belgian cryptographers, Vincent Rijmen and Joan Daemen. The algorithm, originally named Rijndael, is faster than DES and supports key sizes of 128, 192, and 256 bits. It was adopted by the US government as the Advanced Encryption Standard in 2001.

For many years after it was adopted, AES-128 was the preferred use of AES because it ran significantly faster than the more secure AES-256. That extra security is in fact the reason that AES-256 was slower. The design of AES is based on a function that is repeated a certain number of "rounds" for every block of data that the algorithm encrypts. AES-128 has 10 rounds, AES-256 has 14.[64] Today those differences are less significant than they were in 2001, as computers are faster and many microprocessors now contain hardware support to make AES run faster still. In most modern computers, encrypting with AES-128 is essentially free. For example, the Apple iPhone contains a chip that automatically encrypts data with AES when it is written from the CPU out to the phone's flash memory, and automatically decrypts the data when it is read back in.

However, absent quantum computing, the differences between AES-128 and AES-256 are inconsequential for most users. That's because 2^{128} is a really big number: in a world without quantum computers, a message encrypted with a 128-bit key will *never* be cracked using a brute-force, key search attack.

---

[64] AES-256 may in fact be more than 2^{128} times stronger than AES-128, as AES-256 has 14 internal "rounds" of computation, while AES-128 has only 10. If there is an algorithmic weakness in the underlying AES algorithm, that weakness should be easier to exploit if there are fewer rounds.

5.3.3 Cracking AES-128 with Grover's Algorithm

Grover's algorithm makes it possible to use a quantum computer to guess the right key with fewer steps than it would take to try all possible keys. To understand why AES-128 is vulnerable to a quantum computer running Grover's algorithm but AES-256 is not, it is necessary to understand more about how Grover's algorithm works in practice.

Although Grover's discovery is frequently described as an algorithm for speeding up "database search," this gives a misleading impression as to what the algorithm actually does. The "database" is not the kind of database that most people are familiar with: it doesn't actually store data. Instead, the database is a database of guesses and whether or not each guess is correct.

In Table 5.2, we have recast the problem of cracking an encrypted message into a database search problem that could then be searched using Grover's algorithm. To perform a brute force search for the correct key, just start at the top and examine each row until the database value is a 1. In this example, a little more than half of the rows need to be examined. If you have a computer that can examine 90 billion rows a second – on par with the speed of the EFF DES Cracker – then you will find the answer in roughly five days.

A key search attack is possible because 2^{56} is not such a fantastically large number after all – that's the point that Hellman made in his letter to the NBS when he urged that 56 bits was just too small. If NBS had gone with a 64-bit key length, then an average search time of 20 hours would become 1280 days. That's better, but it's still not good enough for government work, which requires that national security secrets be declassified after 50 years[65] unless they contain names of confidential intelligence sources, contain information on weapons of mass destruction technology, would "reveal information that would impair US cryptologic systems or activities," or meet a few other specified requirements.[66] Clearly for US government use, an encryption algorithm that might be crackable at any point in the foreseeable future due to the likely advance of computer technology is not acceptable.

---

[65] For an explanation of the origin of this phrase and its corruption, see Lerman, *Good Enough for Government Work: The Public Reputation Crisis in America (And What We Can Do to Fix It)* (2019).

[66] Obama, "Executive Order 13526: Classified National Security Information" (2009).

Table 5.2. To use Grover's algorithm to crack an encryption key, Table 5.1 is recast as a database search problem, where one row has the value of 1 stored and all of the other rows have the value of 0 . In this example the keys are 56-bit DES keys. If this table instead used 128-bit AES keys, the last row would be number $340\,282\,366\,920\,938\,463\,463\,374\,607\,431\,768\,211\,455$ ($2^{128} - 1$).

Row	Row number in binary	Database Value
0	0000 ... 0000	0
1	0000 ... 0001	0
2	0000 ... 0010	0
3	0000 ... 0011	0
4	0000 ... 0100	0
...		
38 326 038 678 974 151	1000 ... 0111	1
...		
72 057 594 037 927 935	1111 ... 1111	0

As we have stated above, AES-128 doesn't have this problem, because 2^{128} is fantastically larger than 2^{56} – unless the attacker has a functioning quantum computer that's large enough to compute AES-128.

Cracking AES-128 with Grover's algorithm is surprisingly straightforward. First, it is necessary to construct an implementation of AES-128 on a quantum computer with at least 129 qubits, such that when the first 128 qubits have the correct decryption key, the 129th qubit has the value of 1 . Additional qubits are required to implement various details of Grover's algorithm and to properly implement AES-128 (we won't go into the details here).

AES-128 has 10 rounds, which means there is an inner algorithm that is repeated in a loop 10 times. Quantum computers don't have this kind of loop, so it is necessary to *unroll* the rounds, meaning that the circuits for the inner AES function need to be repeated 10 times. Additional circuitry is required to detect when the correct decryption key has been found.

It's relatively straightforward to imagine how the AES-128 circuit might be run on the kinds of superconducting quantum computers being developed by IBM and Google. On these computers, the qubits are "artificial atoms" made up of superconducting circuits operating at close to absolute zero, while the quantum gates and circuits are implemented by precisely timed and aimed pulses of radio waves. The speed of the quantum computation is determined by how quickly the quantum computer can cycle through a specific combination of radio waves that it sends into the artificial atoms. When the computation is finished, the qubits are measured with other radio wave pulses.

To run Grover's algorithm, each of the unknown bits (here, the 128-bit AES key) starts off as a superposition of 0 and 1. The algorithm is then cycled $\sqrt{2^N}$ times, where N is the number of unknown bits. At the end of these cycles, the unknown bits are measured, and they are overwhelmingly likely to have the answer to the problem. Superposition must be maintained for the entire time: if it is lost, the computation is ruined.

It turns out that $\sqrt{2^N} = 2^{N \div 2}$. So when cracking AES-128, only 2^{64} iterations are required, rather than 2^{128}. Because 2^{64} is not a fantastically large number, the mere existence of Grover's algorithm and the possible future existence of large-enough quantum computers was enough for cryptography experts to recommend discontinuing the use of AES-128 when these results became generally understood. However, AES-256 is still fine, because even with Grover's algorithm reducing the security parameter from 2^{256} to 2^{128}, that's okay because 2^{128} is a fantastically large number. All of this was clear from the theory, without the need to create an actual working quantum implementation of AES to actually try out Grover's algorithm.

In 2016, quantum computing theoreticians in Germany and the US carried out the hard work of actually building "working" quantum circuits of AES-128, AES-192, and AES-256 – at least, in theory. They found that implementing cracking a single AES-128 encryption key with Grover's algorithm requires *at most* 2953 qubits and on order of 2^{86} gates. For AES-256 the estimate was 6681 qubits and 2^{151} gates.

"One of our main findings is that the number of logical qubits required to implement a Grover attack on AES is relatively low, namely between around 3000 and 7000 logical qubits. However, due to the large circuit depth of unrolling the entire Grover iteration, it seems challenging to implement this algorithm on an actual physical quantum computer, even if the gates are not error corrected," the authors write. The authors conclude "It seems prudent to move away from 128-bit keys when expecting the availability of at least a moderate size quantum computer."

The word "prudent" requires additional explanation, as even a work factor of 2^{86} is likely to be beyond the limits of any human technology for the foreseeable future. For example, a quantum computer that could sequence quantum gates every *femtosecond* (that is, 10^{15} times per second) would still require 2451 *years* to crack a single AES-128 key using the implementation described in the 2016 publication. And a femtosecond clock would be a big deal – it would be 250 times faster than the clock speed of today's 4 GHz microprocessors. Chemical reactions take place at the femtosecond scale; the time is so short that light only travels 300 nanometers.

Of course, given a cluster of 1024 quantum computers, each running with a femtosecond clock, each one attempting to crack AES-128 with a different 10-bit prefix, an AES-128 message could be cracked in less than a year. So if mass-produced femtosecond quantum computers with a thousand qubits that can compute a single calculation error-free for a year is a risk that you consider relevant, then you should not be using AES-128 to protect your data!

But remember – the 2016 article describes an *upper bound*: it might be possible to create AES-cracking quantum computing circuits that require fewer gates. In fact, two 2019 efforts[67] lowered the upper bound on the work factor to crack AES-128 to 2^{81} and 2^{79} respectively by developing better quantum gate implementations for the AES oracle (the quantum code that determines when the correct key has been guessed). It has long been the case that hand-tuning algorithms to squeeze out the last few cycles of performance has been something of a parlor game among computer scientists.[68] So instead of looking for upper bounds, it might be more productive to look for theoretical lower bounds.

The absolute lowest bound for a circuit that could crack AES using Grover's algorithm would be a circuit that executed a single gate over a large number of qubits: such a perfect implementation would require a minimum of 2^{64} cycles to crack AES-128, and 2^{128} to crack AES-256. We (the authors) do not think that such a circuit is possible. However, this "perfect" quantum AES implementation would be able to crack AES-128 in 5.12 hours using our fictional

---

[67] Jaques et al., "Implementing Grover Oracles for Quantum Key Search on AES and LowMC" (2019); Langenberg, Pham, and Steinwandt, "Reducing The Cost of Implementing AES As a Quantum Circuit" (2019).

[68] For example, in 2010, a group of researchers at the Naval Postgraduate School that included one of us published a high-speed implementation of AES for the Sony PlayStation. See Dinolt et al., *Parallelizing SHA-256, SHA-1 MD5 and AES on The CellBroadbandEngine* (2010).

quantum computer with the femtosecond clock; even this perfect implementation would require $10\,782\,897$ *billion years* to crack a single AES-256 encryption.

To push the absurd hypothetical even more, there's no fundamental reason why we should limit our fictional quantum computer to a femtosecond clock. What if we had a smaller, more compact quantum computer that could fit in a nanosphere – perhaps two thousand packed atoms in a blob just $10\,\mathrm{nm}$ across. The maximum cycle time of this computer would be roughly $\frac{1}{30}$ of a femtosecond, the time it takes light to move from one side of the sphere to the other. With this computer and the (fictional) perfect Grover AES circuit, you could crack AES-128 in just 10 minutes, but it would still take 360 billion years to crack AES-256. Here parallelism finally begins to help: with a billion of these computers, you could crack an AES-256 encryption in at most 3.6 years. Of course, if you have the kind of technology that can make and control a billion of these computers, there are probably far more productive things you would be able to do than to go after AES-256 keys from the 2020s.

So to summarize, although it's conceivable that AES-128 might one day fall to a futuristic quantum computer, there is no conceivable technology that could crack an AES-256 encryption using exhaustive key search. What's more, AES-128 is sufficiently close to the boundary of what a quantum computer might be able to crack over the next 20 or 30 years that it is indeed "prudent" to stop using AES-128 in favor of AES-256. In part, this is because the cost increase of using AES-256 instead of AES-128 is quite minor: on a 2018 Apple "Mac Mini" computer, encrypting a $7\,\mathrm{GiB}$ file took $7.1\,\mathrm{s}$ with AES-128 running in "cipher block chaining" mode; with AES-256 it took $9.1\,\mathrm{s}$. For the vast majority of applications this 28 percent increase in encryption time is simply not significant.

But remember – all of the analysis above assumes that AES-256 is a perfect symmetric encryption algorithm. There might be underlying vulnerabilities that make it possible to crack AES-256 encrypted messages with significantly less work than a full brute-force attack. To date, no such attacks have been published that offer speedup greater than Grover's algorithm,[69] but there's always tomorrow. Certainly, if computer scientists discover that P=NP, then

---

[69]There is one classical attack against AES-256 that lowers the work factor from 2^{256} to $2^{254.4}$; Grover's quantum algorithm lowers the work factor to 2^{128}.

attacking AES-256 could become the stuff of high school science fairs soon thereafter.

5.3.4 Grover's Algorithm Today

The impact of the square-root speedup offered by Grover's algorithm has been systematically misrepresented in the popular press over the past two decades. Recall that although Grover's algorithm speeds up *search*, it is not the kind of search that we do with Google looking for a web page or using an accounting system when we are looking for a specific transaction. Those kinds of searches involve the computer scanning through a database and looking for a matching record, as we discuss in Section 3.5.1 (p. 102). Although Grover's algorithm *could* be applied to such a search, it would require storing the entire database in some kind of quantum storage – a system that has only been well-specified in works of science fiction – playing the entire database through the quantum circuit, a process that would eliminate any speedup provided by Grover's algorithm in the first place.

To date, scientists have accomplished only limited demonstrations of Grover's algorithm. Beit, a quantum software company with a lab in Kraków, Poland, released two unpublished papers in 2020 reporting state-of-the-science accomplishments in applications of Grover's search. A September 2020 paper from the group demonstrated a Grover implementation in IBM hardware, where the team performed an unstructured search among a list with just 16 elements. The goal of such a search is to identify one element in the list successfully, but the system was able to do so on average only 18–24 percent of the time.[70] A subsequent study employed Honeywell's 6-qubit Model H0 ion trap, which is commercially available. In June 2020, Honeywell hailed the device as the world's most powerful quantum computer, claiming that it has a quantum volume of 64.[71] The Beit team, using Honeywell's API, tested Grover's search in 4, 5, and 6-qubit implementations. Respectively, the team could select the right result 66

---

[70]Gwinner et al., "Benchmarking 16-Element Quantum Search Algorithms on IBM Quantum Processors" (2020).

[71]*Quantum volume* (QV) is a metric that IBM created that measures the square of the number of quantum circuits that a quantum computer can implement. According to IBM, QV combines "many aspects of device performance," including "gate errors, measurement errors, the quality of the circuit compiler, and spectator errors" (Jurcevic et al., "Demonstration of Quantum Volume 64 on a Superconducting Quantum Computing System" (2020)).

percent of the time with a 4-qubit circuit (selecting from a list with 16 elements), 25 percent of the time with a 5-qubit circuit (using a list with 32 elements), and just 6 percent of the time using all 6 qubits in a circuit (using a list with 64 elements).[72]

Some articles in the popular press incorrectly describe quantum computers as machines that use superposition to simultaneously consider all possible answers and select the one that is correct. Such machines do exist in the computer science literature, but they are called "nondeterministic Turing machines" (see Section 3.5.3, p. 107). And while such machines do exist *in theory*, they do not exist *in practice*: the conservation of mass and energy makes them impossible to build in this universe.[73]

Quantum computers use superposition to simultaneously consider a multitude of solutions, which *does* allow them to compute the answers to *some kinds of problems* faster than computers that are not based on superposition and entanglement. But they don't do this by coming up with the single, best answer to those problems. Instead, modern quantum computers are like a carefully designed collection of dual-split experiments (see Section B.1.3, p. 490): they have a distribution of possible answers – like an interference pattern on the screen – with the more probable answers coming up more often and the less probable answers coming up less often. The trick to programming the machines is to set up the computer so that the answers that are correct are significantly more probable and that incorrect answers are significantly less probable. This is done, ultimately, with constructive and destructive interference at the quantum level, in the machine's Schrödinger wave equation.

Another source of confusion might be that quantum computers can solve particular kinds of problems in polynomial time that are thought to be harder than the complexity class known as P (polynomial). The key example here is factoring. Because NP (nondetermin-

---

[72]Hlembotskyi et al., "Efficient Unstructured Search Implementation on Current Ion-Trap Quantum Processors" (2020).

[73]Such machines are not even possible if you subscribe to the many-worlds interpretation of quantum physics: it may be that a computer facing an NP-hard problem with a quantum-mechanical random number generator splits the universe 2^N times and that in one of those universes a computer immediately finds the correct answer. The problem is that in all of the *other* $2^N - 1$ universes the computers all discover that their answer is incorrect, and there is no inter-universe network to allow the computer that guessed correctly to inform its clones of the correct choice.

The Limits of Quantum Computation

"The manipulation and transmission of information is today carried out by physical machines (computers, routers, scanners, etc.), in which the embodiment and transformations of this information can be described using the language of classical mechanics," wrote David P. DiVincenzo, then a theoretical physicist at the IBM T.J. Watson Research Center, in 2000.[a] "But the final physical theory of the world is not Newtonian mechanics, and there is no reason to suppose that machines following the laws of quantum mechanics should have the same computational power as classical machines; indeed, since Newtonian mechanics emerges as a special limit of quantum mechanics, quantum machines can only have greater computational power than classical ones."

"So, how much is gained by computing with quantum physics over computing with classical physics? We do not seem to be near to a final answer to this question, which is natural since even the ultimate computing power of classical machines remains unknown."

For example, DiVincenzo wrote, we know that quantum computing does not speed up some problems at all, while some are sped up "moderately" (in the example of Grover's algorithm), and others are "apparently sped up exponentially" (Shor's algorithm).

DiVincenzo notes that, on purely theoretical grounds, quantum computing also could result in a "quadratic reduction" in the amount of data required to be transmitted across a link between two parties to complete certain mathematical protocols. But such a reduction requires the data is transmitted as quantum states – over a quantum network – rather than as classical states. "The list of these tasks that have been considered in the light of quantum capabilities, and for which some advantage has been found in using quantum tools, is fairly long and diverse: it includes secret key distribution, multiparty function evaluation as in appointment scheduling, secret sharing, and game playing."

---

[a]DiVincenzo, "The Physical Implementation of Quantum Computation" (2000).

istic polynomial) is the class that most people think is harder than *P*, and *NP* is the class solved by nondeterministic Turing machines, some people jump to the conclusion that quantum computers can solve NP-hard problems.

There are several problems with this line of thinking. First, just because mathematicians haven't found an algorithm that can factor in polynomial time doesn't mean that such an algorithm doesn't exist: it wasn't until 2002 that mathematicians had an algorithm for primality testing that ran in polynomial time. So factoring might be in *P*, and we just haven't found the algorithm yet. Or, more likely, factoring might be harder than *P* and still not in *NP*. Or, it might be that *P = NP*, which would mean factoring in both *P and NP*, because they would be the same. As we discussed in Section 3.5.6 (p. 116), computer scientists use the complexity class called *BQP* to describe the class of decision problems solvable by a quantum computer in polynomial time. Just as we don't know if *P* is equal to *NP*, we don't know if *BQP* is the same as or different from *P* or *NP*. This can be written as:

$$P \overset{?}{=} BQP \overset{?}{=} NP \tag{7}$$

For further discussion of this topic, we recommend Aaronson's article "The Limits of Quantum."[74]

Similar to the situation with the NP-hard and NP-complete problems, there is no proof that quantum computers would *definitely* be faster at solving these problems than classical computers. Such a mathematical proof would put theoreticians well on their way to solving the whole *P ≠ NP* conjecture, so it is either right around the corner or it is a long way off. It is simply the case that scientists have discovered efficient algorithms for solving these problems on quantum computers, and no such corresponding algorithms have been discovered for classical computers.

5.4 Conclusion

Whereas the electromechanical and early electronic computers of the 1940s were transformative, allowing the United Kingdom to crack the German Enigma code and the United States to create the hydrogen bomb, the main use of quantum computers today in 2021 is by researchers who are developing better quantum computers, better quantum algorithms, and students who are learning about quantum

---

[74]Aaronson, "The Limits of Quantum" (2008).

The Quantum Algorithm Zoo

Stephen Jordan, a physicist at Microsoft Research who works on quantum computing, maintains a database of quantum algorithms – the Quantum Algorithm Zoo. Jordan categorizes today's quantum algorithms into four types:[a]

1. **Algebraic and number theoretic algorithms,** which use properties of quantum computers to solve number theory problems. An example is Shor's algorithm for factoring.

2. **Oracular algorithms,** which depend upon an *oracle* that can provide an answer to a question. An example is Grover's algorithm for speeding up search.

3. **Approximation and simulation algorithms**, such as would be used to simulate the process of nitrogen fixation as discussed in Section 5.1.1, "Nitrogen Fixation, without Simulation" (p. 181).

4. **Optimization, numerics, and machine learning algorithms,** which could be used for improving systems based on so-called neural networks, including speech, vision, and machine translation.

[a]You can find the list of algorithms at Jordan's website, http://quantuma lgorithmzoo.org/, which is based on his May 2008 MIT PhD thesis (S. P. Jordan, *Quantum Computation beyond The Circuit Model* (2008)).

computers. The main output of today's quantum computers is not military intelligence and might, but papers published in prestigious journals.

Nevertheless, it would be a mistake to dismiss this research as quantum navel gazing. Unlike the limits that have impacted Silicon Valley's efforts to make increasingly faster electronic computers, we may be a far way off from hitting any fundamental limit or law of nature that will prevent researchers from making larger and faster quantum computers – provided that governments and industry continue to invest the necessary capital.[75]

---

[75]If it turns out that we can never make machines that work at large scale, then it

This may be why some governments continue to pour money into quantum computing. Although promoters speak about the benefits in terms of simulation and optimization, they are surely also driven by that darker goal of being able to crack today's encryption schemes used to secure the vast majority of information transmitted over the Internet and through the air. And because information transmitted in secret today might be useful if decrypted many decades from today, *the mere possibility* that powerful, reliable quantum computers might exist several decades in the future is a powerful influencer today.

Today's quantum computers are not nearly powerful enough to break the world's cryptography algorithms (or do anything else), but each year they improve, as quantum computing engineers become more adept at precisely controlling fundamental quantum processes. For this reason alone, our society should seek to rapidly transition from today's quantum-vulnerable encryption algorithms like RSA and AES-128 to the next generation of post-quantum encryption algorithms. If our understanding of quantum mechanics is correct, it is only a matter of time until the machines are sufficiently powerful.

We are still at the beginning of quantum computing, and very basic questions of technology and architecture still have to be worked out. The next chapter canvasses the research groups that are wrestling with different physical substrates for representing quantum information, different ways of organizing those physics packages into computing platforms, and different languages that programmers can use to express quantum algorithms. Much research in quantum computing is so preliminary and theoretical that an idea can have a major impact years before it's been reduced to practice and demonstrated. What's concerning is that the field hasn't had a mind-blowing discovery since the breakthroughs of Shor and Grover in the mid-1990s.

---

is likely that there is something fundamentally wrong about our understanding of quantum physics. Many advocates say that this alone is worth the study of quantum computers. And while some funding agencies might disagree, the amount of money spent on quantum computing to date appears to be significantly less than the $10–$20 billion that the US high energy physics community proposed spending on the Superconducting Super Collider in the 1990s, or even the $4.75 billion that Europe spent on the Large Hadron Collider between 1994 and 2014.

Quantum Computing Today

A
T the 25th Solvay Conference on Physics in 2011, John Preskill asked a question about quantum computing for which we still have no answer:

> Is controlling large-scale quantum systems merely **really, really hard**, or is it **ridiculously hard**?[1]

Preskill, who is the Richard P. Feynman Professor of Theoretical Physics at the California Institute of Technology, was asking if building ever larger quantum computers of the kind we envisioned in the last chapter is merely a matter of better engineering, or if there are fundamental limits about the nature of physics, computation, and reality itself that will get in the way. That is, are we likely to have working quantum computers "going beyond what can be achieved with ordinary digital computers" – what Preskill called "quantum supremacy" – after "a few decades of very hard work"? Or are we likely to come up short after even centuries of effort?

Preskill didn't have an answer, but he was enthusiastic about the quest: even if efforts to build a working large-scale quantum computer failed, humanity would still learn important fundamental truths about the fundamental nature of the universe.

---

[1]Preskill, "Quantum Computing and The Entanglement Frontier" (2012), emphasis in the original.

In the last chapter we discussed the first three great applications that have been envisioned for quantum computers: simulating quantum mechanical systems (Feynman), factoring large numbers (Shor), and speeding the search for solutions to any mathematical problem for which it is possible to construct a quantum oracle (Grover). All of these applications were developed by theoreticians working with nothing more than the metaphorical pencil and paper, and the ability to discuss ideas with their collaborators. Actually realizing these applications requires something more: a large-scale, reliable quantum computer.

Companies and research labs are racing to answer Preskill's question. Some are large, established technology powerhouses, like Google, IBM, and Microsoft. Others are well-funded emerging players, such as ColdQuanta, D-Wave and Rigetti. Most are building actual physics packages, with super-cooled superconductors and parts that are literally gold-plated. In most but not all cases, the results of these quantum computers can be reliably simulated using clusters of conventional computers. However, in a few cases, machines have been constructed that can solve problems beyond the capacity of today's digital computers – even when millions of those computers are networked together.

"I proposed the term 'quantum supremacy' to describe the point where quantum computers can do things that classical computers can't, regardless of whether those tasks are useful," Preskill wrote in 2019.[2] "With that new term, I wanted to emphasize that this is a privileged time in the history of our planet, when information technologies based on principles of quantum physics are ascendant."

After gaining traction, Preskill's term *quantum supremacy* has been somewhat supplanted by the term *quantum advantage*. Some researchers prefer this term, because it rightfully implies that quantum computers will be working alongside classical computers to literally confer advantage, just as a modern computer might offload some computations to a graphics processing unit (GPU).

Quantum computers have not scaled up at the same rate as their electronic computing predecessors. We have yet to experience a quantum form of Moore's Law (see Section 3.5, p. 98), in part because quantum engineers have not found a suitable quantum mechanism equivalent to the digital discipline that allows creating ever-larger

---

[2]Preskill, "Why I Called It 'Quantum Supremacy'" (2019).

digital circuits without ever-increasing amounts of systemic error (see Section 3.3 (p. 84)). Although quantum error correction schemes exist, it is unclear if they can scale to allow for meaningfully complex computations, because these schemes themselves require higher quality qubits operational for longer timescales than are currently possible. Without resolving this issue, we will still likely be able to create analog *quantum simulators* for solving questions in physics, chemistry, and biology, but the goal of using quantum computers to crack codes may remain forever out of reach. Nevertheless, researchers at both Google and the University of Science and Technology of China created quantum computing systems that clearly meet Preskill's requirement for quantum supremacy.

In this first section of this chapter we will describe in abstract the basics of how the current generation of quantum computers work. Next, in Section 6.2.2 (p. 237) we discuss the hardware efforts of today and the near future. We discuss what will need to be overcome in Section 6.3 (p. 242). Finally we conclude this chapter with Section 6.4 (p. 253).

6.1 How to Build a Quantum Computer

In Chapter 4 we introduced the basic idea of the Fredkin and Toffoli gates, and in Chapter 5 we discussed the two quantum algorithms that started serious money flowing into the creation of actual quantum computers. In this chapter we'll briefly look at a simple quantum circuit and discuss the barriers to creating quantum circuits of the size necessary to accomplish the computational goals set out in the previous chapter.

In a now classic article, David P. DiVincenzo, then at the IBM T.J. Watson Research Center, formulated five requirements for quantum computing:[3]

1. There needed to be something that could "hold data and perform computation." For simplicity, scientists have focused systems that have two precise states, which we call qubits. Whereas a classical bit can only have two values, `0` and `1`, quantum bits are a superposition of these two states. This superposition is typically written using Paul Dirac's Bra-ket notation as $a|0\rangle + b|1\rangle$, where a and b are taken to be complex numbers such that $|a|^2 + |b|^2 = 1$ during the course of the computation,

---

[3]DiVincenzo, "Topics in Quantum Computers" (1997).

but which become either 0 or 1 when they are measured at the end of the computation.[4] This measurement corresponds to "opening the box" in Schrödinger's famous thought experiment (see p. 523).[5]

2. The ability to initialize the qubits to a known "fiducial starting quantum state." This requirement is akin to resetting all of the bits in a classical computer to 0 . In his 1997 article, DiVincenzo wrote "I do not think that this 'initial state preparation' requirement will be the most difficult one to achieve for quantum computation." Three years later in his follow-up article, DiVincenzo was less sanguine: "The problem of continuous initialization does not have to be solved very soon; still, experimentalists should be aware that the speed with which a qubit can be zeroed will eventually be a very important issue."[6]

3. The ability to interact with each other using some form of *quantum gate*. This is where the Feynman and Toffoli gates from Section 4.5 (p. 151) become relevant. Each gate mixes the quantum state of two, three or more qubits together to perform some sort of simple computation. The physical construction of the quantum computer determines which qubits can be connected together. Ideally, the quantum gates are *universal*, so that they can be used to describe any computation (provided that you have sufficient qubits and time).

As we will see in Chapter 3, this design makes the construction and programming of quantum computers fundamentally different from the way we have built classical computers. In classical computers the bits represented by the presence or absence of an electric charge move through the electronic circuits, which are fixed at the time the computer is manufactured. In a quantum computer, it is the qubits that are fixed when the computer is manufactured, and the system is programmed by playing a sequence of circuits through the qubits to perform

---

[4]With two qubits, the systems state is described by a four-dimensional vector: $a\,|00\rangle + b\,|01\rangle + c\,|10\rangle + d\,|11\rangle$.

[5]Qubits must be physically isolated from the universe such that there is no external energy that would bias the qubit towards being 0 or 1 on measurement. This is why qubits do not need to be isolated from gravity: both the $|0\rangle$ and the $|1\rangle$ states have the same mass.

[6]DiVincenzo, "The Physical Implementation of Quantum Computation" (2000).

the desired computation. Thus, the computing speed of the quantum computer fundamentally depends on the number of qubits that it has and the speed at which the circuits can be constructed; this speed is exactly analogous to the clock speed of a modern microprocessor.[7]

4. The ability to keep the qubits in their *coherent, entangled* state for an extended period of time. This period of time is not measured in seconds, but in terms of how many gates can be played through the qubits. In his article, DiVincenzo suggested that it would be necessary to execute between a thousand and ten thousand gates in order to be able to perform meaningful computations with sufficient quantum error correction.[8]

   An added complication is how error propagates as the quantum computer begins to lose its coherency: if errors are correlated rather than randomly scattered through the system, it may adversely impact the ability to perform meaningful quantum error correction.

5. The ability to measure each qubit at the end of the computation.

We show what this looks like in Figures 6.1 through 6.3. This adder, which would be a small part of a much larger quantum circuit, takes two numbers between 0 and 15 and adds them together. The key difference between this adder and the 4-bit adder that you might find in a classical computer (such as Figure 3.5) is that this adder is reversible. The adder in Figure 6.3 uses 13 qubits and requires 30 gates. The design in Figure 6.3 also requires 30 cycles to operate because none of the gates execute at the same time. However, this algorithm can be optimized (Figure 6.4) by having many of the gates acting simultaneously. This optimized algorithm can run in just 7 cycles.

By *reversible*, we mean that this adder needs to be able to run in reverse. That is, it needs to be able to take the result of the addition, a single number between 0 and 15, and provide the two specific input numbers that were used to create it. This may seem like a magic trick! If we told you that the number 9 is the sum of

---

[7]In his 1997 and 2000 articles, the requirement of "a 'universal' set of quantum gates" is presented as the fourth DiVincenzo criterion.

[8]Long decoherence time was originally presented as the third DiVincenzo criterion.

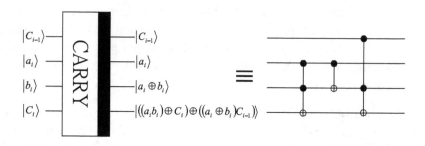

Figure 6.1. A 2-bit quantum carry gate, from Cheng and Tseng, "Quantum Plain and Carry Look-Ahead Adders" (2002), used with permission. The gate reversibly determines whether adding two bits produces a carry operation.

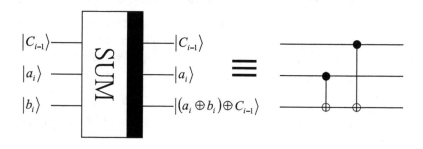

Figure 6.2. A 2-bit quantum sum gate, from Cheng and Tseng, "Quantum Plain and Carry Look-Ahead Adders" (2002), used with permission. The gate reversibly determines whether adding two bits produces a sum.

two numbers and asked you what they were, you would be unable to tell us: the answer might be 0 and 9, or 1 and 8, or 2 and 7, and so on. As a result, the quantum 4-bit adder needs more than 4 bits of output: besides the 4-bit sum, it also preserves half of the input bits. The adder also has an additional input bit called z and an output bit that combines z with the carry bit. Such additional qubits are sometimes called an *ancillary* or *ancilla qubits*; designing efficient quantum circuits that use a minimum number of ancilla qubits is one of the current challenges of quantum computer programming, due to the small number of qubits and the short decoherence times. Programming quantum computers at the circuit level in this manner is exactly analogous to the way that computing's pioneers in the 1940s and 1950s modified the hardware of their computers to add new instructions and programmed the machines using machine code.

In summary, in order to compute at the quantum level, one must

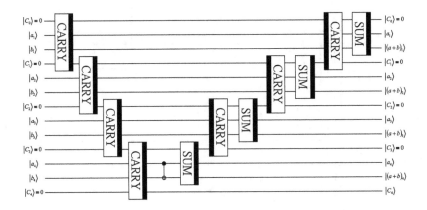

Figure 6.3. A 4-bit quantum adder circuit, from Cheng and Tseng, "Quantum Plain and Carry Look-Ahead Adders" (2002), used with permission. The inputs on the left are the nibbles $a_4a_3a_2a_1$ and $b_4b_3b_2b_1$ and the carry bit C_0. The output bits on the right are the sum $(a + b)_4(a + b)_3(a + b)_2(a + b)_1$, the input value $a_4a_3a_2a_1$, and the carry bit C_4. Time flows from left to right. Compare this with Figure 3.5, the 4-bit classical adder.

be able to generate, maintain, manipulate, and measure quantum states. Thus, quantum sensors are a precursor technology for quantum computing, and this is why this book presented quantum sensing first. In many ways, today's quantum computers are really just large-scale quantum sensor arrays.

6.2 The Quantum Computer Landscape

Preskill's 2019 article argues that the question he posed in 2012 is all but answered, and that we have moved from the era of quantum computing's first steps and into the era of noisy intermediate-scale quantum devices – NISQ – another term that he coined.

Unlike classical computers, which are nearly all based on silicon semiconductors, today's NISQ computers are not dominated by a single physical substrate. Instead, we are in a period of experimentation – one that might stretch out for decades. Today's quantum innovators are experimenting with different approaches to creating and managing the quantum states necessary for computation. To date, no one has realized the scale required for solving meaningful problems outside the world of experimental physics. The different media are promising in different ways, with some offering longer coherence times and greater interconnection, while others lack the need for specialized cooling or have engineering characteristics that might

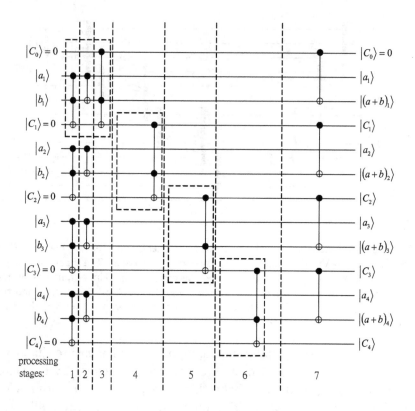

Figure 6.4. The 4-bit quantum adder from Figure 6.3, optimized to execute in fewer cycles. From Cheng and Tseng, "Quantum Plain and Carry Look-Ahead Adders" (2002), used with permission.

make a large-scale computer possible. We don't know which will be the winner.

6.2.1 *Comparing Quantum Media*

Understanding the quantum computing landscape is challenging because virtually every device that's been produced has different characteristics and capabilities. Some competitors claim to have relatively large-scale qubit devices, yet these may not be as interconnected as smaller devices, and large devices' size and architecture may be noisier and less stable than smaller devices. One cannot evaluate today's quantum computers simply by comparing the number of qubits they possess.

Adding to the difficulty, companies claims' on quantum computers may be strategically shaped to capture para-hardware markets, such as software and services. Companies have created vocabularies and software frameworks that are explicitly helpful to them and their business model. Even when claimed to be neutral and universal, these vocabularies and frameworks cannot help but seek to establish a software ecosystem that is favorable to their creators.

Competitors in the field all seek the *logical qubit*, a qubit that can overcome the problems of gate errors, environmental noise, and decoherence long enough to perform quantum operations. Understandably, competitors have chosen different paths for the construction of a stable quantum computer. The paths chosen reflect a deeper design approach philosophy where some innovators are focused on small devices with high levels of interconnectivity and stability, while others are focused on building the largest device possible. The philosophy of the large devices is that with many *physical qubits*, the device can manage its own error.[9]

We've seen this behavior before repeatedly over the 70-year history of computing. Computer engineers in the 1950s experimented with a variety of computing and storage media before settling on silicon for switching, core memory for short-term storage, and a combination of hard drives, magnetic tape and punch cards for long-term storage. Similar technology competitions and selections took place in the world of high-performance supercomputers in the 1970s and 1980s. This fight played out once again during the emergence of cloud computing in the 2000s, with the surprising (to some) discovery that vast computing clouds built from commodity hardware could outperform specialized high-performance systems on a wide variety of tasks, once companies like Amazon and Google developed approaches for overcoming the challenges with scale.

6.2.2 Five Kinds of Quantum Computers

The word "quantum" is attached to a range of devices, and terminology in the field sometimes takes a functional approach. That is, the category of the device is cast by its use rather than its underlying architecture and capabilities. The lines between different categories of quantum computers blur. When it comes to computing, the word *quantum* can describe:

---

[9]Doug Finke, the publisher of the Quantum Computing Report, maintains the most comprehensive and up-to-date summary and categorization of hardware and software approaches by competitors. Finke's site carefully tracks claims of device size, quality, and construction (Finke, "Qubit Count" (2021)).

- **Simulations of quantum computers.** On the most basic level, classical computers can be optimized to simulate quantum effects. The fundamental problem with using classical computers to simulate quantum systems it that today's algorithms require exponentially more steps to simulate a quantum system as the number of quantum particles increases; quantum computers do not have this problem (see **Section 5.1.2, "Modeling Chemical Reactions"**). However, we do not know if this exponential scaling is fundamental or not; an answer to that question would likely also result in an answer to the question of whether or not $P = NP$.

- **Quantum annealers.** Quantum annealers achieve quantum effects in specially prepared materials. D-Wave System's quantum annealer is the most well-known device in this category. A quantum annealer uses a metal material that exhibits quantum properties as it is cooled to temperatures close to absolute zero. Unlike a general purpose quantum computer, which uses gates to process qubits, the annealer is analog. The annealing process directly manipulates qubits.

  Quantum annealers are limited in function. Although D-Wave's machines have literally thousands of qubits,[10] the numbers cannot be compared with other kinds of quantum computers because the D-Wave qubits are not universal: they can only be used to solve a limited range of quantum problems. Specifically, the D-Wave can only solve problems phrased as quadratic unconstrained binary optimization (QUBO) calculations. When it comes to QUBO problems, D-Wave can solve problems that are significantly larger than almost all private companies in the field. D-Wave also hopes that its ability to solve optimization problems will make the system commercially attractive today to companies not interested in learning about quantum computing, but interested in actually using quantum computing to solve other problems. At this point, however, there is no clear

---

[10]D-Wave Systems scaled its annealer from 128 qubits, the D-Wave "One" released in 2011, to the D-Wave 2,000Q, a 2000-qubit annealer, in 2017. The 2,000Q has been commercially available since 2017; popular reporting suggests a $15m price tag (Temperton, "Got a Spare $15 Million? Why Not Buy Your Very Own D-Wave Quantum Computer" (2017)). The D-Wave advantage (2020) has 5000 qubits.

evidence that D-Wave's systems are more cost effective at optimizing than existing commercial optimizers such as CPLEX and Gurobi, run on traditional electronic computers.

- **Quantum simulators.** The Feynman vision that quantum computers would simulate quantum interactions is being pursued in the form of quantum simulators, devices that use, "entanglement and other many-particle quantum phenomena to explore and solve hard scientific, engineering, and computational problems," as described by a report signed by 37 attendees of a 2019 workshop organized by the National Science Foundation. According to the workshop report, there are now more than 300 quantum simulators operating around the world based on a wide variety of underlying platforms. Those working in the field are pursuing a two-phase strategy: in the first phase, early prototypes are built that are research curiosities in themselves. These early devices are intended to bridge to a second phase where a broader set of researchers can employ quantum simulation, with a goal of moving second-generation devices out of quantum computing applied research laboratories and into other fields such as botany, chemistry, materials science, astronomy, and in the creation of other quantum devices, including quantum internet technologies (discussed in Chapter 7). That is, the goal is to stop doing research on quantum simulators, and to start doing research *with* quantum simulators.

Quantum simulators are similar in design to quantum computers, but as with quantum annealers, quantum simulators are not universal: simulators are constructed with a single goal of simulating quantum mechanical systems, and often on a single scientific problem, such as understanding photosynthesis. By taking the complexities involved in the pursuit of universality off the table, some see quantum physics simulators as the most compelling near-term strategy for quantum computing. The NSF group predicted: "Scaling existing bottom-up quantum simulators to hundreds or even thousands of interacting, entangled, and well-controlled quantum elements is realistically within reach."[11]

---

[11] Altman et al., "Quantum Simulators: Architectures and Opportunities" (2019).

- **Noisy Intermediate-Scale Quantum Devices (NISQ).**
  NISQs represent the state-of-the-science in programmable digital quantum computing. Universities, research labs, and private companies are pouring untold sums of money into developing an "intermediate-scale" device that could lend insights into the building of larger devices. That is, a mid-scale quantum computer with 50–100 qubits might reveal characteristics of materials or engineering that make creation of a 500-qubit device possible, and so on.

  NISQs are being built with several technology substrates, all familiar to readers of **Chapter 2, "Quantum Sensing and Metrology"**. Several large companies such as Google and IBM are betting on the superconducting circuit approach, where Josephson junctions form the basis of the architecture. This is the same underlying approach as superconducting quantum interference devices discussed in Section 2.2 (p. 40).

  Others, such as Honeywell, are experimenting with ion trap approaches (see Figure 6.5), where charged electronic particles are held in position with lasers, magnetic fields, or even in a physical substrate, such as the nitrogen-vacancy approach discussed in Section 2.2 (p. 41). Ion traps do not require super-cooling and enjoy long coherence times, but to date have been very limited in their number of qubits.[12]

  Photons are another option for NISQs. Photonic approaches also avoid supercooling and have good stability, and can be implemented using existing materials, like silicon and optical devices from commercial providers such as ThorLabs. As of this writing, the largest quantum computer is a photonic interferometer in China, but the device is limited to a single scientific application (see Figure 6.6).

  Microsoft is pursuing a cutting-edge approach known as "topological qubits," which involves splitting an electron in order to store information redundantly and thus manage noise problems

---

[12]In June 2020, Honeywell announced that it had created "the world's highest performing quantum computer," bench-marking it with IBM's notion of a "quantum volume" of 64 (Honeywell, "The World's Highest Performing Quantum Computer Is Here" (2020)). The computer had only six qubits, yet its interconnection and low noise led the company to make dramatic performance claims (Crane, "Honeywell Claims It Has Built The Most Powerful Quantum Computer Ever" (2020)).

Figure 6.5. The device on the left is a vacuum chamber that houses four trapped ytterbium ions (on right) from Sandia National Laboratory. These ions can be measured using single-photon-sensitive media and are hoped to be a substrate for quantum computing and quantum memory. Photo courtesy US Air Force.

that cause decoherence. This approach is promising, but it is not nearly as developed as other approaches.

Despite their cutting-edge engineering, The National Academies of Sciences (NAS) characterizes NISQs as having "primitive" gate operations and as being plagued by error and decoherence. NAS' 2019 report concluded that today's NISQs will never scale to become the large-scale, general purpose quantum machines so desired.[13]

- **Large-scale quantum computers.** For many of the above-described efforts, the goal is to create a large, stable, universal digital quantum computer with millions of error-corrected qubits. Such a device would be similar to a modern high-performance computer. Stored in its creator's cloud warehouse, its universal functionality could be leased out to users to solve all manner of interesting problems. The question is now to realize that goal.

---

[13]Grumbling and Horowitz, *Quantum Computing: Progress and Prospects* (2019).

One path is through fundamental discoveries in materials science, chemistry, or physics that can be applied to manage qubits. Indeed, while cryptanalysis grabs the news headlines, companies in quantum computing identify chemistry and materials science as their research focus. This is because with a mid-scale quantum computer, one might discover fundamental insights in materials design and in chemistry that elucidate strategies to build a larger quantum computer. Thus, like classical computers before it, quantum computer strategy is to trigger a virtuous cycle of growth. This insight also foreshadows an innovation policy issue: groups that can make those fundamental observations are likely to pull ahead of the pack, building ever-larger computers with teams that were trained over decades, using discoveries that competitors cannot obtain. In this large-scale scenario, quantum computing could be a *winner-take-all technology*, suggesting that the first innovator might well become the most successful one.

Alternatively, the path to the large-scale quantum computer may be just a matter of scaling up existing approaches. This appears to be the strategy of several reputable companies in the quantum computing field that are creating ever-larger devices based on superconducting circuits. Perhaps the manufacture of densely produced, well connected and controlled Josephson junctions will yield room-sized quantum computers with millions of qubits.

When will a large-scale quantum device be built? Even scientists at companies known to enthusiastically promote their technologies say that it will take a decade. Some say several decades. Others say this task is impossible. The next section turns to the reasons why building a quantum computer is so difficult.

6.3 Skeptics Present Quantum Computing's Challenges

Almost 20 years ago, physicists Jonathan P. Dowling and Gerard J. Milburn wrote that humankind had entered a new stage of quantum information science: the second quantum revolution. In the first quantum revolution, scientists used quantum mechanics to better understand our reality. Truly a scientific revolution, the first period of QIS started with theory and expanded over the century as more insights were gained (see Appendix A and Appendix B). The second

Figure 6.6. In 2020, Jian-Wei Pan and Chao-Yang Lu at the University of Science and Technology of China built a large-scale interferometer to solve the "boson sampling" problem, a task insoluble with classical computers. With 25 laser sources and 100 single-photon sensors, the Jiuzhang Device demonstrates the link between quantum sensing and computing. Image courtesy of Jian-Wei Pan.

quantum revolution is a technological one, where scientists actively employ "quantum mechanics to alter the quantum face of our physical world."

Dowling and Milburn canvassed the exciting state-of-the-science developments of this second revolution. Finally they warned that, "A solid-state quantum computer is probably the most daunting quantum technological challenge of all and will require huge advances in almost all the areas of quantum technology we have discussed."[14]

Significant progress has been made since then. Nevertheless, quantum computing still depends on realizing a number of technical feats. Until now we've presented the challenges as significant but surmountable. However, a significant number of well-credentialed experts maintain that general purpose quantum computing is simply not achievable with physics as we understand it today. This section details those challenges.

6.3.1 Scientific Challenges

A 2019 National Academies of Sciences (NAS) report characterized quantum computing as consisting of creating small, proof-of-concept,

---

[14]Dowling and Milburn, "Quantum Technology: The Second Quantum Revolution" (2003).

demonstration devices.[15] This is because quantum computing requires a mastery of quantum superposition and entanglement, development of software and control systems, and management of costly, difficult physical conditions. But more than that, breakthroughs in quantum computing may also require fundamental breakthroughs in basic physics – or at very least, transitioning phenomena that have only been observed in a laboratory setting (and only in the last decade) into engineering prototypes.

To get an idea of the gap between theoretical advance and engineering realization, consider that Microsoft's approach, the "topological qubit,"[16] is based on a 1937 theoretical prediction that single electrons can be split into subparticles.[17] Now Microsoft hopes to use the phenomena to create a working quantum computer. But it took 75 years between the theory's discovery to produce evidence that the subparticles exist.[18] Microsoft collaborated with the Delft University of Technology (TU Delft), the oldest and largest Dutch public technological university in the Netherlands to substantiate the existence of the particles. In 2018, Microsoft published a paper with more evidence but the paper was retracted in 2021.[19]

Some argue that quantum computing will never be achieved; indeed, some claim that modern quantum computing research efforts are reaching the end of what they can accomplish. Physicist Mikhail Dyakonov wrote a short book about the challenges and reprinted a warning that Rolf Landauer urged scientists to include in their papers and talks: "This scheme, like all other schemes for quantum computation, relies on speculative technology, does not in its current form take into account all possible sources of noise, unreliability and manufacturing error, and probably will not work."[20]

A chorus of other commentators have downplayed quantum computing as an overhyped phenomenon. In 2015, a US Air Force advisory board found that technology advocates "herald[ed]" imminent

---

[15]Grumbling and Horowitz, *Quantum Computing: Progress and Prospects* (2019).

[16]Microsoft Corp., "Developing a Topological Qubit" (2018).

[17]Majorana and Maiani, "A Symmetric Theory of Electrons and Positrons" (2006).

[18]Mourik et al., "Signatures of Majorana Fermions in Hybrid Superconductor-Semiconductor Nanowire Devices" (2012).

[19]H. Zhang et al., "Quantized Majorana Conductance" (2018).

[20]Dyakonov, *Will We Ever Have a Quantum Computer?* (2020); Dyakonov, "When Will Useful Quantum Computers Be Constructed? Not in The Foreseeable Future, This Physicist Argues. Here's Why: The Case against: Quantum Computing" (2019).

breakthroughs but nevertheless, "no compelling evidence exists that quantum computers can be usefully applied to computing problems of interest to the Air Force."[21]

The most specific critique comes from the 2019 NAS report of the field that made both economic and technological assessments. On the economic front, the NAS group observed that there are essentially no economically advantaged uses for quantum computers for the foreseeable future (and obviously no consumer ones either).[22] This is directly different from the history of computing, in which spending money on computing was advantageous from the very first dollar spent. From the beginning, spending money on computing – be it mechanical, electromechanical, or electronic – made it possible to do something that wasn't otherwise possible, or to do it faster, or for less money overall. Although quantum computing might one day make it possible to train large-scale artificial intelligence machine learning models faster and with far less electricity than is currently the case, this does not seem to be a breakthrough that is plainly visible on the short-term horizon.

6.3.2 Engineering Challenges

Without uses that produce big savings or profits in the near term, funding for quantum computing is likely to be limited to governments and the largest technology companies. As such, quantum computing lacks the "virtuous cycle," like what was enjoyed with classical computers, with increasing commercial and consumer utility driving demands and willingness to pay for fantastic technological innovations.

The NAS survey's core technological critique relates to the difficulty of scaling up today's quantum systems into larger systems that can be used to solve meaningful problems. As a result of these challenges, the survey found it too uncertain to predict when a scalable quantum computer would be invented and that existing devices could never scale into general-purpose machines.

Quantum computers are characterized by the integration of multiple qubits. Thus, for a quantum computer to work, one needs to be able to encode, entangle, manipulate, and maintain an array of qubits, raising the challenges visited in Chapter 2. The challenges inherent in quantum computing are thus different from the obstacles

---

[21]US Air Force Scientific Advisory Board, *Utility of Quantum Systems for The Air Force Study Abstract* (2016).

[22]Grumbling and Horowitz, *Quantum Computing: Progress and Prospects* (2019).

encountered by engineers building and then scaling digital computers. Classical computers went through an evolution of mechanical, to relay, to tube, and to discrete transistors, and finally to integrated circuits. Each improvement produced systems that were smaller, faster, and required less overall energy to perform a computation. Semiconductors enjoyed their own virtuous cycle, providing chip makers with tools for designing and manufacturing computers that were ever more complex yet less expensive. Quantum computing has not realized a scaling breakthrough on the level of the transistor. Perhaps more to the point, there is no such breakthrough lurking in the future of any realistic technology road map. In many ways this is similar to the days of mechanical, electromechanical and tube-based computing, when larger computers might be faster than smaller ones, but they were also dramatically more expensive and less reliable.

Different technologies can be used to create qubits, but for each, quantum scientists must be able to master and control events at quantum scales (see Appendix A). Mastery and control require substantial technical expertise, reflected in the multidisciplinary nature of quantum computing teams (engineers, physicists, mathematicians, computer scientists, chemists, materials science). This is also a difference from the last 70 years of computing, which generally required mastery of fewer technical domains, and where modularization and isolation between technical domains meant less interdisciplinary work.

Quantum computers require that their qubits be entangled, cohered into a group that can be operated upon. But at the same time, quantum computers must be shielded from the universe, lest noise in the environment cause those qubits to decohere. This makes the quantum computer challenge fundamentally different from the classical computer. The transistor allowed scale with intricately managed stability. However, with quantum computers, scale requires the management of additional, exquisitely fragile quantum states.

When qubits decohere, they lose information. Thus, quantum algorithms have to be crafted to be efficient enough to execute before coherence is lost. As of this writing, some state-of-the-science devices have coherence in the hundreds of *micro*seconds, a time too short for the quantum gates of today to process significant numbers of qubits. This is a time period so short that human physical experience has no analogue for it. A blink of the eye takes about 100 000 microseconds.

The longer quantum computers run, the more performance degrades. In classical computing, extra bits are used to correct ordinary errors that occur in processing. This approach works because of all the engineering performed in classical computers to avoid quantum effects like tunneling. In quantum computing, many of the qubits employed are dedicated to error correction, so many that it creates significant overhead and degrades computing performance. Current thinking is that to emerge from the era of NISQ machines, as many as 90 percent of a quantum computer's qubits might have to be dedicated to error correction.[23] Initially, one might suggest just adding more qubits to achieve reliability, but as more qubits are added, system complexity increases, and quantum devices become more prone to both random environmental interference and to noise from the computer's own control system.

Quantum computers are not fault tolerant. In addition to temperature, vibration and electromagnetic interference can easily destabilize quantum computers. Conventional electronic computers rely on the digital discipline to smooth out errors so that they effectively do not matter.[24] In quantum devices, by contrast, errors are not rounded out, but instead compound until the conclusion of the computation.

To shield quantum computers from environmental noise that triggers decoherence, many quantum computer architectures require supercooling. This cooling is *super* because it is colder than even the background temperature of the universe. Extreme frigidity is needed both to elicit quantum properties from materials (for instance, in analog quantum annealers) but also because heat increases the chances that random energy collisions will generate noise that will interfere with quantum states or cause decoherence.

Keeping quantum devices at 15 millikelvin ($-273\,°C$, $-459\,°F$) means that quantum computer scientists need liquid helium, an increasingly rare and valuable element, of which there is a finite supply on Earth. There are currently no limits on the usage of Earth's helium supply.[25] Unlike quantum computing, many other quantum

---

[23] Möller and Vuik, "On The Impact of Quantum Computing Technology on Future Developments in High-Performance Scientific Computing" (2017).

[24] In classical computing, bits of data are either a `0` or `1`. In that environment, error appears as a decimal value such as `0.1` or `0.9` that can be easily rounded to `0` or `1`. For more information, see p. 84.

[25] Some hope that early quantum computers will solve fundamental challenges in fusion. If that happens, we could create helium via hydrogen fusion.

technologies do not require supercooling. This means that some sensing and communications technologies can be miniaturized, commercialized, and deployed in many more challenging contexts (in outer space, underwater, in missiles) than quantum computers.

6.3.3 Validation Challenges

It will be necessary to validate quantum computers to make sure that the answers they produce are correct. Ironically (and annoyingly), validation is easy for many of the hard, long-term applications for quantum computing, but likely to be harder for the more probable, near-term applications.

For the algorithms like factoring with Shor's algorithm and search with Grover's, validation is easy: just try the answer provided by the quantum computer and see if it works. That is, if the quantum computer says that the 2227 are 131 and 17, one need merely multiply 131×17 to determine if the factorization is correct or not. The same logic applies to using Grover's algorithm to crack an AES-128 key: just try to decrypt the encrypted message: if the message decrypts, the AES-128 key is correct.

On the other hand, approaches for both error correction and validation are less developed for analog quantum simulators. One approach suggested in the 2019 NSF report is to run simulations forward and backwards (theoretically possible, since the computations should be reversible) to see if the simulator retraces its steps. Another approach is to see if different systems that should have equivalent outcomes do indeed have similar outcomes.

6.3.4 Ecosystem Challenges

A final challenge is not technical, but organizational. Significant work still needs to be done to create a rich ecosystem of quantum software. Beyond basic programming languages and compilers, which exist today, there is need for documentation for people at multiple levels of expertise, programming courses, systems on which to run those programs, and finally organizations willing to pay for training and to hire quantum programmers.

On the software front, many teams are developing languages to make interaction with quantum computers more routine and standardized. As of 2021, a growing "zoo" of quantum algorithms included 430 papers.[26] But the overwhelming number of these algorithms are expressed as *papers* in *scientific journals* or on *preprint servers*; they are not code on sites like GitHub that can be downloaded, incorporated into other, larger quantum programs, and run. Recall that Ed Fredkin got himself hired without a college degree to write programs for BBN's first computer in 1956 (and which he convinced BBN to purchase – see Section 4.4.1 (p. 146)). We have not yet reached the point where it is possible to teach yourself quantum programming and get a job at a company that needs someone to write quantum algorithms to run on their quantum computer.

6.3.5 Quantum Supremacy and Quantum Advantage

Quantum Supremacy is an awkward term. As Preskill defined it in 2012, the goal is to perform a computation – any computation – that cannot be performed with a classical computer. But the term is misleading, because quantum engineers in China and the US have clearly achieved "supremacy" as defined by Preskill, but quantum computers are not supreme: for the vast majority of computations performed on planet Earth, you would not be able to use one of today's quantum computers. And even if reliable, large-scale quantum computers are available in the future, it is hard to imagine that these machines will be used for more than a tiny fraction of the world's computing problems. And even in these applications, quantum computers are likely to be co-processors that depend on classical computers for many functions. For these reasons, we prefer the term "quantum advantage" to describe the achievement of solving a problem with a quantum device that cannot be solved with a classical computer.

In December 2020, Jian-Wei Pan and Chao-Yang Lu made the most compelling claim of quantum advantage to date.[27] Their team built a large-scale interferometer to compute a specific problem, Gaussian Boson Sampling (GBS). The team named their device Jiuzhang, for the ancient Chinese manuscript focused upon applied mathematics, *Nine Chapters on the Mathematical Art*. But as exciting as the Jiuzhang development is, the device can perform just one computation. However, it's really fast!

Previously, Google researchers announced in October 2019 that they had achieved quantum supremacy using their 54-qubit Syca-

---

[26]Montanaro, "Quantum Algorithms: an Overview" (2016); S. P. Jordan, "Quantum Algorithm Zoo" (2021).

[27]Zhong et al., "Quantum Computational Advantage Using Photons" (2020).

Figure 6.7. Computing a specific distribution of photons that would have taken 600 million years to solve on the fastest existing classical supercomputer in 2020 was done in 200 seconds with a reported 99 percent fidelity by Jian-Wei Pan and Chao-Yang Lu at the Hefei National Laboratory, University of Science and Technology of China. However, turning the device into a "fault-tolerant universal quantum computer, is a very long-term goal and requires many more challenges to tackle, including ultra-high-efficiency quantum light sources and detectors, and ultra-fast and ultra-low-loss optical switch," Lu told us. Image courtesy of Jian-Wei Pan.

more superconducting approach.[28] Google's researchers programmed their computer to create and then evaluate random quantum circuits. IBM, a chief rival to Google, quickly disputed the supremacy claim, arguing on its research blog that "ideal simulation of the same task can be performed on a classical system in 2.5 days and with far greater fidelity."[29] In March 2021, two Chinese scientists claimed that they replicated the Google approach with higher fidelity using classical GPUs.[30] The scientists concluded with a humble brag that their "proposed algorithm can be used straightforwardly for simulating and verifying existing and near-future NISQ quantum circuits" and helpfully posted their approach on GitHub. These quick retorts

---

[28] Arute et al., "Quantum Supremacy Using a Programmable Superconducting Processor" (2019).

[29] Pednault et al., "On 'Quantum Supremacy'" (2019).

[30] Pan and P. Zhang, "Simulating The Sycamore Quantum Supremacy Circuits" (2021).

The Helium Challenge

Helium's stability, non-reactivity, and phase as a fluid at near absolute zero makes it useful for cooling both quantum computers and the magnets in Magnetic Resonance Imaging machines. And while helium is abundant in the universe, on Earth it is a non-renewable resource. The small amount of helium that our planet has is the result of underground radioactive decay. Helium is rendered along with natural gas; if it is released and not captured, it is no longer financially viable to collect from the air.

The US and Qatar are the largest producers of helium, with the US supply provided by a storage and enrichment facility in Amarillo, Texas, run by the US Bureau of Land Management. Russia's Gazprom and China are building plants in order to reduce their reliance on US sources. Because of helium's many uses, limited availability, and strategic relevance, conservationists have called for an international helium agency to preserve supply and prevent a crisis in availability, and to expand extraction of helium from existing natural gas plants.[a] But don't feel guilty about helium balloons: such consumption is inconsequential compared to industrial and medical uses.

Today the biggest consumers of helium are MRI machines and devices used at border crossings to detect dirty bombs and other nuclear devices. Quantum computers use less helium, and modern cryogenics equipment attempts to conserve and recycle it. D-Wave explicitly markets its annealer as recycling helium to avoid the need to continuously resupply the machine's local store of helium.

Some quantum computers require light helium, Helium-3. This is extracted from nuclear reactors, and is somewhat controlled. IBM's plans for a 1000-qubit superconducting device caused the company to develop a custom dilution refrigerator. Others are building supercooling capacities that do not use a cryogen like helium or liquid nitrogen. These non-cryogen coolers have a major disadvantage: they require much more electricity for cooling. However, as nations signal an interest in decoupling their technology stacks, nations without access to helium sales may simply turn to electric cooling.

---

[a]Nuttall, Clarke, and Glowacki, "Stop Squandering Helium" (2012).

to Google's claim demonstrate how scientists value their quantum computing bragging rights, even if the bragging is only about the ability to solve otherwise meaningless random quantum puzzles.

The Jiuzhang device is a clear demonstration of quantum advantage, but the device has no practical application. Whereas Google's claim of advantage stands on contested ground, its Sycamore device can be programmed to solve problems other than random puzzles, so it is probably more important from a commercial point of view.

For computer scientists, achieving quantum advantage was long seen as a kind of Rubicon. But for most organizations, the real quantum computing Rubicon will be the moment that quantum computing can perform some useful commercial, defense, or intelligence application. Competitors strive to make the case that they have some advantage to sell from quantum computing. Perhaps the most promising in the near term are proposals that use quantum computers to solve part of a problem or those that apply "low-depth algorithms" that promise some quantum speedup with practical payoff. For instance, Goldman Sachs proclaimed that by optimizing algorithms, there will be a quantum advantage in derivatives pricing from even small quantum computers by 2025.[31] If they are correct – or even if other financial services firms believe that Goldman Sachs is correct – the development could create a gold rush in quantum computing.

How can one make sense of quantum computers' power when they rely on different physical media (ranging from photonics to trapped ions to annealing) and when innovators claim to have more qubits than competing devices? Quantum computers cannot be evaluated simply by the number of qubits they have, otherwise D-Wave's 2000-qubit system would be leagues ahead of teams at IBM, Google, and Microsoft – even when those systems can clearly perform computations that the quantum annealer can't. To evaluate quantum devices, IBM created its own metric called *quantum volume*.[32] A computer's quantum volume is "the largest random circuit of equal width and depth that the computer successfully implements." Thus, quantum volumes are necessarily perfect squares: 2, 4, 9, 16, and so on. Unfortunately, the largest quantum volume that IBM measured was 16,

---

[31]Giurgica-Tiron et al., "Low Depth Algorithms for Quantum Amplitude Estimation" (2020).

[32]Cross et al., "Validating Quantum Computers Using Randomized Model Circuits" (2019).

on a machine with 4 qubits running a circuit with a depth of four gates. "We conjecture that systems with higher connectivity will have higher quantum volume given otherwise similar performance parameters," the authors state.

Despite all these challenges, governments and large technology companies (e.g. Fujitsu, Google, IBM, Microsoft, Toshiba) have devoted major resources to quantum computing, and several startups (e.g. IonQ, Rigetti, Xanadu) are betting the company on it. Competition has produced wonderful resources to learn about and even experiment with quantum computing. For instance, IBM and others have made instructional videos, extensive, carefully curated explanatory material, and even made rudimentary quantum computers available through the Web at quantum-computing.ibm.com for anyone who wants to try their hand at programming the machines.

Quantum computing efforts are either basic or applied research. Basic research projects, like the Large Hadron Collider (LHC) at the European Organization for Nuclear Research (CERN), can be huge impressive projects that reveal fundamental truths about the nature of the universe: at a cost of approximately $9 billion, the LHC is one of the most expensive scientific instruments ever built, and it is responsible for the "discovery" of the Higgs boson, but it is hard to draw a line from the LHC to improvements in day-to-day life of anyone except for several thousand construction workers, physicists, and science journalists. On the other hand, nuclear fission was discovered in December 1938 by physicists Lise Meitner and Otto Frisch,[33] which led to the creation of a working nuclear bomb within just seven years and the first nuclear power plants in 1954. Such is the unpredictability of research.

6.4 The Outlook for Quantum Computing

The long-term outlook for quantum computing may be hazy, but the near-term outlook for quantum computing companies appears to be quite bright.

As we saw in the last chapter, although it was the potential for quantum computers to crack codes that led to the initial burst of enthusiasm, interest in quantum computing is likely being sustained by the promise of using quantum technology as an advanced scientific instrument for learning more about quantum physics and quantum

---

[33]Tretkoff, "This Month in Physics History: December 1938: Discovery of Nuclear Fission" (2007).

chemistry. The payoffs may be directly in these fields, or they may simply be the development of superior quantum sensors that are usable throughout the military industrial complex.

As such, there are many practical regulatory implications at least in the short term:

1. Because of their expense and complexity, only large firms and governments are likely to be able to afford quantum computers for some time. This means that governments have a relatively small number of players to police in quantum computing, and that the technologies may be easier to monitor and control. This period of large-organization exclusivity may continue for decades. Consider that classical computers were the domain of universities, governments, and large companies until the personal computer revolution of the 1970s.

2. Because of their complexity, quantum computers require teams of multidisciplinary experts. This means that one cannot simply sell a quantum computer and expect a user to make sense of it. Sellers will be on the premises of buyers and will probably know about the buyers' intended uses of the devices. The business model may be selling services as much as selling the device itself.

3. Because of their sensitivity to interference of all types, quantum computers are likely to be placed in low-noise environments. For instance, the D-Wave system occupies a $10 \times 10 \times 10$ foot housing plus three auxiliary cabinets for control systems. The cabinet is part of a system to produce quantum effects in D-Wave's annealer, where the chip is the size of a thumbnail. This requires a vacuum environment, a low-vibration floor, shielding to 50 000 times less than the Earth's magnetic field, and cooling to 0.0012 K.[34] Such devices are unlikely to be installed in jets for forward-deployed use, although they might be deployable in a suitably outfitted ship.

4. Finally, large firms that build the first quantum computers are likely to offer services through the cloud until the engineering becomes easier and medium-sized enterprises can purchase their own devices. Until then, quantum computing is likely to

---

[34]R. Copeland, "The International Quantum Race" (2017).

be offered as an enhanced service, one optimized for specific problems.[35,36]

Taken together, these limits will shape the trajectory and offerings of quantum computers.

Despite the lack of a practical demonstration, many scientists believe that sufficiently large quantum computers will be much more powerful than classical computers for solving certain kinds of problems. We lack *proof* that quantum computers will be innately more powerful for the same reason that we lack proof that factoring is fundamentally more difficult than primality testing, or that mixed integer linear programming is fundamentally harder than linear programming. That is, we don't have a proof that $P \neq NP$.

---

[35]Ibid.
[36]Gibney, "The Quantum Gold Rush" (2019).

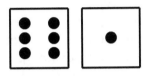

Quantum Communications

"QUANTUM communications" refers to two related applications: first, the use of quantum states to ensure true randomness in number selection and to communicate encryption keys to other parties, known respectively as quantum random number generation and quantum key distribution; second, the use of quantum effects themselves, such as the spin of photons, to encode a message, which is known as quantum internet or quantum networking.

There are four reasons to be excited by quantum communications and all are strategically relevant:

1. Properly implemented, quantum communications applications enjoy *information-theoretic security*, which means that no adversary, regardless of their computing resources or background knowledge, can decipher communications that have been covertly intercepted. Not even a quantum computer can decrypt such communications! This is because the security is a property of the underlying mathematics and quantum physics, rather than the putative "hardness" of a particular math problem.

   Quantum security guarantees to protect institutions against the future. Those continuing to use *computationally secure* post-quantum classical alternatives for distributing their keys rely on assumptions that may be proven incorrect. For instance, a mathematician may discover a new algorithm that unscrambles post-quantum encryption.

2. Quantum communications systems, unlike classical ones, reveal when a communication has been intercepted. That interception could be a surveilor, or it might be ordinary environmental interference, such as electronic noise or malfunctioning hardware. (Users of such systems typically cannot determine if the message failure was an accident of the environment or the actual presence of an eavesdropper.) The detection of interception capability results from the nature of quantum states. The act of interception interferes with quantum states, and this interference can be detected, unlike in classical communications, where interception is both easy and stealthy.

   For this reason, properly implemented quantum communications systems are not susceptible to proxying attacks. (You may also see these attacks referred to as "machine-in-the-middle" or "man-in-the-middle" attacks.) That's because if an attacker does intercept a photon carrying a particular quantum state, it is impossible for the attacker to both measure the photon's quantum state and retransmit a photon with the same quantum state.

3. In a fully quantum network that uses quantum states themselves to communicate, communication security becomes end-to-end. Users no longer have to rely on network trust, and can shut out eavesdroppers from both the content of their communications *and the metadata about those conversations.* Because governments extensively use metadata to study adversaries, this metadata-denying affordance of quantum internet schemes may be what is driving quantum network investments in Europe and China.

4. Just as Grover's algorithm speeds up some kinds of computations when performed on a quantum computer, some kinds of multi-party mathematical protocols enjoy a similar speedup when the parties communicate over a quantum network.

These benefits of quantum communications – information-theoretic security, awareness of message interception, the possibility of metadata secrecy, and certain kinds of optimizations – are driving both interest in quantum communications and its early commercialization. Indeed, the first quantum key distribution systems reached the market in 2005.[1]

Although quantum communication was discovered before quantum computing, another way to think about quantum communications systems is as a quantum computer with a "flying qubit" that travels from one party to the second, or with two flying qubits that travel from a common sender to two different receiving parties.

Quantum communications builds upon the technologies of quantum sensing discussed in Chapter 2, including single-photon detectors, the ability to perform low-noise measurements of quantum states, and even superconducting quantum devices.[2]

This chapter sets the stage for interest in quantum communications by briefly explaining the rise of signals intelligence (SIGINT) (Section 7.2 (p. 264)) capabilities of governments and the proliferation of these powers to nongovernmental actors. SIGINT is information derived from communications systems, radars, and weapons systems.[3] The chapter continues by explaining three quantum communications technologies, all of which can contribute to the confidentiality and integrity of communications.

First, quantum random number generation techniques use quantum uncertainty to create truly random numbers. Computer systems use high-quality random numbers in security, in simulations, and statistical models.

Second, quantum key distribution techniques use randomness to make secure encryption keys and ensure their confidentiality and integrity when they are transmitted to multiple parties. Although these protocols are called *quantum* key distribution, they are ultimately used to secure *classical* communications, for instance over the regular Internet or even the telephone.

Finally, a quantum internet would preserve quantum states and allow quantum computation between parties in different physical locations – possibly over great distances. This would provide both security against interception and secrecy of metadata. If the quantum

---

[1]Garfinkel, "Quantum Physics to The Rescue: Cryptographic Systems Can Be Cracked. And People Make Mistakes. Take Those Two Factors out of The Equation, and You Have Quantum Cryptography and a New Way to Protect Your Data" (2005).

[2]Takemoto et al., "Quantum Key Distribution Over 120 km Using Ultrahigh Purity Single-Photon Source and Superconducting Single-Photon Detectors" (2015).

[3]Director of National Intelligence, "What Is Intelligence?" (2019).

networking necessary to achieve the ideal of a quantum internet were achieved, one could likely use the technology to connect disparate, small quantum devices into a larger cluster computer, or connect multiple quantum computers together to create a larger quantum computer.

7.1 Information-Theoretic Security

To understand the power of information-theoretic security is to understand the sublime attraction of quantum methods for protecting communications. Because many readers will not be familiar with the concept of information-theoretic security, we present below three math problems: one that is easy, one that was hard in 1977 when it was posed but was solved in 1994, and one that is information-theoretic secure, which means that it cannot be solved with the information that we present, even by an attacker who has unlimited computer power.

7.1.1 An Easy Math Problem

Here is an easy math problem. The variables p and q are positive integers and p is less than q $(p < q)$.

$$p \times q = 15 \tag{1}$$

That is, what two numbers multiplied by each other equal 15? The answer is 3 and 5. This is an easy problem.

Recall that 15 is the number factored by IBM's quantum computer in 2001 (Section 5.2 (p. 188)). A simple way to think about this problem is to imagine that you have 15 cubes in a single line and you want to arrange them into a rectangle. If you did that, what would the dimensions of that rectangle be?

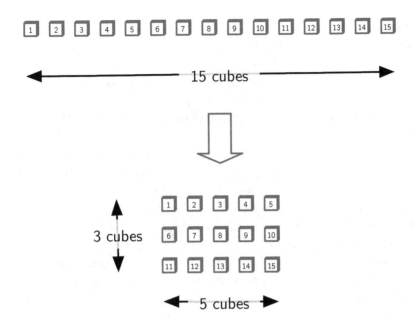

It turns out that there is only one way to make that rectangle, and that's with three rows of five cubes each.[4]

7.1.2 A Hard Math Problem

Here is a math problem that was posed in 1977 but was not solved until 1991, when it was cracked by an international team of 600 volunteers using more than a thousand computers. Instead of trying to factor the 2-digit number 15, try to break this number down to its prime factors p and q:

$$p \times q = 11438162575788886766923577997614661201021829672124236256256184293570693524573389783059 \quad (2)$$
$$71235639587050589890751475992900268795435 41$$

This 129-digit number is called RSA-129. It was chosen by Ron Rivest in 1977 as a puzzle to accompany the publication of a Martin Gardner column in *Scientific American*.[5] Like the number 15 in

---

[4]Turning the rectangle 90° so that it's five rows of three cubes each doesn't count as another "way" in this situation, because we required that the first factor be less than the second.

[5]Gardner, "Mathematical Games: A New Kind of Cipher That Would Take Millions of Years to Break" (1977).

equation 1, RSA-129 has two factors, here called p and q.[6] But what are p and q in this case? That was the problem posed by Rivest.

RSA-129 has a curious property: if you factor the number into its two primes, you can use the result to decrypt a secret message that Rivest wrote and encrypted back in 1977.

Factoring RSA-129 was computationally infeasible in 1977. Rivest didn't know how long it would be until computers were fast enough that it would be feasible. Gardner's column claims that Rivest estimated it would take "40 quadrillion years" to factor such a number. But that estimate was based on a single 1977 computer running with the best factoring algorithm of the day: in the following years computers got faster, factoring algorithms got better; it also became possible to connect many computers together to work on the same number at the same time. This is what we mean when we say that factoring RSA-129 was *computational infeasible* in 1977, or alternatively, that RSA-129 was *computationally secure* then. Finding the factors of RSA-129 is left as an exercise for the reader.

7.1.3 An Impossible Math Problem

Now here is a math problem that you can't solve no matter how much computational power you have:

> There is a line that passes through the points (x_1, y_1) and (x_2, y_2). Find the value of y where the line passes through the y-axis (that is, when $x = 0$), given that one of the points is (3,5).

That is, solve for y in this equation given $x = 0$, knowing that $x_1 = 3$ and $y_1 = 5$:

$$y = mx + b \tag{3}$$

This equation can't be solved to give a unique solution for y: you aren't provided with enough information. The equation $y = mx + b$ describes a line on a graph, where m is the slope of the line and b is y-intercept. It's the y-intercept that you are trying to find. You can't find the y-intercept because you only have one point on the graph.

---

[6]Mathematicians frequently reuse variable names like p and q in different equations, just as lawyers reuse labels like "plaintiff," "defendant," and "the Court" in different lawsuits.

This is an example of a problem that is information-theoretic secure (see the sidebar "Secret Sharing" on page 266).

Today nearly every use of encryption on the planet is protected using ciphers that are computationally secure. As we saw in Chapter 5, these algorithms can be cracked simply by trying every possible decryption key and recognizing the message when it is properly decrypted. Quantum computers promise to make this process faster. Even post-quantum encryption algorithms are still merely computationally secure: we know that with enough computer power, these algorithms can be cracked. There might also be short-cuts to cracking these algorithms that haven't yet been discovered, just as better approaches for factoring were discovered after 1977 that made it easier to factor RSA-129.

Adopters of a properly implemented quantum encryption system do not have to rely on *computationally secure* algorithms for distributing their keys. Instead, they use qubits, safe with the knowledge that if the qubits are intercepted by an adversary, then the legitimate sender and recipient will be able to determine this fact.

There are actually two ways to use quantum cryptography, one that is secure given what we know about quantum computers today, and a second that is secure given our understanding of quantum physics and the physical laws of the universe:

1. With **Quantum Key Exchange**, flying qubits are used to exchange an encryption key that is then used with a conventional quantum-resistant symmetric encryption algorithm, such as AES-256. Because we believe that AES-256 cannot be cracked on a quantum computer, this approach is believed to be secure for the foreseeable future. That is, the key exchange is information-theoretic secure, but the bulk encryption is only computationally secure.[7]

2. With **Quantum networking** or **"quantum internet,"** flying qubits are used to exchange *all* of the information end-to-end between the parties. This approach is information-theoretic

---

[7]Note that AES-256 is only computationally secure against our current notions of quantum computing. It might not be secure against a computer based on quantum gravity, or strange matter, multiverse computation, or some kind of physics that we haven't yet imagined. Specifically, it might not be secure against a device that could solve NP-hard problems in polynomial time.

secure if the laws of quantum computing are correct. Put another way, it is secure as long as it is impossible to predict the future with absolute accuracy.

7.2 Golden Ages: SIGINT and Encryption Adoption

Signals Intelligence is one of the oldest intelligence gathering disciplines (Table 7.1). Many histories of SIGINT start with the use of wireless during World War I by both German and Allied forces: radio offered the advantage of instantaneous communications to troops in the field, potentially anywhere in the world, but suffered from risk that the enemy could be privy to the communications as well. Radio was too powerful to ignore, but too dangerous to use without some mechanism for protecting communications. Military users resolved this conflict by turning to encryption.[8]

In recent years events surely have altered the balance between those who wish to eavesdrop on communications and those who wish to keep their communications private. However, there is no clear accounting as to which side is now ahead.

7.2.1 The Golden Age of SIGINT

On the SIGINT side, many governments have developed audacious, comprehensive, systematic programs to capture communications and personal data in order to identify people, to attribute actions to parties and adversaries, to perform link analysis (the evaluation of relationships among people, adversaries, and others), and to capture communications content. For instance, it is alleged that in 2011 the Iranian government used compromised encryption certificates to access the email accounts of hundreds of thousands of Iranians who used Google's Gmail.[9]

In recent years, there have been repeated accounts in the US media of both Chinese and Russian successes in exfiltrating data

---

[8]In fact, the use of both encryption and cryptanalysis by militaries predates the invention of radio by at least 2500 years. For a history of code-making and code-breaking, we recommend David Kahn's updated classic (Kahn, *The Codebreakers: The Comprehensive History of Secret Communication From Ancient Times to The Internet* (1996)) as well as the more manageable (Singh, *The Code Book: The Science of Secrecy From Ancient Egypt to Quantum Cryptography* (2000)). For a contemporaneous account of code-breaking during World War I, we recommend Yardley, *The American Black Chamber* (1931).

[9]Hoogstraaten et al., *Black Tulip Report of The Investigation into The DigiNotar Certificate Authority Breach* (2012).

Table 7.1. A sampling of the intelligence gathering disciplines (Director of National Intelligence, "What Is Intelligence?" (2019)).

GEOINT Geospatial Intelligence Gathered from satellite, aerial photography, and maps.

HUMINT Human Intelligence Gathered from a person. Includes diplomatic reporting, espionage, interrogation, traveler debriefing, and other activities.

IMINT Imagery Intelligence Analysis of images for their intelligence value. The National Geospatial-Intelligence Agency has primary responsibility for IMINT.

MASINT Measurement and Signature Intelligence Intelligence typically reviewed through the use of scientific measurement instruments. The Defense Intelligence Agency has primary responsibility for MASINT.

OSINT Open-Source Intelligence Analysis of information sources that are generally available, including news media and social media. The Director of National Intelligence's Open Source Center and the National Air and Space Intelligence Center are major contributors to OSINT.

SIGINT Signals Intelligence Intelligence gathered by analyzing "signals," which may include the analysis of intentional communications (COMINT – communications intelligence) and analysis of unintentional electronic emanations (ELINT – electronic intelligence). "The National Security Agency is responsible for collecting, processing and reporting SIGINT."

Secret Sharing

Secret sharing is an information-theoretic approach to splitting a secret into multiple parts. Invented independently in 1977 by G. R. Blakley[a] and Adi Shamir,[b] one primary use of secret sharing is splitting cryptographic keys used for data backups. Doing this renders the backup unusable unless multiple parties receiving the secret shares get together and reassemble the secret, allowing the backup to be decrypted.

Secret sharing works by representing the secret as a mathematical function that cannot be solved with the information present alone in each of the shares. In the example below, the secret is the y-intercept, which is where the straight line crosses the Y axis. Each share is a point on the line. Two points uniquely define a line, so without a second share, there is no way to identify the y-intercept.

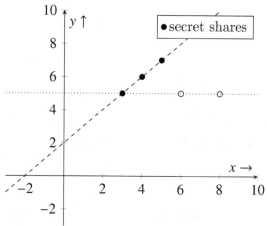

Here we see an example of secret sharing at work. The secret is $y = 2$ (the dashed line). The shares are $x_1, y_1 = (3, 5)$, $x_2, y_2 = (4, 6)$ and $x_3, y_3 = (5, 7)$. Combining any two secrets allows reconstructing the line. Notice that if the shares had been $(3, 5)$, $(6, 5)$ and $(8, 5)$, then the secret would have been $y = 5$. Thus, there is no way for a person receiving the share of $(3, 5)$ to know the value of the secret without combining their share with a share that someone else received.

[a]Blakley, "Safeguarding Cryptographic Keys" (1979).
[b]Adi Shamir, "How to Share a Secret" (1979).

from both public and private US information systems. With respect to China, the breach of the US Office of Personnel Management database resulted in the theft of records on more than 20 million current and past federal employees, including fingerprint records and lengthy, detailed forms used when applying for a security clearance. Chinese hackers are also reported to have stolen the credit reports on over a hundred million Americans. Between these two attacks, China can presumably identify and target people who are both likely involved in intelligence efforts and who are economically vulnerable. This data surveillance has real consequences for US efforts and is believed to have enabled China to identify multiple CIA assets in Africa.[10] Turning to Russia, the former superpower has many satellites, terrestrial assets, and near-shore submarines, all of which can be used for collection of SIGINT. At the end of 2020, the US intelligence stated that a supply chain attack on the US company Solar Winds, which makes software to help organizations monitor their computer systems, was "likely Russian in origin."[11] More than ten thousand US companies and government agencies were compromised as a result of the attack.

Books and reports that synthesize government programs into single readings, like Barton Gellman's *Dark Mirror*,[12] can seem like paranoid science fiction. In that book, for instance, Edward Snowden refuses to reveal whether he has a blender, for fear that the appliance's electrical signal would reveal his location to intelligence agencies. There is no way to know from public sources if Snowden's fears are justified. But we do know that in 2014 a smart refrigerator was taken over by hackers and used to send spam,[13] and that in 2019 the FBI's Oregon office warned that hackers can take over the microphones and cameras in smart TVs and use them for surveillance.[14] More recently, *New York Times* cybersecurity reporter Nicole Perlroth published the bestseller *This Is How They Tell Me the World*

---

[10]Zach, "China Used Stolen Data to Expose CIA Operatives in Africa and Europe" (2020).

[11]Cybersecurity and Infrastructure Security Agency, "Joint Statement by The Federal Bureau of Investigation (FBI), The Cybersecurity and Infrastructure Security Agency (CISA), The Office of The Director of National Intelligence (ODNI), and The National Security Agency (NSA)" (2021).

[12]Gellman, *Dark Mirror: Edward Snowden and The American Surveillance State* (2020).

[13]Starr, "Fridge Caught Sending Spam Emails in Botnet Attack" (2014).

[14]Steele, "Oregon FBI Tech Tuesday: Securing Smart TVs" (2019).

Ends which details decades of offensive hacking efforts by China, Iran, Israel, North Korea, Russia, and the US to access information and booby-trap information protection systems.[15]

Peter Swire, who served under two presidential administrations and was responsible for reviewing intelligence community activities after the Snowden documents were dumped, argues that we live in "The Golden Age of Surveillance."[16] Not only do nation states like China, Russia, and the US have well-funded institutions with technically gifted employees searching for new ways to monitor, but important other factors have also begun to enhance surveillance powers.

As information traverses the Internet, operators of servers can log *metadata* about activity. US law currently makes it much easier for law enforcement to obtain metadata than content. Perhaps this is because the content/metadata distinction was in part driven from the days when a telephone's content was recorded with a pair of alligator clips onto a reel-to-reel tape recorder and metadata was captured with a *dialed number recorder* that literally recovered the numbers that a person dialed *and nothing else.*

Metadata is commonly believed to be less sensitive than content. However, there is a good argument to be made that metadata is more revealing than content. Metadata is easier to structure in computer databases and analyze. Consider the act of watching and interacting with a YouTube video. The *content* of the session includes:

- The visual content of the video, including the individual frames, the images of the people in the frames, the images of the buildings, etc.

- The audio content of the video, including the sounds, music, and other information.

- The text of any comments left on the video.

But if you were an analyst, consider the knowledge that could be derived from the same video's metadata:

- The video's unique identifier and its title.

- The time that the video was recorded, uploaded, and edited.

---

[15]Perlroth, *This Is How They Tell Me The World Ends: The Cyberweapons Arms Race* (2021).

[16]Swire, "The Golden Age of Surveillance" (2015).

- The unique identifiers of each person that watched the video, their geographic location, their internet protocol (IP) address, and the time that it was watched.

- Whether the viewers clicked "thumbs up" or "thumbs down" on the video.

- Whether the viewers shared the video with friends and, if so, whom.

- The identifiers of any individuals in the video found with face recognition software.

The additional information available from metadata – particularly surrounding the identity of the community of users interested in the video and the people to whom they send it, might be far more important than the video's actual content.

The lines between content and metadata are not sharp. A transcript of the video might be considered content, but keywords extracted from the transcript might be considered metadata. While we classify the comments as content, the timings between individual keystrokes when the comments were left might be considered metadata – even if software can recover the actual typed words using those timings.

Metadata can thus indicate location, the identities of friends, and provide many hints about the content of communications and actual activities online. In many cases, the metadata/content distinction is functionally irrelevant, because operators of servers and services directly examine the content of our email, photographs, and other communications in the dual interests of security (anti-spam) and commercialization (behavioral-based advertising). The private sector plays a critical role by assembling dossiers of both proprietary company data and open source information on people; such products can then be sold to both marketers and (even foreign) government agencies.

The move to the "cloud" means that governments can obtain troves of data about people that previously would have been confined to a home or a business with legal process (or simply by guessing or otherwise obtaining the user's password). Individual users of technology also contribute to surveillance power by documenting their lives on social networks, and by carrying mobile trackers and dutifully storing contact books in them, which give companies and intelligence agencies alike access to location data and fodder for link analysis.

As much as technological trends have benefited nation states, these capabilities have devolved to many private sector actors as well.[17]

Especially concerning to some is the use of state collection capabilities to support domestic industries and silence critics living abroad. In the 1990s, for example, France was accused of using its intelligence apparatus to spy against Boeing, Textron, and Bell.[18] More recently businesses have raised concerns about intellectual property exfiltration by China, which then shares the information with commercial rivals in China. Businesses are concerned about China and other nations using a range of surveillance capabilities to collect information on dissidents, regime critics, and refugees who live outside of the country. For example, in 2010 Google revealed that its Gmail system had been hacked by China and that information from the email accounts of human rights activists had been pilfered.[19] Businesses are also concerned about the convergence of organized crime and government in Russia, which not only directly engages in financial fraud but also creates platforms and even a market for others to do so.[20]

7.2.2 The Golden Age of Encryption

The Golden Age of Surveillance is accompanied by a corresponding golden age of encryption *adoption by default*. Since 1991, users with significant technical ability have been able to use strong encryption in the form of Phil Zimmerman's Pretty Good Privacy,[21] although even later versions that were heralded as being easy to use were

---

[17]Weinbaum et al., *SIGINT for Anyone: The Growing Availability of Signals Intelligence in The Public Domain* (2017); Koller, *The Future of Ubiquitous, Realtime Intelligence: A GEOINT Singularity* (2019).

[18]Doyle, "Business Spy War Erupts between US and France: Paris Forced to Come Clean on Hi-Tech Dirty Tricks" (1993); Greve, "Boeing Called A Target Of French Spy Effort" (1993).

[19]Zetter, "Google to Stop Censoring Search Results in China After Hack Attack" (2018).

[20]Organized Crime and Corruption Reporting Project, "The Russian Laundromat Exposed" (2017); US Agency for International Development, Bureau for Africa, "Government Complicity in Organized Crime" (2019).

[21]Garfinkel, *PGP: Pretty Good Privacy* (1994).

still too difficult for most people.[22] Since then, technologists have sought to change the security landscape by implementing encryption by default in seamless ways. Perhaps most notable is the shift of addresses on the World Wide Web from being prefixed by `http://` to `https://`, which provides users greater confidentiality and integrity in their web browsing. Prior to this change, users' web browsing was sent over the Internet without encryption, allowing adversaries and telecommunications providers alike to monitor users' website visits or even change the content of web pages as they were being viewed.[23] Email likewise has moved from communications where most messages sent over the Internet backbone were sent entirely in plain-text to a system where such messages are largely encrypted (although email encryption is not generally end-to-end – see "Is Your Email Encrypted?" on page 272). Likewise, the popular messaging app WhatsApp offers end-to-end encryption. When WhatsApp was acquired by Facebook, the creators left to support Signal, another messaging application offering end-to-end encryption. Likewise, Apple's iPhone and its newest laptops and desktops use encryption for storage and for text messages sent between Apple users. Although such techniques can be defeated through the use of so-called 0-day attacks,[24] companies like Apple are typically quick to fix such vulnerabilities when they become public.

Central to this rise in encryption is that the user need not understand, configure, or even activate it because encryption is on by default. This offers a lesson for the confidentiality and integrity gains possible in quantum communications: for these innovations to be realized, they must not only be easy to use, they must be secure and integrated into the fabric of communications systems and consumer-facing applications.

7.3 Quantum Random Number Generation (QRNG)

All of these encryption systems we discussed in the last section are based on more-or-less the same technology stack: the AES encryption algorithm to encrypt the messages, a secure random number

---

[22]Whitten and Tygar, "Why Johnny Can't Encrypt: A Usability Evaluation of PGP 5.0" (1999).

[23]The advent of free encryption certificate services and a policy from Google that sites with TLS would get higher rankings in search results caused a rush to adopt the `https://` prefix.

[24]Perlroth, *This Is How They Tell Me The World Ends: The Cyberweapons Arms Race* (2021).

Is Your Email Encrypted?

Much email sent today is between two Gmail users. These messages are encrypted by the Transport Layer Security (TLS) as they travel from the sender's web browser to Google's web-mail service. Although the messages are not encrypted in the memory of Google's servers, they are encrypted when they are written to Google's disks where the messages are stored.[a] Likewise, the email messages are encrypted when they are sent from Google's servers to the Gmail recipient.

Mail that gets sent from Gmail to other mail providers, such as Microsoft's Office 365 cloud platform, are frequently encrypted using the SMTP STARTTLS protocol.[b]

This kind of protection is not as strong as the so-called *end-to-end* encryption offered by the S/MIME and PGP encryption systems. However, STARTTLS is significantly easier to use because each user does not need to create or otherwise obtain a public/private keypair.

---

[a]Google LLC, "Encryption at Rest" (2021).
[b]Rose et al., *Trustworthy Email* (2019).

---

generator to create the AES key, and public key cryptography to get the per-message key from the message sender to the recipient. Earlier in this book we discussed the role of the AES and public key cryptography algorithms. In this section we will discuss the role of random numbers.

Cryptography depends on strong random numbers. For instance, a RSA-2048 key is generated from prime numbers that are over 300 digits long: these prime numbers are found by guessing random numbers and checking them to see if they are prime. (Unlike factoring, there are mathematical tricks that are used to rapidly determine if a number is prime or not.) Likewise, the AES-256 keys are themselves random numbers.

Random numbers thus form the very basis of the security provided by encryption. If a 256-bit key is random, then that means every key is equally probable. But if an attacker can somehow interfere with the randomness of the number generation process, it can dramatically reduce the possible number of encryption keys. For such an attack, the strength of AES-256 with a key that is not very random might not be strong at all.

The NIST Randomness Beacon

In 2013, the US National Institute of Standards and Technology deployed its "Randomness Beacon," a web-based service that posted random numbers in blocks of 512 bits every minute. Like an electronic lottery machine, the bits posted to the NIST website are unpredictable.

The randomness service is an endless source of numbers that can be used in situations where a random choice needs to be made, and the person making the choice wants to demonstrate that they made the choice fairly. In football games, for example, the receiving team is chosen by a coin toss – but how do we know the coin is fair? In this and similar situations where a decision must be made on a random choice, the NIST service can be relied upon by both parties to ensure a selection that is unbiased.

Example applications that NIST proposed included selection for random screening at security checkpoints, selection of test and control groups in scientific trials, selection of people for random tax audits, assignment of judges to cases, and so forth. Because the beacon is public, and because each bitsream is added to a hash chain (or blockchain), the system can be audited by any party. Of course, being public comes with a risk as well: the bits should not be used in cases were both randomness and secrecy are required. To drive in this lesson, the NIST website states:[a]

WARNING:
DO NOT USE BEACON GENERATED VALUES
AS SECRET CRYPTOGRAPHIC KEYS.

---

[a]See beacon.nist.gov/home

Modern computers generate random numbers by using an initial *random seed* which is then used with a deterministic random bit generator, also called a pseudo-random number generator (PRNG). Typically, the random seed is created by combining many events that, if not completely random, are at least unpredictable. For example, the early PGP program instructed users to type on the keyboard and used the inter-character timing as a source of randomness. Other sources of randomness include the arrival time of packets at a network interface, inputs to digital cameras, and even seismic sensors. In practice, the quality of random numbers is determined by the samples taken from the "random" source, the quality of the mixing, and the quality of the PRNG. If any of these produce output that is somewhat predictable, or for which there is correlation between successive values, then a knowledgeable adversary can gain advantage when attempting to decrypt a message that was encrypted with such "poor quality" randomness.

Concerns about the strength of random number generators has been raised many times in the past. One such case from the US involves the Dual Elliptic Curve Deterministic Random Bit Generator (Dual_EC_DRBG).[25] When Dual_EC_DRBG was proposed, security professional Bruce Schneier and others raised concerns that the algorithm might include a "secret backdoor" that would allow the US government to predict the algorithm's "random" outputs.[26] These concerns were confirmed in 2013.[27] Following the disclosure, NIST issued guidance stating "NIST strongly recommends that, pending the resolution of the security concerns and the re-issuance of SP 800-90A, the Dual_EC_DRBG, as specified in the January 2012 version of SP 800-90A, no longer be used."[28] In 2015, the Director of Research at the National Security Agency said that the agency's "failure to drop support for the Dual_EC_DRBG" after vulnerabilities were identified in 2007 was "regrettable."[29,30]

In 2019 cryptographers stated that two Russian-designed encryption systems, Streebog and Kuznyechik, might also contain a secret backdoor that would give an advantage to a knowledgeable attacker trying to decrypt a message protected with the algorithm. In this

---

[25] Barker and Kelsey, *Recommendation for Random Number Generation Using Deterministic Random Bit Generators (Revised)* (2007).

[26] Schneier, "Did NSA Put a Secret Backdoor in New Encryption Standard?" (2007).

[27] Perlroth, "Government Announces Steps to Restore Confidence on Encryption Standards" (2013); Buchanan, *The Hacker and The State: Cyber Attacks and The New Normal of Geopolitics* (2020).

[28] Information Technology Laboratory, "Supplemental ITL Bulletin for September 2013" (2013).

[29] Wertheimer, "Encryption and The NSA Role in International Standards" (2015).

[30] This story and others surrounding the quest to produce high-quality random numbers at scale is discussed in Garfinkel and Leclerc, "Randomness Concerns When Deploying Differential Privacy" (2020), from which this story and its references are taken.

case, the weakness was not in the random number generator, but in the algorithms' so-called "substitution boxes."[31]

Quantum states provide the best source for strong, unbiased randomness. Scientists have developed several different methods to derive strong randomness from quantum events, including the path that photons take when light is split, the polarization of individual photons, and the phase of quantum states and processes.[32] A notional device bears similarity to the dual-slit experiment discussed in Section B.1.3, "Light: It Acts Like a Wave" (p. 490). The device works by cycling a particle or photon in and out of superposition. Measurement disturbs the superposition, causing decoherence and the production of a random bit. That bit is then used as a basis to generate random numbers. One way to think of these machines is as a quantum computer with a single qubit that is constantly computing the answer to the question "is the qubit 0 or 1?"

Number generation in such a scheme faces two sets of challenges. The first is the cycle speed of the prepare-superposition process and the speed of the measurement-decoherence process, which together determines how fast these systems can produce random bits. These machines may also be impacted by errors produced by classical noise and the reliability and tolerances of the quantum source and of the measurement mechanism, which can bias the results.

Properly implemented, QRNG produces strong randomness.[33] In fact, it probably produces the strongest possible random numbers, since modern physics holds that quantum processes are the ultimate source of all nondeterminism that we observe in the universe. QRNG has also been commercially available for years. In fact, after scientists created a QRNG system at the Australian National University in 2011,[34] the investigators found they had more random numbers than they would ever need for experiments. So they created a free QRNG service on the web.[35] In 2020, IBM and Cambridge Quantum Computing offered QRNG as a cloud service. And NIST is deploy-

---

[31] Perrin, "Partitions in The S-Box of Streebog and Kuznyechik" (2019).

[32] X. Ma et al., "Quantum Random Number Generation" (2016).

[33] Acin and Masanes, "Certified Randomness in Quantum Physics" (2016); Bierhorst et al., "Experimentally Generated Randomness Certified by The Impossibility of Superluminal Signals" (2018).

[34] Symul, Assad, and Lam, "Real Time Demonstration of High Bitrate Quantum Random Number Generation with Coherent Laser Light" (2011).

[35] See qrng.anu.edu.au/

ing Entropy as a Service (EaaS), a public, quantum-based source of random numbers.

Using these remote, cloud-based services requires some reliance on the provider, but there are measures that can be taken to reduce the risk. Instead of using the source directly, it can be combined with a secret key and then used in a cryptographically strong PRNG – a CSPRNG! This approach works as long as the secret key is kept secret and as long the PRNG is really a CSPRNG. That's the use case that NIST envisions for its EaaS. The EaaS project is explicitly designed to serve Internet of Things (IoT) devices by providing random numbers that these devices can use to create strong encryption keys. The idea is that IoT devices will be small and inexpensive, so much so that even high-end brands will cut corners on security, thus the chances that the market will produce QRNG for IoT devices is particularly unlikely. NIST is in effect substituting the market with security fundamentals for anyone to use. NIST is also upgrading its Randomness Beacon to use QRNG, as currently it uses two classical generators to prevent guile.

Higher levels of assurance require implementing the QRNG locally, so that the high-quality random bits are generated where they are needed, and not by some third party. For instance, ID Quantique has long sold QRNG hardware that plugs into a standard personal computer or server. In 2020, the company announced a QRNG chip that could fit into mobile phone handsets.[36] This device uses the random "shot noise" from a light-emitting diode (LED) to generate numbers. Every time the LED fires, the number of photons emitted fluctuates randomly. A CMOS sensor array sensitive to single-photon events detects the number emitted and their positions (see Figure 7.1).

7.4 Quantum Key Distribution

When Rivest, Shamir, and Adleman wrote their article introducing the RSA encryption system, they explained it with a woman, "Alice," who wanted to send a secret message to a man named "Bob."[37] Since then, Alice, Bob, and a whole cast of other characters have been used to help scientists analyze and explain security protocols. There is Eve, the eavesdropper, who attempts to "intercept" (a strained metaphor)

---

[36]Quantique, "Quantis QRNG Chip" (2020).
[37]Ronald L. Rivest, Adi Shamir, and Len Adleman, "A Method for Obtaining Digital Signatures and Public-Key Cryptosystems" (1978).

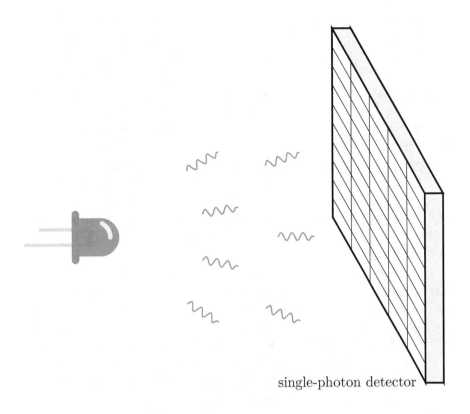

single-photon detector

Figure 7.1. A mechanism for QRNG designed by ID Quantique fits into a mobile phone handset and pairs an LED and single-photon sensor array to derive randomness from photonic noise.

this conversation. And there is Mallory, a malicious attacker, who can modify the message or inject new ones.

Quantum Key Distribution (QKD) describes an approach where Alice and Bob can exchange an encryption key guaranteed to enjoy *unconditional* security. No computer available today or in the future can compromise this system, because the attacker does not have enough information to make sense of the ciphertext.

7.4.1 BB84

In 1984, Charles Bennett and Giles Brassard published the BB84 protocol, demonstrating how Alice and Bob could exchange encryption keys using quantum states.[38] Using the protocol, Alice and Bob get the same stream of `0` and `1` bits that they can use for any purpose. For example, they can use the sequence in 8-bit chunks as a *one-time pad* (see Figure 7.2), using each group of 8 bits to encrypt the next byte of the message. Alternatively, they can use the sequence in 256-bit chunks as AES-256 encryption keys.

The one-time pad is the gold standard for communications security because it is information-theoretic secure.[39] Even if the attacker tries every possible key, there is not enough information in the encrypted message to distinguish a correctly decrypted message from an incorrectly decrypted message. The reason is that the key is as long as the message, so every possible key makes the message decrypt a different way. This means that trying every possible key makes the encrypted message decrypt to every possible message.

One-time pads are the stuff of spy thrillers and history books, but they are not used much today because it is too difficult to distribute the pads in advance and then assure that each is used just once. The Soviet Union attempted to use one-time pads for its diplomatic communications after World War II and it failed; the NSA revealed its success in cracking the Soviet codes in 1995 (see Figure 7.6).[40]

BB84 is revolutionary, because Bennett and Brassard's approach deals with two central challenges in communication: how to generate a secure, shared secret, and how to distribute it at a distance. Two other key challenges – usability and the time it takes to generate and transmit the key securely – are up to the companies that create applications using QKD protocols.

However, modern QKD systems cannot generate a stream of bits fast enough to encrypt modern data links. For this reason, QKD systems typically operate in a slightly less secure mode in which BB84 is used to exchange 256-bit encryption keys which are then used with conventional encryption algorithms such as AES-256. With a 256-bit key, each encrypted message will have only 2^{256} possible decryptions, and the likelihood is that all but one of them will be gibberish. As we discussed in Chapter 5, it isn't possible to try all 2^{256} keys, so using BB84 to exchange AES-256 keys is considered secure. However, it is only computationally secure, not information-theoretic

---

[38]C. H. Bennett and G. Brassard, "Quantum Cryptography: Public Key Distribution and Coin Tossing" (1984).

[39]Shannon, *Communication Theory of Secrecy Systems* (1949).

[40]National Security Agency and Central Security Service, "VENONA" (2021).

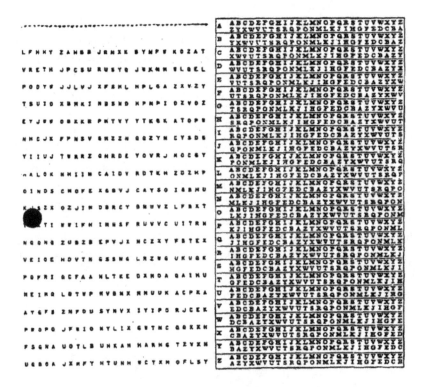

Figure 7.2. This table from the NSA's DIANA program illustrates how one-time pads produce messages with keys the same length of ciphertext. The key is on the left-hand side. The right-hand side is the table used to convert plain text to ciphertext (and vice versa). This key starts with the letter "L," so the user encrypting a message would use the L row on the table to choose the first letter of ciphertext. Assume that Alice wants to say "The Magic Words Are Squeamish Ossifrage" to Bob. To encrypt, Alice notes the first letter from the key, left-hand pane, which is L. Turning to the table, row L, and then to the letter T, the corresponding ciphertext underneath the T is a V. To encrypt the next letter, Alice would then use F from the key to locate the letter H and choose the ciphertext N, and so on. Alice and Bob must have identical cards and must destroy them after the process.

secure. As a compromise, these systems might change their AES-256 keys every few seconds, to minimize the amount of ciphertext that has been encrypted with any given AES-256 key.

7.4.2 How QKD Works

Most QKD systems are based on the idea of sending a stream of photons from a sender (Alice) to a recipient (Bob). For more background on polarized light, see Appendix B.3, "Quantum Effects 2: Polarization".

Here we provide a simplified explanation for how BB84 operates. The first thing to know is that actually using BB84 in a production system requires considerable mastery of the quantum realm and engineering cleverness not explained here.

In modern QKD systems, the photons either travel down a fiber-optic strand, or they are created in pairs in a satellite and sent to two independent ground stations.[41] In the first case, Alice prepares a stream of photons by sending each through a polarizing filter that is either polarized horizontally (H), vertically (V), at a 45° angle, or at a 135° angle. Alice makes this choice at random, recording both the number of the photon and the orientation of her polarizing filter. Sending with a H or a 45° is tentatively sending a 0 , while sending with a V or a 135° is tentatively sending a 1 . (Alice can't actually number each photon, so instead she will encode each photon's value in the light stream itself.)

Let's say Alice sends 10 photons:

Photon #	Alice Filter orientation	Tentative bit
0	45°	0
1	45°	0
2	45°	0
3	H	0
4	V	1
5	135°	1
6	45°	0
7	45°	0
8	H	0
9	135°	1

When Bob receives the photons, he also passes them through a filter that is also randomly oriented at either V or at 135°. He then measures the presence or absence of the photon with a single photon detector:

[41]The protocol involving a pair of entangled photons is called E91, after its inventor Artur Ekert (Ekert, "Quantum Cryptography Based on Bell's Theorem" (1991)).

Photon #	Bob Filter orientation	Photon detected?	tentative bit
0	135°	NO	0
1	135°	NO	0
2	V	YES	1
3	V	NO	0
4	V	YES	1
5	V	YES	1
6	135°	NO	0
7	V	NO	0
8	135°	NO	0
9	V	YES	1

Now Alice and Bob need to compare notes to see if the measurement that Bob made of the photon was compatible with the photon that Alice prepared and sent. If Bob measured with his V filter, then he will detect light if Alice sent the light with her V filter, but not if she used her H filter. But if Alice sent it with her 45° or 135° filters, the measurement that Bob made is meaningless: there's a 50–50 chance that a photon polarized with the 45° filter will pass through a V filter.

To compare notes, Bob can reveal which filter he used to measure each photon. Alice then tells Bob which of his measurements he should keep and which he should throw out:

Photon #	Bob to Alice	Alice to Bob
0	135°	KEEP
1	135°	KEEP
2	V	–
3	V	KEEP
4	V	KEEP
5	V	–
6	135°	KEEP
7	V	–
8	135°	–
9	V	–

At this point, Alice and Bob know that photons 0, 1, 3, 4, and 6 were sent and received with compatible polarizing filters. Alice looks

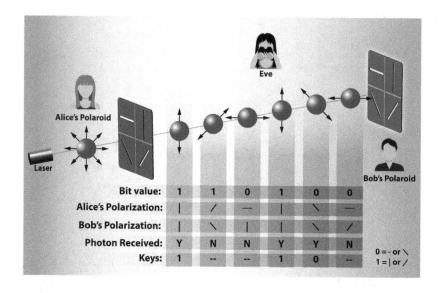

Bit value:	1	1	0	1	0	0
Alice's Polarization:	|	/	—	|	\	—
Bob's Polarization:	|	\	|	|	\	/
Photon Received:	Y	N	N	Y	Y	N
Keys:	1	--	--	1	0	--

0 = - or \
1 = \| or /

Figure 7.3. The BB84 protocol illustrated. Adapted from Aliberti and Bruen by Twitter user farooqumer89.

at her table and discovers that the tentative bits corresponding to those numbers are 0 0 0 1 0 . Bob looks at his table and gets the same sequence of bits.

To determine that the system is operating properly, Alice and Bob can now decide to reveal every even bit of the resulting sequence. Alice says that even bits are 0 , 0 , and 0 . Bob notes that his are the same. Alice and Bob then use the remaining bits (0 , 1) as their secret key.

If Alice and Bob do not reveal to each other the same bits, then either the system is not operating properly, or else an attacker is intercepting the beam and injecting a photon sequence of their own. In either case, Alice and Bob know not to use that key.

Because of measurement error, the sequence of bits that Alice and Bob recover are not exactly the same. A variety of error correction techniques exist that can be used to account for these errors, at the cost of using even more bits.

The two-photon system is similar, except that a pair of entangled photons are sent from the satellite to both Alice and Bob, who then both measure the polarization and compare notes. In this design, the satellite cannot determine the key that Alice and Bob agree upon, nor can anything else in the universe: each photon can only be measured once. Of course, once Alice and Bob agree upon a key, a suitably skillful attacker might be able to steal it from either Alice or Bob if their QKD device does not properly protect the key after it has been created.

7.4.3 Why QKD Is Secure

What makes QKD secure is the fact that the actions of Alice and Bob measuring the photon are independent, but the measurements are correlated *if and only if Alice and Bob choose compatible measurements*. If Alice measures the photon with a horizontal polarizing filter and Bob uses a filter that is polarized vertically, their measured results are linked and they have now agreed on a common bit. But if Bob uses a filter at 45°, the measures are incompatible and there is no correlation between them. This is the essence of Einstein's "spooky action at a distance," the paradox of entanglement. Because Alice and Bob chose their measurements at random, only 50 percent of them will be compatible: the remaining measurements will be thrown out.

Now let's say an attacker, Eve, tries to crash the party. Eve attempts the well-known "man-in-the-middle" attack: she catches the photons headed for Bob, measures them, and then prepares a new photon and sends it to Bob. Can Eve get away with this deception? In a properly implemented QKD system, the answer is no. That's because when Eve receives, measures, and retransmits the photon, she doesn't know how Bob is going to measure it. By chance, she will only measure the photon in a compatible manner 50 percent of the time. The other 50 percent of the time, she will measure the photon in a way that is incompatible. When she sends each of those incorrectly measured photons to Bob, Eve has a 50 percent chance of sending them in the correct state, and 50 percent chance of sending them in the wrong state.

When Bob compares notes with Alice, they first reveal how the photons were measured and throw out the photons for which Alice's and Bob's measurements were incompatible. But after this step, they intentionally reveal a certain percentage of the remaining photons. When Bob and Alice discuss these intentionally revealed photons, they will discover that their measurements disagree roughly half of the time. This indicates either that their equipment is not working properly, or that Eve is attempting to perform a man-in-the-middle attack.

Quantum Computing and Bitcoin

Cryptocurrencies such as Bitcoin are speculative investment and value transfer mechanisms that are based on a *distributed ledger*, a kind of shared database, that is difficult to corrupt. Bitcoin, the first cryptocurrency, relies on SHA-256 to build its ledger.

The Bitcoin ledger consists of many transactions, each of which is basically an electronic check that is signed with a private key. The check transfers some amount of Bitcoin from the user's corresponding public key (a Bitcoin "address") to another public key. These transactions are grouped into blocks. In addition to these electronic checks, each block contains the hash of the previous block, a signature by the block's "miner," and a block of random values placed there by the miner. The random values are manipulated such that the SHA-256 hash of the new block begins with a large number of zeros. To do so, the Bitcoin "miner" takes the block of transactions and makes systematic changes to that random block until the hash has enough zeroes.

Because the hashes generated by SHA-256 appear random, with each bit having an equal chance of being a `0` or a `1`, finding hashes with a large number of leading zeros is computationally intensive. In March 2020, Bitcoin blocks had 76 leading binary `0`s, followed by 180 bits of `0`s and `1`s; the number of leading `0`s is automatically adjusted to be longer and longer as more and faster Bitcoin miners join the network; each additional leading `0` requires roughly twice as much computational power to find.

In 2019, the National Academies estimated that a large quantum computer could attack Bitcoin's ledger system but the attack requires 2403 qubits and 180 000 years. Given that the ledger gets a new block every 10 minutes, attacking the ledger itself in order to obtain free Bitcoin appears unlikely.

Bitcoin holders may still be vulnerable because a quantum computer could be tasked with cracking the public key of an individual Bitcoin user's wallet and then stealing that user's money. Alas, the victim would have little recourse owing to the social contract underlying cryptocurrencies.

Quantum Money

Stephen Wiesner's idea of using the entanglement of two particles to create unforgeable banknotes (see p. 137) led Bennett and Brassard to come up with the idea of quantum cryptography in the first place. Since then, many scientists have proposed systems that rely on quantum effects to store and transmit value, now broadly called *quantum money*. These schemes vary in their implementation. Some provide information-theoretic security while others rely on public key systems.[a] But given current constraints in quantum memory, computing, and networking, hopes for quantum money systems are far off.

If they ever do arrive, some of the affordances promised will be contested by parties with interests in transactions. Cryptocurrencies like Bitcoin and most if not all envisioned quantum currencies contain mechanisms to ensure that a purchaser actually has sufficient funds and to prevent "double spending." Beyond that, however, most of these mathematical monies are quite spartan.

Conventional value transfer mechanisms such as checks and credit cards are complex for many reasons. For instance, policy decisions must be made to reconcile the different, conflicting interests held by ordinary consumers, merchants, banks, and governments in payments. A consumer might want the ability to repudiate a value transfer, in case of fraud, coercion, or because of poor-quality goods received, while merchants might want to block repudiation. Governments typically want the ability to unmask all parties in a transaction. Such mechanisms are missing – intentionally – from cryptocurrencies.

Yet, as Bitcoin has become more mainstream, the original vision of a bank-free, anonymous, peer-to-peer payment system has ceded to something more akin to a commodities market, one mediated by exchanges that are regulated by governments and that follow taxation and anti-money-laundering rules to identify market participants.

---

[a]Hull et al., "Quantum Technology for Economists" (2020).

Of course, Eve could go further, and pretend to be Bob to Alice and to be Alice to Bob. To prevent this, Alice and Bob need to have a way of authenticating the open messages that they send to each other. Today the easiest way to do this authentication is with public key cryptography. This use of public key cryptography is considered acceptable in QKD systems, because even if an attacker records the authentication messages and cracks the private keys behind them at some point in the future, that won't change the fact that the messages were properly authenticated when they were sent. No secret information is revealed if the authentication keys are cracked in the future.

Eve can prevent Alice and Bob from communicating securely by using electronic warfare approaches. Eve could inject noise to deny or degrade the quantum channel and cause Alice and Bob to have to revert to other, less secure communication, but she can't decipher the messages sent. (Indeed, risks of denial of service are among the reasons the NSA has spurned QKD in favor of quantum-resistant (or post-quantum) cryptography.[42]) And once the key is exchanged between Alice and Bob, the duo do not need a "quantum internet" or quantum states to talk securely. Alice and Bob can use the quantum key to communicate on existing classical channels, encrypting their communications with a conventional quantum-resistant symmetric algorithm such as AES-256.

7.4.4 QKD Gains Momentum

Since BB84 was proposed, new protocols and even implementations have emerged. For instance, in 1991, Arthur Ekert proposed the satellite entanglement protocol described above.[43] Recall that Alice and Bob receive correlated photons from a split-beam laser. Using Bell tests (see Section B.4, p. 513), Alice and Bob compare the correlations of their photons to ensure that Eve has not intercepted them. Under Ekert's proposal, even if Eve is operating the laser, she cannot determine the states of Alice and Bob's photons without interfering with the Bell correlations, thus revealing her attack. Ekert's proposal anticipates the possibility of a QKD-as-a-service approach – a satellite delivering entangled photons from space to the ground, allowing

---

[42]National Security Agency, "Quantum Key Distribution (QKD) and Quantum Cryptography (QC)" (2020).

[43]Ekert, "Quantum Cryptography Based on Bell's Theorem" (1991).

Quantum Submarine Communication

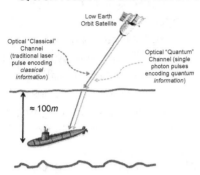

Low Earth
Orbit Satellite

Optical "Classical"
Channel
(traditional laser
pulse encoding
classical
information)

Optical "Quantum"
Channel (single
photon pulses
encoding *quantum*
information)

≈ 100*m*

Optical channel provides some advantage over the cumbersome and bandwidth limited VLF (and the now defunct ELF) submarine communications

We have shown that biologically-inspired quantum photodetectors could allow efficient classical and quantum communications in the optical window of sea water.

VLF and the now defunct ELF provide low bandwidth (300 bits/s for VLF and a few characters per minute for ELF) and require cumbersome buoys or towed antenna arrays, and require specific course and speed alterations.

Our theoretical models predict an *unconditionally secure* key generation rate of 170 kb/s at 100 m deep in Jerlov Type I waters (about 600 times improvement over VLF).

Figure 7.4. In a 2018 address to the National Academies, Dr. Marco Lanzagorta, explained how quantum communications might enable new forms of secure, satellite-to-submarine communication. Image courtesy US Naval Research Laboratory.

any two parties to communicate securely, and not even the satellite can decipher their shared key.

Scientists have also proposed BB84 protocols to improve communications with satellites directly. In one scheme, a submarine equipped with a photosensor or towing a small buoy can exchange photons with a satellite, even while submerged (see Figure 7.4). The submarine would have to make speed versus depth tradeoffs, that is, at a depth of about 60 meters, data could be exchanged at 170 kilobits per second, but this rate drops in murky waters and at deeper levels. Nonetheless, the approach is stealthy and has advantages over existing submarine communication approaches.[44]

Long-distance quantum channels for key distribution require special ingenuity to overcome a variety of technical challenges. Chinese scientists, led by that nation's "father of quantum," Jian-Wei Pan, demonstrated entanglement at 1200 kilometers by using a satellite

---

[44]Marco Lanzagorta, "Envisioning The Future of Quantum Sensing and Communications" (2018); Marco Lanzagorta, *Underwater Communications* (2013).

nicknamed Micius.[45] The satellite beamed photons between distant base stations that were in the coverage area of the Micius for just five minutes.[46] Pan's team pointed to the use of the entangled photons for an Ekert-protocol secure exchange, at a distance currently impossible to achieve with terrestrial, fiber-optic connections (the quantum states degrade in the glass fiber after a distance of around 100 km without taking special measures). Yet, the approach still faces many challenges as revealed in the paper's methods. Pan's team had to beam millions of photons a second to maintain the link, and only a handful reached the base stations because of atmospheric and other interference.

Pan's achievement is part of a $100 million project in China, the Quantum Experiments at Space Scale program (QuESS). The entangled distribution over such a great distance demonstrated a substantial goal of the program. Key exchange was realized later the same year, using a mixed fiber-optic/satellite path of over 7000 km.[47] Pan's team demonstrated the key exchange by holding a videoconference between Beijing and Austria. However, this demonstration did not use end-to-end entanglement between Alice and Bob, as described by Ekert. In this initial experiment, Pan's team used the BB84 protocol, and the satellite operated as a trusted relay. Micius exchanged separate keys with each of the different ground stations.

With a relay, the implementation is not fully quantum – it's not a quantum internet – and the parties must trust the satellite's security. That's a concern. Governments will probably trust their own satellites, but this trust should not be absolute, as the computers in satellites are vulnerable to cyber attack just like computers down here on the ground.

In 2020, Pan's team announced a satellite-terrestrial quantum network covering 4600 km. The network has over 150 users, and achieved a transfer rate of 47 kilobytes a second, more than sufficient for exchanging 256-bit AES keys.[48]

---

[45]Launched in 2016 at the low-earth orbit of 500 km, Micius travels in a Sun-synchronous path. Micius is named for the fifth-century BCE Chinese philosopher Mozi, founder of Moism, who wrote original works on optics.

[46]Yin et al., "Satellite-Based Entanglement Distribution Over 1200 Kilometers" (2017).

[47]Liao et al., "Satellite-Relayed Intercontinental Quantum Network" (2018).

[48]Y.-A. Chen et al., "An Integrated Space-To-Ground Quantum Communication Network Over 4,600 Kilometres" (2021).

In the US, fewer than ten QKD networks have been implemented in recent years. The first, DARPA's QKD network, was implemented by Raytheon BBN, at Harvard and Boston Universities in 2003.[49] The team used dark fiber (unused fiber-optic cables) in Cambridge, Massachusetts to connect the almost 30 km long network. The network, which had trusted optical point-to-point systems and untrusted, relaying infrastructure, operated for four years. Here "untrusted" means that the relaying infrastructure could not impact the security of the data sent over the fiber.

At Los Alamos National Laboratory, scientists created a hub-and-spoke quantum network.[50] In the implementation, a central, trusted server performs the key exchange, which then enables nodes in the spokes to communicate among each other with authenticated quantum encryption. This sort of trust model works when all of the networks have a some reason to trust the central node; in the LANL demonstration, their model was a power distribution network.

Major challenges still exist for QKD implementation. The point-to-point nature required to preserve quantum states between Alice and Bob makes QKD networks more like the early telegraph than the telephone or Internet. Quantum states decohere in long fiber runs, so some networks require repeating, which, like the Micius satellite demonstration, requires trusting the repeater. Alice and Bob also need sophisticated equipment: lasers, single-photon detectors, interferometers and the like. These are now packaged in commodity QKD systems that communicate over fiber-optics, although systems that communicate in free space or using satellites are still basic science endeavors. Even so, QKD is among the most mature quantum technologies, and solving these limitations is receiving significant attention. The next section turns to such commercialization.

7.4.5 QKD Commercialized, Miniaturized

As early as 2009, three companies (ID Quantique, Switzerland; MagiQ Technologies, US; and Smartquantum, France) offered working QKD devices.[51] According to the Quantum Computing Report, at least a

---

[49]Elliott and Yeh, *DARPA Quantum Network Testbed* (2007).

[50]Hughes et al., "Network-Centric Quantum Communications with Application to Critical Infrastructure Protection" (2013).

[51]Scarani, Bechmann-Pasquinucci, et al., "The Security of Practical Quantum Key Distribution" (2009).

Figure 7.5. In 2019, Air Force Research Laboratory scientists demonstrated daylight QKD using this rig at the Starfire Optical Range, located at Kirtland Air Force Base in Albuquerque, New Mexico. This is important because stray daylight entering the collector causes substantial noise that interferes with the measurement, limiting long-distance QKD during the daytime. (The Air Force's Directed Energy Directorate, which developers lasers and optics, was identified for transfer to the US Space Force in 2020.) Image by US Air Force photographer Todd Berenger.

dozen private firms are working on QKD offerings, along with a few large public companies.[52]

Despite the growing competition in QKD, adoption of QKD has been weak. For starters, without large, encryption-breaking quantum computers, there is no demonstrated need for the technology. In 2015, an unclassified summary of the US Air Force advisory board report threw cold water on QKD, apparently finding that QKD significantly increases system complexity while providing "little advantage over the best classical alternatives."[53] The USAF's full report is not publicly available, but perhaps the board meant that as system com-

---

[52] ArQit, InfiniQuant, KETS Quantum Security, Phase Space Computing, QEYnet, Qrate Quantum Communications, Quantropi, Quantum Xchange, Qubit Reset LLC, Quintessence Labs, QuNu Labs, SeQureNet, and VeriQloud; larger firms include Nippon Telegraph and Telephone Corporation (NTT), Raytheon BBN Technologies, and Toshiba.

[53] US Air Force Scientific Advisory Board, *Utility of Quantum Systems for The Air Force Study Abstract* (2016).

4. New York - Moscow 1340 [753], 21 September
[20 September] 1944:

--149P-- detained VOLOK (?who is?) working at the
ENORMOZ plant. He is a fellow countryman [U.S. Communist].
--1U-- (?recognition?) (?of? ?from?) his work they
dismissed (?him?). The cause of the dismissal was
his active work in the past in progressive organizations.

According to --1U-- of the fellow countrymen
[U.S. Communists], LIBERAL (?is in touch with CHESTER
he --2P-- cutter **ERCESE? [this part very dubious])
once a month. CHESTER is interested in whether we
are satisfied with the cooperation and whether there
are not any misunderstandings. About concrete details
of the work he does not inquire. Inasmuch as CHESTER
knows about the role of LIBERAL's group we beg consent
to inquire of CH. through LIBERAL about (?sketches
(drafts)?) from (?the milieu?) of persons working on
ENORMOZ and other spheres of technical science.

Here the subject changes; in the new section, there is
some mention of a person named LARIN, but the text is un-
intelligible. The signature is MAY.

5. New York - Moscow 1699 [conclusion of 940], 2 December 1944
(the preceding part or parts of this message cannot be located):

Conclusion of telegram no. 940

Stated to be (?participants?) --1U-- (?research?)
on the problem are HANS BETHE, NIELS BOHR, ENRICO FERMI,
JOHN NEUMANN, BRUNO ROSSI, GEORGE KISTIAKOWSKI, EMILIO
SEGRE, G.I. TAYLOR, WILLIAM PENNEY, ARTHUR COMPTON,
ERNEST LAWRENCE, HAROLD UREY, HANS (?STAN? ?STROGN?)
AR(?K? ?L? ?M?), EDWARD TELLER, PERCY BRIDGEMAN,
WERNER EISENBERG[a], --1F-- AS --4F-- [There follows
a repetition of all these names.] --5F-- (?of?) our
country turned [or "applied"] to NAPOLI the latter
(?did not?) --2P-- him [or "his"] --2P-- BEK [Beck?]
--7F--. When he tried to see RULEV, he was not admitted
to see him by the latter's secretary.

(?ANTON?)

a. Mistake for WERNER HEISENBERG? It has been known for some
time that Heisenberg was working for the German Reich
throughout the war.

Figure 7.6. Richard Hallock, an analyst at the US Army's Signal Intelligence Service, discovered that Soviet spies were reusing portions of one-time pads. The revelation allowed the Service, a forerunner of the National Security Agency, to decrypt them. This summary of intercepted communications shows that the Soviets had identified the main scientists involved in the Manhattan Project (Soviet cryptonym "ENORMOZ"; "LIBERAL" is Julius Rosenberg). The American analysts also ponder whether the Russians thought that Werner Heisenberg was working on the American fission project; alas he was working for the Germans. The decryption project, code name VENONA, ran from 1943 through 1980. (National Security Agency and Central Security Service, "VENONA" (2021))

plexity increases, so do attack surfaces. A more complex system gives attackers more opportunities to interfere with communications, and perhaps the side channel attacks possible on quantum devices will be more difficult for network operators to understand. Aside from device problems, there remains the old problem that users can be fooled into granting access. Perhaps the USAF report's skepticism reflects that the US government has a decades-old system of using trusted human couriers to transport high-value key material.

In October 2020, the NSA released a statement clarifying that it would not use QKD to secure the classified and sensitive-level networks it is responsible for protecting, and this NSA statement articulated the likely reasons why QKD has not been more commercially successful. Calling out the hype, the NSA statement recognized that QKD advocates "occasionally state bold claims based on theory" but that in reality, the technology is "highly implementation-dependent rather than assured by laws of physics." The NSA's specific objections related to the need to install new, more complex and expensive infrastructure that itself may have vulnerabilities.[54] Indeed, Russian scientist Vadim Marakov has elucidated a series of attacks on QKD *systems* (but not the underlying BB84 protocol).[55] The NSA concluded that whatever confidentiality QKD offers "can be provided by quantum-resistant cryptography, which is typically less expensive with a better understood risk profile."[56] As with the NSA, many companies probably see little reason to adopt a technology that will require infrastructure changes, require more training, introduce new complexities, and all for limited benefits against attackers many years in the future.

Nevertheless, QKD vendors are trying to overcome the skepticism. Four recent developments paint a path for greater QKD adoption in both the private sector and in governments. First, QKD devices have been miniaturized. ID Quantique and MagiQ both market rack-mounted QKD systems. Second, the general upset caused by the Snowden documents caused policymakers in other regions to make stronger communications security a priority and to make large

---

[54]Scarani and Kurtsiefer, "The Black Paper of Quantum Cryptography: Real Implementation Problems" (2014).

[55]Anqi et al., "Implementation Vulnerabilities in General Quantum Cryptography" (2018).

[56]National Security Agency, "Quantum Key Distribution (QKD) and Quantum Cryptography (QC)" (2020).

vertical industrial policy investments in quantum technologies. This policy commitment may overcome the natural resistance to a switch to QKD. For instance, the European Union's quantum technologies strategy makes wide dispersal of QKD (and QRNG) a priority, even for consumer devices. The European Union's OpenQKD project, a three-year €15 million program (2019–2022), explicitly seeks standardization and other objectives to kick start a Continental QKD industry. Third, progress is being made on technical challenges, such as increasing the length of fiber over which QKD can operate: in 2018 scientists demonstrated QKD over a 400 km fiber run.[57] These ultra-long runs cause signal attenuation, and key acquisition slows to a crawl (as much as 24 hours for a key block), but improvements are steady. Finally, concerns about the privacy and security of 5G telecommunications networks is driving international concern and an unprecedented search for technical security measures.

On this last point, the security of 5G, consider the activity of South Korea Telecom (SK Telecom). Operating in the shadow of North Korea, with its active, audacious intelligence activities, SK Telecom officials must contemplate that their own employees might be forced into revealing telecommunications data to North Korea. In 2016, SK Telecom started implementing QKD in some back-haul operations of their LTE network. This effort expanded in later years to 5G infrastructure. As QKD is implemented in SK Telecom's stack, the number of employees who could be coerced into revealing information to North Korea presumably winnows.

QKD or quantum networking to a consumer handset will probably never be a reality, but QRNG may be on the threshold of widespread adoption: In May 2020, ID Quantique announced that its system-on-a-chip QRNG had been implemented in a handset offered by SK Telecom. In September 2020, as part of South Korea's $133 billion "digital new deal" program, the country will pilot QKD implementations in several critical infrastructures.

7.5 Quantum Internet

What's colloquially called "quantum internet" could be thought of as the attempt to bring quantum computing to an infrastructure reminiscent of the Internet. With a quantum internet, any two parties on a large network could communicate over some kind of quantum

---

[57]Boaron et al., "Secure Quantum Key Distribution Over 421 Km of Optical Fiber" (2018).

circuit made up of flying qubits, just as the conventional Internet allows two parties to communicate using a virtual circuit built using packet switching. With a quantum network, Alice and Bob could communicate using quantum states, allowing them to enjoy the protection of quantum cryptography, and also giving them the ability to engage in quantum protocols or compute with quantum algorithms.

There are three non-obvious advances that follow from the resilient management of quantum states across distance and devices: first, mastery of quantum networking would make it possible to assemble a quantum computing cluster. Thus quantum networking could change the strategy by which organizations plan to build large quantum computers. Instead of mastering the management of a single device with many qubits, a quantum network would allow organizations to connect together several smaller, perhaps less expensive and easier-to-manage devices into a cluster that has more qubits and volume than any competitor. Such a quantum network might reside within a single building. But while companies such as IBM, with its research lab full of quantum devices, seems well poised to do this, there is (as of yet) no public evidence that IBM or others are taking this tack.

Second, a quantum network could enable *blind* quantum computing. Recall that quantum computing, because of its expense and complexity, is likely to be available as a cloud service rather than as on-premises devices. Currently, users of cloud-based quantum computers offered by Amazon and its competitors access those devices through classical communication-and-control computers. In a world with a functioning quantum internet, that cloud access could become end-to-end quantum intermediated. At that point, the owner of the cloud-based quantum computer would be blind to the user's action. Being blinded would limit policy options because the quantum computing owner might not be able to detect and deter unwanted uses of the device, such as cryptanalysis or currently unimagined noisome behavior.

Depending on how it is implemented, a quantum internet might deny adversaries the ability to spy on metadata. Currently metadata, the data about data in the communications network, such as who calls whom and when, is a key tool of intelligence agencies. Metadata is well structured and relatively easy to analyze. Most people can be identified by their metadata (because most people do not constantly obtain new, clean communications devices) and even though metadata lacks information about the content of communications, metadata often hints at individuals' activities. If a quantum internet is used to set up quantum circuits between the endpoints so that the flying qubits properly travel from Alice to Bob, then such a setup might be susceptible to surveillance. But if the quantum internet is itself controlled *inband* with its own quantum signaling, then it will be difficult to track who is talking to whom. Although this would be a real "going dark" problem that might have intelligence agencies and advertising agencies alike worried, such a possible network seems decades in the future.

Indeed, the challenge of realizing a large-scale quantum network is related to the very attributes that give quantum communications so much privacy: the no-cloning property. Jian-Wei Pan's team demonstrated quantum communication over short distances, extending networks on optical fiber over a distance of about 100 kilometers in 2008.[58] In traditional fiber-optic networks, light becomes diffused from the twists and turns of the fiber and needs to be periodically "repeated," or boosted, to travel to its final destination.[59] But the act of repeating requires copying, which is something that quantum networks can't do. Thus, a repeater on a quantum network breaks the end-to-end guarantees that users of a quantum network would want the network to provide. Although an approach may be developed to address this problem, in the near term, quantum networks will likely involve some sort of trusted repeater that catches the flying qubit, performs a classical computation, and then transmits a brand-new flying qubit down the fiber.

Repeater node trust could be seen as a blessing or a curse: depending on one's perspective, it either can enable lawful access to otherwise unbreakable key exchange, or it represents a problematic security loophole. Still, even a classically relayed quantum network is advantageous, in that if one controls the relay points, one could detect interception and still enjoy lawful access when needed.[60] For instance, the political attributes of China probably fit neatly with

---

[58] Yuan et al., "Experimental Demonstration of a BDCZ Quantum Repeater Node" (2008).

[59] Briegel et al., "Quantum Repeaters: The Role of Imperfect Local Operations in Quantum Communication" (1998).

[60] Farrell and Newman, "Weaponized Interdependence: How Global Economic Networks Shape State Coercion" (2019). Consider the rise of "Weaponized Interdependence," state use of networked infrastructures to leverage panoptic capabilities and use chokepoints for control.

the limits of classical repeaters. Those nodes could be operated by state-controlled companies, and surveilled when desired by domestic law enforcement and intelligence, while denying that same ability to foreign adversaries. Jian-Wei Pan himself boasted, "China is completely capable of making full use of quantum communications in a regional war ... The direction of development in the future calls for using relay satellites to realize quantum communications and control that covers the entire army."

A *quantum repeater* or *quantum memory router* can overcome the trust problem. The first re-transmits the flying qubit, and the second allows the flying qubit to fly off in one of several possible directions. Such devices are still in their infancy.[61] Quantum internet routers are in effect small quantum computers. One approach uses atomic vapor technologies, specifically Electromagnetically Induced Transparency (EIT), introduced in Section 2.2, "Atomic vapor technologies" (p. 41). Scientists are working on the fidelity of copying and storage time; as of 2019, EIT memory loses fidelity in just microseconds.[62]

Quantum "teleportation" is a mechanism being explored to build quantum networks. Teleportation in science fiction is as unexplained as it is exciting. What exactly do teleporters do? How they work seems to change from season to season and among different series. The most well-developed fictional teleportation system appears in *Star Trek*, but the fictional "transporter" was originally created by the series writers to save the cost (in terms of special effects and screen time) of needing to use the ship's shuttle craft to send the crew down to the planet.[63] Over time, the transporter became a useful plot device for creating and then exploring psychological situations, but similar to the show's "warp drive," the underlying physics were never satisfactorily explained.[64]

---

[61]Yan and Fan, "Single-Photon Quantum Router with Multiple Output Ports" (2014); Pant et al., "Routing Entanglement in The Quantum Internet" (2019); Korzeczek and Braun, "Quantum-Router: Storing and Redirecting Light at The Photon Level" (2020).

[62]Yunfei Wang et al., "Efficient Quantum Memory for Single-Photon Polarization Qubits" (2019).

[63]Whitfield and Roddenberry, *The Making of Star Trek* (1968).

[64]In both the original and Next Generation Star Trek series, transporters caused accidents and created doppelgangers: a good and evil Captain Kirk, and a copy of Commander Riker. In Star Trek Voyager, a teleporter accident fused a Vulcan (Tuvok) with a Talaxian (Neelix), creating the unfortunate Tuvix. In Spaceballs (1987), President Skroob's head materialized backwards, so that he faced his pos-

In contrast to mythical teleportation devices, *quantum teleportation* is an effect that is well understood and has even been demonstrated. Quantum teleportation moves the *quantum state* from one particle to a second, irrevocably changing the state of the first particle in the process. Because the state is moved and not copied, quantum teleportation violates neither the Heisenberg uncertainty principle nor the "No Cloning" theorem, which holds that quantum states cannot be precisely copied.

One possible way to construct a quantum router is to use quantum teleportation to transmit data to some point in the distance, in effect creating a point-to-point communication between Alice and Bob. Teams at TU-Delft led by Stephanie Wehner and Ronald Hanson have impressive accomplishments in advancing entanglement and in teleportation. In a TU-Delft demonstration of quantum teleportation, Alice and Bob share a classical communication channel and an entangled particle. The entangled particle is a nitrogen-14 spin inside a diamond. Known as a "nitrogen-vacancy" chamber, this imperfection in a synthetic diamond isolates and insulates the nitrogen atom from the outside environment (see Chapter 2, Section 2.2, "Nitrogen vacancy" (p. 41)). That isolation makes the nitrogen spin more resilient to unwanted interference. With the nitrogen atoms entangled over a distance, Alice takes a second atom, the information bit, and performs a so-called "Bell measurement" between her entangled atom and the second atom. The measurement causes a corresponding change to Bob's entangled qubit. Bob can then extract the information – the state that Alice sent – by communicating with Alice over a classical channel. Alice tells Bob the transformations she made; by performing these same steps, Bob can extract the value of the original state.[65] Because this process uses both quantum entanglement and classical channels as a medium, teleportation protocols do not support faster-than-light communication, as is sometimes claimed.[66]

---

terior, to the delight of the crew. An earlier transporter appeared in the movie "The Fly" (1958), in which a teleporter affixed a fly's head atop a smart scientist's body. The scientist kept his mind, but was under siege from the fly's entomic instincts. See Rzetelny, "Is Beaming Down in Star Trek a Death Sentence?" (2017) for contemporary examination regarding the philosophical implications of creating a perfect copy of a person while destroying the original.

[65] Pfaff et al., "Unconditional Quantum Teleportation between Distant Solid-State Quantum Bits" (2014).

[66] J. G. Ren et al., "Ground-To-Satellite Quantum Teleportation" (2017).

Figure 7.7. xkcd #465: Quantum Teleportation. Used with permission. xkcd.com/4 65/

(See the sidebar "Alas, Faster-than-light Communication Is Not Possible" on page 301.)

Quantum teleportation was first conceived by an international team that included Charles Bennett and Gilles Brassard.[67] In 1997, scientists at the Austrian Institut für Experimentalphysik demonstrated teleportation in a laboratory setting using photons and their spins. Jian-Wei Pan was part of that team, then training under Austrian physicist Anton Zeilinger. Since then, teleportation has been demonstrated at greater distances. The TU-Delft team demonstrated teleportation at 3 meters in 2014 and by 2017, Jian-Wei Pan's team demonstrated teleportation at 1400 km using entangled photons between a base station in Ngari, Tibet (elevation 4500 m) and the Micius satellite.

To enable teleportation over greater distances, and indeed in a quantum internet, scientists are experimenting with entanglement *swapping*. In entanglement swapping, communication between Alice and Bob is made possible even if they lack a point-to-point path. The process works with a device, operated by a third party (here called Faythe), close enough to Alice and Bob to receive an entangled photon separately from each of them.[68]

The European Union has identified a quantum internet as a central goal in its €1 billion investment in quantum technologies,[69] and scientists there have already achieved several key steps towards the creation of a quantum internet. The most synoptic expression of this vision, written by the German physicist Stephanie Wehner, makes it

---

[67]Charles H. Bennett et al., "Teleporting an Unknown Quantum State via Dual Classical and Einstein-Podolsky-Rosen Channels" (1993).

[68]Halder et al., "Entangling Independent Photons by Time Measurement" (2007).

[69]European Commission, High Level Steering Committee, DG Connect, "Quantum Technologies Flagship Intermediate Report" (2017b).

clear that a quantum internet is seen as a special purpose network to exist alongside the conventional Internet.[70] The quantum internet is intended to maintain a channel capable of special functions, such as quantum key distribution, secure identification, and others.

If nations decided to invest in creating a quantum internet, network paths would become a key focus. From a technical perspective, all paths would have to be fully quantum mechanical, or the quantum state would collapse and the technology would fail. Strategically, adversaries along those paths could easily interfere with the quantum state, causing it to collapse. These attacks on availability need not be at the router or even that sophisticated. Anything that degrades the light will work, meaning that these attacks might be easily deniable, and attributable to accident and so on.

Going back to the time of the telegraph, communications find their way along wires on specified routes. If a telegraph pole fell in a storm, that path would be interrupted, and the pole would have to be replaced or a new path set into place. One major advance of the Internet was packet switching, the conversion of communications into datagrams that could take multiple routes. The sender and recipient need not specify these routes. But this lack of specificity comes with a downside: because the communications' paths change dynamically, an attack can intentionally interfere with one route and force the communications to travel over another route with lower legal or technical protections.[71] Recently, the risk that internet communications take unnecessarily circuitous routes through other legal jurisdictions has become a concern of some nations. A 2019 study focusing on path-based risks evaluated tens of thousands of likely paths a user's browser might take when visiting popular sites. The group found that 33 percent "unnecessarily expose network traffic to at least one nation state, often more."[72] Some nations are building local internet exchange points to keep more communications domestic, and out of paths that traverse China, Russia, the US, or its "five-eyes" allies.

A quantum internet would almost certainly require that nations and sophisticated companies create dedicated fiber links for a quan-

---

[70]Wehner, Elkouss, and Hanson, "Quantum Internet: A Vision for The Road Ahead" (2018).

[71]Woo, Swire, and Desai, "The Important, Justifiable, and Constrained Role of Nationality in Foreign Intelligence Surveillance" (2019).

[72]Holland, J. M. Smith, and Schuchard, "Measuring Irregular Geographic Exposure on The Internet" (2019).

tum network, making it more like a separate, dedicated private network. The infrastructure for communication is likely to become much more state-specific. Already, sophisticated users are able to choose the paths that their conventional internet communications travel; the same will likely be true of quantum networks, if they are ever created. Already the Dutch telecom provider KPN has built a fiber-optic, quantum channel network backbone between Leiden, Delft, Amsterdam, and The Hague. (The KPN network does not require repeating, because of the short distances among these cities.[73])

Another option comes from satellites. It seems less likely that a satellite could be manipulated by an adversary than an underwater repeater. At least a half a dozen countries are pursuing satellite-based QKD programs.[74] Either physical or cyber manipulations could be impactful. Thus, initiatives such as Elon Musk's SpaceX/Starlink satellite network, which intends to populate the sky with internet-providing satellites, could also form the backbone of a tamper-resistant network that is mostly classical but could include quantum elements: perhaps two quantum-enabled ground-stations on opposite sides of the planet would communicate with a message passed from satellite to satellite.

Similarly, one might imagine businesses that place point-to-point servers connected by quantum channels in physically inaccessible places, for instance submerged in containers that if opened would fail.

7.6 Conclusion

Quantum communications can be binned into two categories: first, the related applications of quantum random number generation and key distribution, and second, technologies that enable a quantum network or quantum internet. While quantum random number generation and key distribution are both maturing technologies, early systems have been commercialized and are in use today. These technologies meet two central requirements for secure communications technologies: they are information-theoretically secure and enable distribution of keys at a distance. Those who adopt QKD will never have to worry that the keys they use today in encryption systems based on the RSA or Elliptic Curve public key cryptography systems might be cracked by some powerful quantum computer in the future

---

[73]Baloo, "KPN's Quantum Journey, Cyberweek 2019, Tel Aviv, Israel" (2019).
[74]Khan et al., "Satellite-Based QKD" (2018).

Alas, Faster-Than-Light Communication Is Not Possible

Experiments in entanglement show that entangled particles somehow "know" the quantum state of their twin. One might think of entangled particles as parts of a connected system. Scientists do not know how they are connected, but scientists can show through Bell tests (see Section B.4 (p. 513)) that they are.

Quantum teleportation takes advantage of the linkage between distant particles to teleport a state from Alice's entangled particle to Bob's. Because Bob's particle reacts instantly, even when separated by great distances, some have speculated that teleportation could somehow enable faster-than-light (superluminal) communication. Alas, quantum teleportation does not enable faster-than-light communication.

Superluminal communication is impossible because quantum teleportation protocols depend on classical channels to extract the meaning from the entangled qubits. After teleporting a state to Bob, Alice and Bob communicate over a classical channel. Bob determines the teleported state by applying transformations that correspond to Alice's instructions.[a] This is the basis of the BB84 and E91 protocols.

So as one can see, the reversion to a classical channel, and the complexity of the information exchange and discovery, makes it impossible to communicate faster than light speed.

---

[a]Pfaff et al., "Unconditional Quantum Teleportation between Distant Solid-State Quantum Bits" (2014).

– although adopters of today's QKD systems still need to verify that the QKD systems themselves are still secure against traditional vulnerabilities, such as electromagnetic radiation or cyberattack.

Yet, if experience with other privacy-enhancing technologies holds, only entities with the most to lose will affirmatively adopt them. Banks, militaries, intelligence agencies, and other entities with the awareness and budget are likely adopters. But for everyone else, three other requirements must be met: the system has to be fast, it has to be usable by anyone, and it has to be on by default. The coming availability of classical encryption that is quantum resistant will be satisfactory for many actors. Unless some economic interest arises and militates strongly in favor of quantum encryption, most consumers and businesses will rely on classical alternatives.

The quantum internet's best use in the future – aside from its ability to procure funding for prestigious science projects – seems to be the interconnection of existing, small quantum computers into a cluster of unprecedented power. The other benefits, relating to time synchronization and astronomy, seem so tethered to scientific and technical users that it is difficult to see how they would inspire a commitment to outlay the money to make a quantum internet happen. In the nearer term, the quantum internet's potential to make communications end-to-end secure and eliminate metadata surveillance may be the driving factor for nation states to invest in the technology.

Part 10

Shaping the Quantum Future

Part I introduced the functional capabilities of quantum technologies. This part focuses on the policy issues that emerge from these technologies. For example, the possibility of improved quantum computing tomorrow means that we must start upgrading encryption algorithms *today*. Quantum computing will provide speedups in certain kinds of computations that are important for scientific discovery broadly. And while the most assured and lowest-risk approach to avoid quantum cryptanalysis is to use quantum key distribution, another approach is to use improved mathematical algorithms that are believed to be quantum resistant. Meanwhile, quantum sensors that can measure gravimetric and magnetic fields more precisely and accurately may enable visibility into secret compounds or even private homes.

Part II builds on the implications and continues our discussion of the technological possibilities that flow from quantum technologies, which we discuss in Chapter 8. Once the likely paths of the technology are understood, the next chapter proceeds to the legal and policy issues raised by the special affordances of quantum metrology and sensing, communications, and computing technologies in Chapter 9. We conclude the book in Chapter 10.

Quantum Technologies and Possible Futures

W HAT are the most likely paths for quantum technologies? Are we facing a future where quantum technologies are the domain of governments, with asymmetric powers to collect information about us and to make sense of it? Or might the future bring some other landscape, where quantum technologies protect the communications of the average person and quantum sensing helps us diagnose and treat illness?

This chapter uses scenario analysis to seed a policy discussion for quantum technologies. We envision four likely outcomes of the quantum technology race, and these different visions provide motivation for contemplating the strategic, political, and social dimensions of quantum technologies. The next chapter considers how different policy measures could address these risks.

8.1 Do Quantum Artifacts Have Politics?

Langdon Winner, in his seminal 1980 article, "Do Artifacts Have Politics?",[1] argued that "technical things have political qualities." This is different from the popular notion that "technologies are seen as neutral tools that can be used well or poorly, for good, evil, or something in between," he wrote.

The notion of technology neutrality is a powerful one, adhered to by many. Such adherents observe that technologies, what Winner calls *artifacts*, are just tools wielded by individuals who decide

---

[1]Winner, "Do Artifacts Have Politics?" (2018).

how to use them. A hammer could be used to build your home or
to break your neighbor's windows. But Winner's argument is more
nuanced and strikes at a deeper level. It is not that the individual is
blameless or without control, it is that the tool shapes the possible
and the broader social landscape. Winner argued that some technologies are "inherently political" in two senses. First, a technology can
be adopted to settle a contested issue. For instance, internet users
may value anonymity at times, but an advertising company that
develops web browsers might deploy its software so that users are
always identified and no real chance of anonymity is possible. The
advertiser's web browser settles the debate between anonymity and
perfect identification in favor of its own preferred outcome.

Second, and more problematically, a technology might require
a certain political, economic, or social order. These are *inherently
political technologies*. To press the point, Winner contrasts forceful
examples: nuclear power and solar energy. A society with the power
of nuclear fission or fusion cannot allow the technology to devolve
to ordinary citizens. Instead, only powerful institutions secured with
military-like safeguards can possess these technologies. Indeed, federal rules specify that sites with special nuclear material must have
trained, qualified, ballistic-armor wearing guards in possession of as-
sault rifles, shotguns, and handguns.[2] Even with these safeguards,
civilian technologies such as nuclear power present fantastic risks.
Just imagine if the September 11, 2001 hijackers crashed a jet into
the Indian Point nuclear power plant, just 36 miles from Manhattan,
instead of the city's World Trade Center. Atomic energy requires cen-
tralized political, economic, and social power arrangements because
of the risk of misuse, accident, and disaster.

Consider solar power as a counterexample. Solar power is distributed, often on the roofs of homeowners or in community-clustered
solar farms. Solar power has its disadvantages and its own costs, of
course. But Winner's point about its politics still holds: solar power
leads to different political, economic, and social orders. A world that
invests billions in solar energy is one where communities and even
individuals can have both policy and technical control over energy
generation and storage. There is no need for armed police, secrecy,
or worry about widespread disaster. In fact, because it is distributed
widely and to individual citizens, solar power may be resilient against

---

[2]See 10 C.F.R. Part 73.

the very attacks we are so concerned about with regard to ordinary power stations.

Nuclear power – a quantum technology – was identified by Winner as inherently political. But what about quantum sensing, computing, and communications? There is an obvious path to quantum technologies becoming inherently political. In this path, quantum technologies are shaped by the small elite who understand and can use them for purposes that are political, such as military and intelligence uses. For a historical comparison, consider early computing, which was dominated by military and industrial applications (see Chapter 3). Renowned MIT computer scientist Joseph Weizenbaum characterized the computer as fundamentally a conservative force, one that allowed institutions to maintain and centralize their power.[3] Alas, democratization with the personal computer revolution changed public perceptions and the political possibilities of computing. Today, the personal computer is seen as a tool of creative expression and entertainment and few remember its early uses for artillery tables. But quantum technologies will not necessarily see a personal computing revolution. Today, only an elite few from powerful institutions can understand and use quantum technologies. Quantum technologies might become associated with the needs of this military-intelligence elite, perhaps even earning a "taboo" or taint as did mainframe computing.

8.1.1 Threat Modeling

Threat modeling is a technique for understanding the different ways technology can be used to attack, be attacked, or fail, and helps prepare organizations to mitigate these threats in a systemic way. Threat modeling can be used in software development to understand the complex dependencies and vulnerabilities in enterprise systems and, as a result, develop software that is more secure and resilient. Adam Shostack created a straightforward, four-step model for security threat modeling[4] which we have adapted to anticipate the likely ways that adversaries could use quantum technologies.

In Shostack's model, one begins by defining the problem being analyzed. Quantum technologies, as a field, are too broad for analysis. Some reductionists might argue that most modern technologies must be viewed as quantum technologies – even classical electronic

---

[3]ben-Aaron, "Weizenbaum Examines Computers and Society" (1985).
[4]Shostack, *Threat Modeling: Designing for Security* (2014).

computers – because their functions are best described using concepts from quantum mechanics such as electrons, photons and atoms.
Such reductionist approaches are unhelpful. Instead, here we cordon
off quantum technologies from others by restricting our analysis to
those technologies that specifically leverage quantum effects in order
to perform some useful function.

As discussed in Part 01 , our tripartite categorization decom-
poses "quantum technologies" into quantum sensing, quantum computing, and quantum communication. These three share the characteristic of gaining utility from harnessing quantum effects, but each
presents challenges and uses so different that they are recognized as
separate fields.

Drawing from our previous chapters, we assume the following in
this chapter's analysis:

- All sectors will continue to adopt quantum sensing, resulting
  in sensors that are less expensive, smaller and more powerful. Some sensors will be mounted on satellites, some will be
  mounted on unmanned aerial vehicles, while others may be in
  ground vehicles, handheld, or even in fixed locations.

- Intelligence and military agencies, particularly in countries with
  space programs, will implement quantum sensing devices to
  detect both hidden matériel and to understand adversaries' infrastructure, as discussed in Chapter 2.

- Programmable quantum computers that are large enough to
  solve useful problems (as discussed in Chapter 4 and Chapter 5)
  will be built within 10 years.

- Quantum Key Distribution will be *selectively* adopted to secure data in transmission; most users will be content using
  post-quantum-computer encryption schemes for the majority
  of uses, as discussed in Chapter 7. These algorithms will be
  standardized, broadly deployed, and become the default encryption technology for key exchange.

8.1.2 *Future Quantum Technology Scenarios*

In Shostack's framework for threat modeling, analysts define a problem and then ask broadly, "what could go wrong?" In the computer
security context, the most relevant risks are known by the mnemonic

STRIDE: **S**poofing, **T**ampering, **R**epudiation, **I**nformation disclosure, **D**enial of service, and **E**levation of privilege.[5]

Turning to quantum technologies, the dynamics go far beyond STRIDE. Quantum technologies could alter world order, with certain nations gaining important advantages over others. For example, quantum sensing might impart such a dramatic advantage that it causes nations to focus their initial attack on each other's satellites. Competition for advantage could also alter innovation strategies, with some nations racing ahead in hopes of being the first to achieve benefits, while others might realize that their optimal strategy is to copy – or steal – the innovations of first movers.

To explore what could "go wrong" – and go right – this chapter explores four high-level scenarios[6] for quantum technologies:

- **Government Superior and Dominant**: where a government possesses more capabilities than all others and can deny others the ability to acquire or use quantum technologies;

- **Public/Private Utopia**: a landscape where companies and governments share different levels of prowess in quantum technologies;

- **Pubic/Private East/West**: a version of the public/private landscape colored by East/West bloc competition;

- **Quantum Winter**: the possibility that quantum technologies ultimately fail to be consequential, similar to the "AI winters" that chilled the field of artificial intelligence in the 1970s and 1980s, where hype cycles were followed by disappointment and dormancy.

8.2 Scenario 1: Government Superior and Dominant

One possible future scenario is a world where a major government – likely the US or China – achieves superiority in quantum technologies, and uses that superiority both to maintain their technological dominance and as an enabler to take actions without significant interference by others governments.

This scenario is based on the concepts of *deterrence theory*. Nations mostly seek superiority not to win conflicts, but to prevent

---

[5]Shostack, "The Threats to Our Products" (2009).
[6]Heuer Jr. and Pherson, *Structured Analytic Techniques for Intelligence Analysis* (2015).

conflicts from happening. For example, for decades the US military
strategy has been to create a war-fighting force that is so superior to
other nations, and so omnipresent throughout the world that other
nations dare not attack. This level of military supremacy, in theory,
produces an alignment that makes conflict less likely. Two pieces of
historical evidence in support of the theory are the post-World War
II peace in western Europe – the longest in history – and the fact
that all US conflicts since 1945 have either been conflicts of choice,
or (in the single case of Afghanistan) the result of an attack by a
non-state actor.

As a definitional matter, superiority only means that one actor
is stronger than all others. Left unchecked, competitor nations will
start nipping at the heels of a superior state until they reach technological parity. Thus, to maintain technological superiority, a nation
must pursue *dominance*: a level of superiority reaching supremacy,
where one both enjoys freedom of action and can (at will) deny freedom of action to others.

What would the path to quantum technology dominance look
like? Is dominance even possible? We believe that the possibility for
dominance depends on whether quantum computing is a winner-take-
all (or winner-take-most) technology.

8.2.1 Winner Take All

At first, quantum technologies would appear not to present a winner-
take-all opportunity. Consider quantum communications. American,
Chinese, and Dutch scientists have all demonstrated major achievements in quantum communications, publishing their work in scientific journals. The underlying hardware for photonic transmission
and capture (such as single-photon emitters and detectors) is commercially available and can be found in many physics labs. But most
importantly, quantum communication technology does not appear
to benefit from a *virtuous circle*: breakthroughs in quantum key distribution do not themselves create new tools for developing better
breakthroughs.

But unlike quantum communications, quantum computing is likely
a domain in which dominance is possible. It's true that competition
is booming in the private sector and companies are experimenting
with an array of different physical systems to create quantum computers. Likewise, none of this research is being kept secret. Instead,
scientists and their corporate backers are apparently competing for

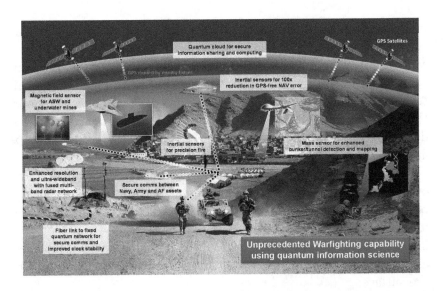

Figure 8.1. A 2018 vision of quantum technology use by the US Army Research Laboratory.

academic glory by publishing their findings in prestigious journals like *Science* and *Nature*.

But while the detailed scientific papers that are appearing may have hundred-page appendices explaining all of the science, they do not come with detailed technical information that is necessary to actually manufacture the underlying scientific apparatus. Such information would easily run to tens of thousands of pages, and in any event would be largely unusable, because using such information requires mastery of manufacturing processes and operational know-how that is built upon years of practice.

Unlike quantum communications, quantum computing does enjoy a virtuous circle, in that advances in quantum computing could almost certainly be used to develop more powerful quantum computers.

Consider this scenario: A nation develops an intermediate-scale quantum computer. Perhaps it does so by carefully observing commercial activities, and uses a different approach that has been less researched but that appears, in light of new discoveries, more promising. Instead of publicizing this achievement, or using it for cracking encryption keys, this nation focuses on understanding materials science. Specifically, that nation would attempt to build a larger quantum computer based on the insights that only it can gain from its more complex view of the underlying physics of materials. Just as classical computers help one build larger classical computers, the same strategy could be important to gaining superiority in quantum computers. In this scenario, quantum computing is a winner-take-all technology. The early winner learns secrets of materials and physics that allow it to race ahead of competitors. This winner might even dangle false leads to competitors – not fake science, but perhaps apparently promising paths that lead to dead ends.

Secrecy will be a key element of winner-take-all dominance. Thus, one signpost of the government-dominant scenario is the public appearance that the government has no quantum computing program in the space at all (perhaps it signals that it has given up), or that inexplicable holes exist in the publicly available literature, but there are indicators of an aggressive quantum program operating below the surface.

An important factor in maintaining dominance is crushing competitors' will to compete. In quantum computing, such a strategy could be accomplished by eventually revealing the existence of a superior quantum program and selling commercial access. Such access would necessarily be subtly restricted. For example, users could be restricted to less powerful machines, or could be prohibited from solving particular kinds of problems. Recall that quantum computers have control systems run by classical ones; these classical computers can function as a filter to prevent certain unwanted uses of the dominant actor's quantum computer. Such access would quench funding for commercial competitors, and would likely cause scientists entering the field to concentrate on applications rather than underlying systems design: why spend billions trying to discover something that has already been discovered elsewhere?

A government that pulls ahead in quantum computing will also likely be superior in quantum sensing. This is because sensing and metrology are antecedent to computing. That is, one must master the management and manipulation of a large ensemble of qubits. That technical ability implies a mastery of smaller systems used for sensing.

To use quantum sensors in a way that helps in a competition with nations, sensors need to be deployed. Nations with sophisticated unmanned aerial vehicle technology and access to outer space will have more ability to sense without restriction.

Space programs are a source of national prestige and scores of nations have one. However, only about a dozen nations have realized a satellite launch capability. The United States has the most satellites in space (1007), followed by China (323) and Russia (164).[7]

Other nations are dependent upon launch-capable states. And these launch-capable states are unlikely to facilitate a competing nation's quantum sensing advances, particularly if they allow the launch-dependent nation to somehow leapfrog others. This is consistent with Henry Farrell and Abraham Newman's theory of *Weaponized Interdependence*.[8] According to duo, nations take advantage of economic and technological choke points for both surveillance and control of adversaries. In the case of quantum technologies, there are not good options to prevent adversaries from building or buying components, but launch-capable states could deter the most powerful implementations of those technologies – by controlling outer space.

Government Dominant

A government enjoys advanced quantum technologies and can operate without significant interference from adversaries.

Key Policy Characteristics

Industrial policy, secrecy, export control, non-proliferation-like strategies.

Key Enabling Factors

Making the right bet on qubit substrates, winner-take-all virtuous cycle, access to outer space.

Strategic Surprise

Sensing technologies that can see adversary matériel, illumination of low-observable (stealth) technologies, cryptanalysis, secretive weapons development.

Outlook

Rich private sector with high-powered incentives to commercialize makes government-exclusive control of quantum technologies unlikely.

Commercial launches might appear to be a promising way to evade the space launch choke point, however, just like seafaring vessels, satellites have a national "flag." Nations regulate such launches

---

[7]Union of Concerned Scientists, "UCS Satellite Database" (2021).

[8]Farrell and Newman, "Weaponized Interdependence: How Global Economic Networks Shape State Coercion" (2019).

with domestic and international law. If a nation sought to use an American company, such as SpaceX, to launch its quantum sensing network, the payload must be licensed and would be subject to review by multiple agencies. Such review explicitly considers whether the satellite would endanger national security, raise foreign policy concerns, or undermine international obligations.[9] Regulations promulgated by the Trump administration require private remote sensing companies to disclose many details about the architecture and capabilities of sensing systems, including resolution and collection rates, and whether the sensor can "look" off-axis. Imaging of other "artificial resident space objects" requires special permission – meaning that the proposed device may not look at other (potentially secret) satellites. It seems unlikely that countries without space launch capability will be able to purchase such capability to achieve quantum parity with those that have it.

The advantages of a space program go beyond sensing. United States Naval Research Laboratory (NRL) scientist Marco Lanzagorta speculates that satellite-based quantum communications systems will enable advances in submarine communication. As long as the water was sufficiently clear and lacking in turbidity, NRL predicts key distribution is possible as deep as 100 m, and at a rate hundreds of times faster than existing very low frequency communication methods.[10] This may change the "lone wolf" strategy of submarine operation.[11] Existing communications require alterations to optimal speeds and paths, ones that might help an adversary track a submarine. Thus, faster and more flexible transmission could enable more communications without detection.

The winner-take-all scenario could also happen in the private sector. For instance, Microsoft has pursued *topological* structures to develop a quantum computer while its competitors have used superconducting systems. If the topological approach turns out to be the winning medium, Microsoft could race ahead in a way its competitors could not, at least for now. Microsoft could also keep important aspects of its engineering a trade secret by selling its quantum computers as a service rather than as standalone devices. Locked in a

---

[9]See e.g. 15 CFR Part 960 (2020).

[10]Jeffrey Uhlmann, Marco Lanzagorta, and Salvador E. Venegas-Andraca, "Quantum Communications in The Maritime Environment" (2015).

[11]Kania and John Costello, *Quantum Hegemony? China's Ambitions and The Challenge to US Innovation Leadership* (2018).

vault-like data center, each Microsoft employee working on the program would only see a small part of the overall project – enough to use it and contribute, but not enough to duplicate a working system. Microsoft would then be able to maintain dominance in quantum computing much as IBM maintained its decades-long dominance in computing, and as Google continues to maintain its dominance in Internet search.

A private-sector winner-take-all outcome is very different from a government one. With the rise of the power and wealth of corporations, private companies with a quantum computer could make far more money selling to other companies than to governments. Furthermore, the sale to governments for military and intelligence purposes can be lucrative, but these activities come with other restrictions and complications that ultimately narrow options for selling one's technology. Thus, a private winner-take-all outcome would drive a great evangelizing of the technology and its uses outside defense and intelligence. A dominant company would want to sell its cloud service to every industry in almost all nations. Access to quantum computing for non-military purposes would likely be democratized, even if the devices themselves were carefully controlled. Military applications would likely remain available to the host country – which in the case of Microsoft, would likely be the US.

Combined quantum technologies may have real and lasting consequence for nation-state conflict. Indeed, some military technology experts refer to quantum sensing and communications as the "atomic bomb" of information theory, and urge us to contemplate a quantum "strategic surprise."[12]

What would strategic surprise look like in a government-superior and dominant quantum technology world? In this section we look at strategic surprise in three areas: cryptanalysis, nuclear weapons, and remote sensing.

8.2.2 Strategic Surprise: Cryptanalysis

Quantum cryptanalysis is the most obvious example of strategic surprise that could be enabled by quantum computing, and it is the motivating example that is primarily responsible for the interest in quantum computing over the past two decades.

---

[12]Marco Lanzagorta, "Envisioning The Future of Quantum Sensing and Communications" (2018).

In order to foresee the implications of quantum cryptanalysis,
it is important to first understand how cryptography is used today.
Here we focus on three purposes of encryption: protecting stored
data ("data-at-rest"), protecting data that is sent over the Internet
("data-in-flight"), and authenticating software ("digital signatures").

The most broadly used encryption algorithm today is the Advanced Encryption Standard (AES).[13] There are basically two versions of AES in use: AES-128, which has a 128-bit secret key, and
AES-256, which has a 256-bit secret key. Both of these algorithms
are considered uncrackable with classical computers for the foreseeable future.[14] However, given that Grover's algorithm can speed up
this kind of search so that it takes only $\sqrt{2^{128}} = 2^{64}$ operations, it
might be possible to crack an AES-128 message using a fully realized
quantum computer. It would still be impossible to crack an AES-256
message.

AES is a secret-key algorithm, meaning that both the sender and
the recipient must agree on the same key. In practice, these keys are
randomly created for every encrypted hard drive, and for every individual web page or email message as it is sent over the Internet.[15]
The keys are then encrypted using a public key cryptography algorithm such as RSA or the Diffie–Hellman key exchange protocol.
The security for both of these algorithms rests on the difficulty of
factoring large numbers, so an attacker with a functioning quantum
computer would be well positioned to decrypt the information sent
over the Internet today provided that three things are true:

- The future attacker has a copy of the information that the attacker wanted to decrypt. Presumably this information would
  be obtained through a search of an office (to get an encrypted
  hard drive), a wiretap or other interception technique.

---

[13] Dworkin et al., *Advanced Encryption Standard (AES)* (2001).

[14] The best approaches for cracking AES-128 typically require on the order of 2^{128}
mathematical operations. If an attacker has a billion computers that can perform
a billion operations per second, then the attacker can perform $10^9 \times 10^9 = 10^{18}$
operations per second. However, $2^{128} = 10^{38}$, so the hypothetical attacker would
require on the order of 10^{20} seconds, or 3168 billion years.

[15] This section only considers encrypted messages sent over the Internet. Native
wireless communication protocols, such as those used to set up LTE cellular
telephone calls, are generally less secure due to the need to have backwards com-
patibility.

- The future attacker knows the protocol that was used to send the information. This is generally not a problem because most information is sent using standard protocols.

- The future attacker has been allocated sufficient time on the quantum computer to actually crack the key.

So clearly, a functioning quantum computer does not mean the total collapse of data confidentiality. Instead, it creates the possibility that a well-positioned attacker could decrypt or forge selected messages.

Encrypted Data-at-Rest

Whether or not a fully realized quantum computer could decrypt stored data has everything to do with the way that the data are encrypted. If the data are encrypted with AES-128, or if they are encrypted with AES-256 and *that* key is encrypted with a *c.* 2020 public key algorithm (that is, one that does not offer post-quantum resistance), then the public key could be cracked. However, a common construction for disk and document encryption systems is to take a user-supplied *passphrase*, compute the *cryptographic hash* using an algorithm such as SHA-256, and use that hash to encrypt the AES-256 key. As near as we can tell, SHA-256 is quantum resistant, and the speedup afforded by Grover's algorithm would be insufficient to achieve a single cracked passphrase within the expected lifetime of the Sun. But 5 billion years is a long time, and it's possible that a flaw will be discovered in SHA-256 or AES-256 that would obviate the need to crack the code using brute-force search before the Sun becomes a red giant and engulfs the Earth.

This means that data-at-rest encrypted *today* might be crackable at some point in the future when quantum computers are available. However, it is relatively easy to design data-at-rest systems to be quantum-resistant, and many of today's systems encryption systems have already been redesigned to take that possible future threat into account.

Encrypted Data-in-Flight

Whereas data-at-rest is a message that you send to yourself in the future, data-in-flight is data that you send to someone else. The fundamental difference between these two scenarios is how the intended user gets access to the decryption encryption key. In the first case, since you are sharing the key with your future self, you have it now – just don't lose it! But when Alice sends her encrypted message to Bob, Bob typically doesn't have the key that was used to encrypt the message. This is the problem for which public key cryptography was invented. The modern solution is that Alice generates a random message key and uses that to encrypt the message, then encrypts the message key with Bob's public key and sends the encrypted message key along with the message. Bob receives both the encrypted key and the message, decrypts the message key with his public key, and uses the decrypted message key to decrypt the message.

As we discussed earlier, technologists are working hard to develop and deploy post-quantum public key cryptography algorithms. If they succeed in developing algorithms that are just as efficient as RSA and Diffie–Hellman, the world will likely transition to them. Such a transition would probably take five to ten years, given the speed of similar cryptography transitions.[16]

If workable quantum computers become available, the data-in-flight with its privacy most likely in jeopardy will not be data being sent at some point in the future, but the data that was sent between 1995 and today that was captured and archived by various national intelligence agencies.

There is no information that is both public and trustworthy regarding the systematic recording of telecommunications in the world today. For example, around 2011 the National Security Agency broke ground on its Utah Data Center, a massive data warehouse costing over a billion dollars.[17] It has been speculated that one purpose of this facility is to warehouse all the data the NSA collects for future analysis. A 2013 article in *Forbes* estimated the capacity of the facility at 12 000 PB stored on 10 000 racks of equipment. To convey the size of this storage, the article notes that all the "voice recordings of all the phone calls made in the US in a year would take up about

---

[16]For example, the first attacks on the cryptographic hash MD5 were discovered in 2006. See Black, Cochran, and Highland, "A Study of The MD5 Attacks: Insights and Improvements" (2006). Yet Microsoft still allowed limited use of the MD5 algorithm for certifying root certificates as late as 2013. See Microsoft Corp., "Microsoft Security Advisory 2862973: Update for Deprecation of MD5 Hashing Algorithm for Microsoft Root Certificate Program" (2013).

[17]National Security Agency, "Groundbreaking Ceremony Held for 1.2 Billion Utah Data Center" (2001).

272 petabytes,"[18] although the likely target of the data center is not US, but foreign communications (as the NSA is generally prohibited from collecting inside the United States). Such data would be prime targets for decryption if they are encrypted and the NSA were to later acquire a quantum computer.

More concerning for US readers than possible surveillance by the US government (which is regulated) may be the electronic surveillance activities of China, Russia, and other governments against US and European targets. Russia and China[19] are also known to have extensive capabilities for Signals Intelligence (SIGINT) and are presumably collecting worldwide, although once again, hard details are somewhat elusive. The Global Signals Intelligence market was said to be $12.8B in 2018 and expected to rise to $15.6B by 2023, according to a market research report,[20] with much of the growth coming from China and India.

In general, it seems prudent to assume that any message transmitted today in any part of the world might be might be captured, indexed and archived by anywhere from two to five governments or non-government organizations, and that the message might be unlocked at some point in the future if sufficient need arises.

Forged signatures

A third application for quantum cryptanalysis will be to crack the keys that are used to sign software updates, electronic documents, and websites.

Digital signatures are an aspect of cryptography that is less publicized than protecting the secrecy of web browsing and email, but it many ways they are more important, because they provide for the underlying security of the computers themselves. Virtually every program that runs on a modern computer is digitally signed by the computer's manufacturer, the operating system vendor, or the software publisher. The computer then verifies these signatures when it boots and as it runs. Companies like Intel also use digital signatures to validate updates for the microcode that runs inside microproces-

---

[18]Hill, "Blueprints of NSA's Ridiculously Expensive Data Center in Utah Suggest It Holds Less Info Than Thought" (2013).

[19]China's SIGINT capabilities go back to the 1950s, see Hagestad, "Chinese IW Capabilities" (2012).

[20]Wood, "Global $15.6Bn Signals Intelligence (SIGINT) Market by Type, Application and Region – Forecast to 2023 – ResearchAndMarkets.com" (2019).

sors. These updates make it possible for Intel and others to fix bugs in microprocessors after they have shipped to customers.

Digital signatures are similar to traditional wet-ink signatures in that they are typically used by an author to sign something that the author has written to demonstrate the author's authorship. However, in practice, an author can sign anything that the author wants. Authors can also be tricked into signing documents unknowingly or be forced against their will. But whereas an ink signature is bound to a particular piece of paper, a digital signature is linked to a specific sequence of bits. If just one bit changes, the signature is no longer valid.

Digital signatures are written with an encryption key that is unsurprisingly called a signing key. These keys are typically certified by organizations that are unsurprisingly called *certificate authorities*. These certifications are also performed using digital signatures. The certifications are verified with the certificate authority's *public key certificate*, which is supplied with the computer's operating system.

To give a palpable example, you rely on these certificates (and thus on these certificate authorities) when you visit the website of your bank or other important services. When the browser visits the putative bank website, the bank sends its certificate to the browser along with a reference to the issuing certificate authority. If the web browser accepts that certificate authority, the browser signals (typically with a lock icon) that the connection is secure, and in some cases, avers the identity of the website as belonging to a certain company. If the certificate is compromised or certificate authority was dishonest, an impostor could masquerade as your bank, and you would be none the wiser.

Although there are different algorithms used for digital signatures than for message secrecy, the algorithms are based on the same underlying mathematics. As a result, a quantum computer that could be used to crack the public keys that are used to encrypt messages, and thus make it possible to decrypt those messages, could also crack the public keys used to verify digital signatures, and thus allow signatures to be forged.

Hashing, Digital Signatures, and Grover's Algorithm

Another way that quantum computers might be able to attack digital signatures is by searching for *hash collisions* in cryptographic hash functions.

A cryptographic hash, sometimes called a digital fingerprint, is a number that results from running an input document through a special kind of one-way digital function. These functions are designed so that no matter the size of the input, the output is a constant size – for example, the US Government's Secure Hash Algorithm #1 (SHA1) always outputs 160 bits (40 hexadecimal characters). Cryptographic hash functions are further designed so that roughly half of the output bits change in an unpredictable manner if a single bit in the input changes.

For example, here we apply SHA1 to the strings *hi* and *hh*, which differ by exactly one bit:

String	Bits	SHA1 (hex)
hi	01101000 01101001	c22b5f9178342609428d
		6f51b2c5af4c0bde6a42
hh	01101000 01101000	d3fc13dc12d8d7a58e7a
		e87295e93dbaddb5d36b

Digital signature systems actually sign hashes of documents, rather than the documents themselves. So if it is possible to find two documents that have the same hash, there is no way to tell if a digital signature from the first is moved to the second.

Quantum computers, using Grover's algorithm, could offer a speedup in finding such collisions, which could be used in attacks to place malware on other computers and otherwise enable attackers to fool recipients about the integrity of files. In order to offer such speedup, however, it might be necessary to implement the entire cryptographic hash function as a set of quantum gates. Grover's algorithm gives a speedup of a square-root, so roughly speaking it would make a 512-bit hash as secure as a 256-bit hash. Since 512-bit algorithms such as SHA-512 and SHA-3-512 are widely deployed today, and since a work-factor of 2^{256} is considered unbreakable, Grover's algorithm is unlikely to have an impact on the security of today's digital signatures absent additional mathematical developments.

False software signatures are valuable because with a single hostile software update, even just a somewhat sophisticated attacker can
take over a computer and capture all information from it instead of
laboriously decrypting individual files and communications.[21] A malicious update allows the attacker to operate the device as a regular
user, and thus avoid the time-intensive requirements of investigating
a suspect through their communication logs or through interviewing
people who conversed with the suspect.[22] Not limited to mere communications surveillance, a hostile update can covertly enable the
computer's microphone and camera, and perform searches on the
user's files. If the computer is used for web-based banking, the update can transfer money out of the user's bank account. If the user
accesses their work computer from home, the work network can be
equally compromised as well.

8.2.3 *Forged Signatures and Our Legal Realities*

Digital signatures are used throughout the digital economy. The ability to forge signatures would render virtually every computerized
system vulnerable to some kind of attack. This includes web servers,
the Internet's underlying domain name system, embedded firmware,
vehicle control systems ... practically everything. A nation with the
capability to create fake software updates could take over the industrial control systems of other nations, and corrupt devices such as
radar systems or targeting systems that are relied upon to compute
properly during a conflict.

Digital signature attacks are real and can have dire consequences
for victims. Consider the attack on the Dutch certificate authority,
DigiNotar, whose certificate authority public keys were relied on by
popular web browsers including Google's Chrome, Microsoft's Internet Explorer and Mozilla's Firefox.[23]

In 2011, intruders thought to be working for the Islamic Republic
of Iran hacked DigiNotar's systems and issued over 500 certificates
in the names of popular web services including Gmail and Facebook.
Combined with the Islamic Republic's control of the Iranian's internet connections, these certificates allowed the holders of the corre-

---

[21]T. Li et al., "Security Attack Analysis Using Attack Patterns" (2016).

[22]Vidas, Votipka, and Christin, "All Your Droid Are Belong to Us: A Survey of
Current Android Attacks" (2011).

[23]Hoogstraaten et al., *Black Tulip Report of The Investigation into The DigiNotar
Certificate Authority Breach* (2012).

sponding private keys to intercept communications between users in Iran and these services, allowing the theft of content such as email messages and postings, as well as passwords and other information. Services belonging to the US Central Intelligence Agency and Israel's Mossad were also allegedly targeted. The DigiNotar attack shows that as individuals in repressive states use the Internet to organize and communicate with the outside world, attackers who can issue false certificates (or crack the private keys of certificates already in use) gain a powerful ability to monitor, change, and block these activities. They can identify participants in communications and masquerade as the activists themselves, all while the users think their communications are protected by advanced encryption.

The DigiNotar incident is a clear demonstration that technologies such as encryption – thought to be the ultimate technical guarantee against spying – often require extraordinary reliance on unknown third parties.[24] We must rely on these third parties to both properly design and to properly operate these systems. This includes anticipating attacks on confidentiality and integrity, and finding ways to upgrade existing systems to be resilient against future adversaries.

Imagine a future where this reliance on encryption deepens by spreading to more contexts, when not only web communication but all sorts of societal functions depend on digital signatures. As governments consider "e-government" services, the most radical approach is to go "digital first" with documents of record. Estonia has done so, meaning that the nation's official document of record is computerized rather than on paper.[25] In Estonia, citizens and businesses can use an electronic identity infrastructure to hold a record of their personal information, and then use this system to avoid the noisome paperwork that major (or even minor) life events trigger. For instance, babies can be registered with the government (i.e. obtain a birth certificate) without paperwork, prescriptions are requested online and filled, taxes can be paid online, citizens can vote online, one can create a corporation online quickly, and one can pay for myriad services, from public transportation to parking fines, all online. Of course many nations provide services like this, but in the US, for instance, there is no single identity architecture and the different

---

[24] Arnbak and van Eijk, *Certificate Authority Collapse: Regulating Systemic Vulnerabilities in The HTTPS Value Chain* (2012).

[25] Heller, "The Digital Republic: Is Estonia The Answer to The Crisis of Nation-States?" (2017).

services tend to be developed and offered by different entities, for better or worse.[26]

As nations implement similar e-government approaches, they become susceptible to integrity attacks that are impracticable in a paper-record society, or even in a society that provides the same services from disparate entities with different systems. As such, quantum computing attacks on signatures could affect the documents that define our legal relationships, spreading uncertainty, allowing people to cheat, and making it difficult to determine what the "ground truth" is. Adversaries could do this by forging signatures and subtly altering important records. Imagine if an adversary changed property lines, changed ownership records or taxes, edited contracts or other negotiated legal instruments, altered voting registrations or actual vote tallies, or even revised another nation's laws by forging the certificates that guaranteed the authenticity of information. We have long lived in a world with fake news,[27] but what if we also lost bearings on the fundamental integrity of legal processes with "fake law"? We have to anticipate that attacks on integrity will alter our fundamental legal relationships, making it easier to cheat and to hide cheating. And technologies such as blockchain may be of no use, since it is the hashes of documents that are typically put on blockchains, rather than the documents themselves.

---

[26]Competitive pressure has prevented a single identity architecture from emerging in the US. In particular, banks have been resistant to a collectivized identity regime, because the process of customer identification and authentication itself helps banks control the customer relationship and prevent churn to competitors. In addition, many retailers have resisted single-sign-on offerings from Google and Facebook, despite the probability that these options are more secure, because single sign-on (SSO) jeopardizes branding and because of the risk that Google or Facebook might use the authentication system to compete against the retailers relying on the system. For instance, imagine using Google's single sign-on to login to a pharmacy. Because the company has access to user email, it knows the user is refilling a prescription for birth control, and so it offers an advertisement for a competing pharmacy, or competing treatment, or perhaps even an issue-advocacy message protesting the use of birth control. Or maybe it decides to enter the pharmacy business based on intelligence from these sources.

[27]Plutarch describes the mob massacre of second-century reform politician Tiberius Gracchus and supporters by patricians who were enraged by false accounts that he sought a crown. Plutarch, *Lives. Vol. 10, Agis and Cleomenes, Tiberius and Caius Gracchus, Philopoemen and Flamninius* (1921).

8.2.4 Attacks on Passwords and Other Authentication Systems

Usernames and passwords are the default security mechanism for most computing services. Developed in the days of the mainframe, usernames identified the account that should be billed for using the computer, and the password prevented one person from accidentally spending from the wrong account. Decades later, passwords are the primary control not for just for billing, but for protecting information. Thus using a quantum computer to attack passwords would also seem to be a more strategic use than decrypting single messages.

An authentication system generally consists of three parts:[28]

1. The *user* who seeks to use it to prove their identity. The user may do this by knowing a password or a PIN, or by participating in a biometric challenge.

2. The computer that receives the password and uses it to identify. (The *relying party*.)

3. The service or database that the relying party uses to verify the identity. (The *identity provider*.)

There are many ways to attack these systems. For example:

Attack 1 The attacker can intercept the communication between the relying party and the identity provider and convince the relying party that the identity provided by the attacker to the relying party is correct. (A *proxy interception attack*, also known as a *machine-in-the-middle* (MITM) attack.) Section 8.2.2 (p. 317) would be applicable here as well.

Attack 2 The attacker can pretend to be the user and repeatedly guess new username/password combinations until one succeeds. (An *online password-guessing attack*.)

Attack 3 The attacker can break into the identity provider's computer and steal a copy of the registration database containing hashes of user passwords. With a password dictionary, the attacker then hashes each password in the password dictionary to see if

---

[28]While this section uses the standardized terminology of the OpenID protocol and the FIDO alliance, the example is intended to be sufficiently general as to apply to any authentication system.

the hashed dictionary password matches a hash in the stolen
database. (An *offline password-guessing attack*.)

In the case of Attack 1, these communications are generally protected by public-key cryptography. Today's recorded communications might be crackable with a quantum computer in the future
(see Section 8.2.2 (p. 317)), so passwords recorded today might be
divulged at some point in the future. Fortunately, there's a simple
mitigation: once quantum computers become available to your adversary, change your passwords.

In case of Attack 2, online password-guessing attacks are limited
by how many passwords can be guessed every second, how many
passwords can be guessed before the user's account locks out, and
the password guessing dictionary used by the attacker. None of these
should be directly affected by quantum computers. Attackers might
be able to use quantum computers to construct better password
guessing dictionaries, but this would be of minor use in an online
attack.

In case of Attack 3, modern identity providers encrypt passwords
with one-way algorithms: there's no way to decrypt the encrypted
password, so attackers try encrypting millions or billions of potential
passwords to see if any of them match the encrypted passwords under
attack. Some algorithms are stronger than others, and increasingly
attackers have enough computer power that they can try all possible
passwords that a person can type. This is the reason that contemporary password systems require you to type a password that includes
uppercase letters, lowercase letters, and symbols: it increases the
number of possible passwords that an attacker has to try (see the
sidebar "Password Complexity Is Complicated!" on page 327).

Quantum computers may offer some quantum advantage to attackers conducting offline password attacks, but the advantage is
likely to be minimal. As modern password encryption schemes do not
rely on number-theory based constructors (see Section 7.1, p. 260)
that would be susceptible to Shor's algorithm, current thinking is
that the maximal quantum speedup would be through the use of
Grover's algorithm – that is, reducing the work for cracking each
password. Like other quantum computing capabilities, this kind of
attack would be dependent on a large device that could implement
the entire function as a series of gates without losing coherence (i.e.
the quantum computer would have to be large enough to store the

Password Complexity Is Complicated!

With an 8-character password comprised solely of lowercase letters, there are $26^8 = 208\,827\,064\,576 \approx 2 \times 10^{11}$ possible passwords. If an attacker can try a billion (10^9) passwords a second, it takes on the order of 200 seconds to try them all.

Password complexity rules attempt to increase the number of possible passwords. For example, any one of those characters can be an uppercase letter, a lowercase letter, or a number, then each character can be one of $26+26+10 = 66$ possible characters, so the total number of possible passwords increases to $66^8 = 360\,040\,606\,269\,696 \approx 3 \times 10^{14}$. The added complexity increases attack time to $300\,000$ seconds or 83 hours.

Unfortunately, such calculations are subverted by the way that people actually guess passwords. Faced with a requirement that an 8-character password must contain an uppercase letter and a number, the typical user will add a single uppercase letter and a single number to their password. An attacker now merely needs to try all passwords containing 6 lowercase letters, 1 uppercase letter, and 1 number. There are $26^6 \times 26 \times 10 = 80\,318\,101\,760$ such combinations. For each of these combinations, the digit can be in any one of 8 positions ($\times 8$) and the uppercase letter can be in any of the remaining 7 ($\times 7$), so an attacker will start by trying these $26^6 \times 26 \times 10 \times 8 \times 7 = 1\,729\,928\,345\,600 \approx 1 \times 10^{12}$ combinations.

While password requirements increase attack burdens, they decrease the usability because of user error. Requiring longer passwords but allowing them to be all lowercase is a viable alternative. A 16-character, all-lower-case password increases the number of potential passwords to at least $26^{16} \approx 4 \times 22$. This is dramatically more secure than eight-character passwords with case restrictions, and is probably easier for most people to remember.[a]

---

[a]For an excellent overview of password security and usability, see Bonneau, Cormac Herley, et al., "Passwords and The Evolution of Imperfect Authentication" (2015). Meanwhile for a comprehensive analysis of alternatives to passwords, see Bonneau, C. Herley, et al., "The Quest to Replace Passwords: A Framework for Comparative Evaluation of Web Authentication Schemes" (2012).

entire set of possible passwords). Thousands of iterations would be required for each password. According to the 2019 National Academy of Sciences report, this process would require 2.3×10^7 *years* to break a single password.[29]

As we write this book in 2021, however, the most valuable passwords are not stolen by brute-force attacks on encrypted databases, but by targeted attacks on key individuals. The fateful email dump of John Podesta, Hillary Clinton advisor and former White House Chief of Staff, illustrates this. Among the most powerful people in America, Podesta used the 10-character password "Runner4567" to protect his Google Gmail account. This password was elicited from Podesta by a phishing attack, so its complexity was not relevant. Podesta's Gmail account was not protected by a second-factor. Thus, once his password was obtained, it allowed a Russian disinformation machine to access and publicize years of archived email messages.[30]

Security incidents where entire user databases are captured by attackers are another source of high-value passwords that does not require quantum computers for analysis or cracking. Cyberintelligence firms estimate that 35 such incidents occur a day, leaving full customer databases online and unprotected.[31] These security incidents provide much simpler means than quantum computing to break into accounts. Indeed, cyberintelligence companies show that many customer databases stolen and circulating online have failed to implement countermeasures and thus the passwords are available in free text. Because users reuse passwords, these databases can be used for online password guessing against individuals, or at scale in what is known as a "credential-stuffing" attack. For instance, in a credential-stuffing attack, if just 1 or 2 percent of users in a compro-

---

[29] Grumbling and Horowitz, *Quantum Computing: Progress and Prospects* (2019), p. 98.

[30] Rid, *Active Measures: The Secret History of Disinformation and Political Warfare* (2020).

[31] 4iQ, "2020 4iQ Identity Breach Report" (2020). Because of the volume of incidents, services such as "have i been pwned?" have in excess of 10 billion credentials that have been aggregated from misconfigured or hacked services. Oftentimes the attacker, or someone who found the database stolen by the attacker, provides this information directly to cybersecurity intelligence companies. Most of this activity is not well known publicly, because losses of customer databases, even if enormous and sensitive, are not always subject to security breach notification laws. As of this writing, haveibeenpwned.com/ makes over 610 million plain-text passwords available for services that wish to prevent users from choosing passwords that are already widely available.

mised database use the same password for Facebook or Gmail, that could result in hundreds or thousands of compromised accounts that can be quickly scanned for the presence of gift cards or other forms of stored value.

Tasking, Targeting, and Deconfliction

Organizations that possess quantum computers will need to carefully consider both their quantum computing capacity and the *key value* of keys that they wish to crack. In all likelihood, each quantum computer will be used to crack a single key at a time. Cracking time will be a major barrier to the widespread use of these machines: the National Academies estimated that a strong RSA key would take 28 hours to crack,[32] while a 2019 Google paper proposed a method that would require 8 hours.[33]

Quantum computing resources will therefore be limited and rationed. Even if the first working machine costs $100 billion to build and each additional machine can be built for the cost of a modern laptop, there will still be far fewer machines than messages to crack. Some process will need to be adopted for allocating the use of these machines.

Military doctrine envisions a process involving *targeting, tasking orders*, and *deconfliction* for making such decisions. Targeting "is the process of selecting and prioritizing targets and matching the appropriate response to them, taking account of command objectives, operational requirements and capabilities."[34] Once targets are chosen, a military command will issue a *tasking order*, which is a "method used to task and to disseminate to components, subordinate units, and command and control agencies projected targets and specific missions as well as general and specific instructions for accomplishment of the mission."[35]

To illustrate why this process is important, consider an organization that is able to intercept wireless messages between a target's phone and a publication service such as Twitter. Each wireless message might contain a tweet destined for immediate publication, a

---

[32]Grumbling and Horowitz, *Quantum Computing: Progress and Prospects* (2019).

[33]Gidney and Ekerå, "How to Factor 2048 Bit RSA Integers in 8 Hours Using 20 Million Noisy Qubits" (2019).

[34]Curtis E. LeMay Center for Doctrine Development and Education, "Introduction to Targeting" (2019).

[35]Joint Chiefs of Staff, *DOD Dictionary of Military and Associated Terms* (2020).

tweet scheduled to be published at some point in the future, a direct message to another user, or perhaps a status check, polling the service for other messages posted by other users. Some of these messages are clearly more valuable than others, but they all require the same level of effort to decrypt. And here's the problem: with a well-designed encryption system, there is no obvious way to tell which message is which before it is decrypted. Encrypted messages are easy to create, so a smart adversary can generate many worthless ones to soak up the capacity of another state to decrypt. This is why obtaining and evaluating external information can be a critical part of the tasking and targeting decisions. Indeed, metadata, which is typically not encrypted, will be important to providing hints about key value.

The term *deconfliction* describes systematic management procedures to coordinate the use of resources by various stakeholders. Quantum cryptanalysis will require multiple layers of deconfliction. At the most basic level, management will need to assure that resources are not used to crack the same key more than once. More strategically, management will need to decide whether the results from cryptanalysis can be directly exploited, or the results will need to be closely held to prevent adversaries from learning the extent of the organization's cryptanalytic capabilities.

Another area that might be of concern is how much information is revealed to adversaries through the use of information gained through quantum cryptanalysis. A nation will change its behavior depending on if it thinks an adversary has possibly one functioning quantum computer, if the adversary is known to have one functioning quantum computer, or if the adversary is known to have a thousand such machines. Countries that have publicly known but nascent quantum cryptographic capabilities might seek to project that they have significantly more capabilities than they in fact do, to keep their adversaries off-balance, while countries that have vast capabilities may seek to keep them secret, in order to lull their adversaries into a false sense of security.

In sum, quantum cryptanalysis is a threat, but one that we consider to be overhyped. Simply put, quantum computers will not magically break all encryption quickly, as sometimes implied by the news media and even by some policy analysts. Instead, attackers will carefully choose and focus their cryptanalysis resources on high-value keys, presumably ones that cannot be attacked using other intelligence trade-craft.

Those other methods of attack also provide context. One tends to look to technology for dramatic intelligence gains, when in reality, simpler approaches may do. For instance, many of the great US intelligence losses have been the result of insiders: John Anthony Walker (1968–1985), Robert Hansen (1979–1981, 1985–1991, 1992–2001), Jonathan Pollard (1984–1985), Ana Montes (1985–2001), Chelsea Elizabeth Manning (2009–2010), and Edward Snowden (2009–2013). Consider Snowden, of whom we likely know the most. Despite his clear technical talents, Snowden's attack was straightforward: privilege escalation, password acquisition, and a mass exfiltration of documents he had access to by virtue of his job.[36] Even in a world with quantum computing, traditional spycraft, including recruitment of insiders and placement of assets, is likely to remain a reliable, effective, and far less costly modality for accessing protected secrets.

8.2.5 Strategic Surprise: Nuclear Weapons

Simulating nuclear physics (presumably for weapons testing) was the existential reason that Feynman proposed quantum computing in the first place. We therefore reason that once governments have functioning quantum computers, they will use them for this purpose – to simulate the action of current and proposed nuclear weapons.

The connection between computing and weapons delivery and design runs deep. The original mechanical, electromechanical, and electronic computers were developed for the purpose of targeting munitions. Later, the design and operation of nuclear weapons drove the development of electronic computers in the 1940s, and supercomputers since the 1960s.

Prohibiting testing was a major diplomatic priority of the Soviet Union, particularly in the last decade of the Cold War. Aside from reducing the overall stockpile of weapons, Soviet strategists were worried that continued testing was a key precursor to President Reagan's anti-ICBM technology, known as the Strategic Defense Initiative (SDI).[37] Mocked as "star wars," SDI made it clear that space

---

[36] US Congress, House Permanent Select Committee on Intelligence, "Executive Summary of Review of The Unauthorized Disclosures of Former National Security Agency Contractor Edward Snowden" (2016).

[37] Hoffman, *The Dead Hand: The Untold Story of The Cold War Arms Race and Its Dangerous Legacy* (2009).

was a new domain for military conflict, and raised military spending to levels that the Soviets ultimately could not afford.

Today there are comprehensive test bans in place prohibiting nuclear testing in outer space, in the atmosphere, and underground. As a result, governments must turn to computers to simulate the "physics package" of nuclear weapons. But more than a simple replacement for testing, computers make it possible to explain many possible designs without producing a blast, radiation or fallout. For this reason, quantum computers might end up significantly accelerating the development of novel physics packages with particular characteristics, such as very-low yield, enhanced radiation, or fallout with particularly short half-lives. As such, quantum computers might paradoxically enable the creation of nuclear weapons with fewer barriers-to-use.

Indeed, with quantum computers, simulations of ICBM flight, the design of warheads, and their destructive potential will all improve, but in the privacy of computing, hidden from satellites and possibly other forms of intelligence gathering.

8.2.6 Quantum Strategic Surprise: Chemical, Biological, and Genetic Weapons

Nuclear weapons occupy a central place in the modern psyche. We all live less than 30 minutes from an attack that could end life on Earth. Not as much attention is devoted to the potential of gigadeaths from chemical, biological, or genomic weapons. This may be because of the worldwide consensus against so-called "weapons of mass destruction" that emerged from World War I. The first international ban on chemical and biological weapons was the Geneva Protocol of 1925, formally known as the Protocol for the Prohibition of the Use in War of Asphyxiating, Poisonous or other Gases, and of Bacteriological Methods of Warfare. In 1972 many countries entered into the Biological Weapons Convention, which prohibited the development, stockpiling, testing, acquisition and retention of such weapons (although the Soviet Union continued to develop and stockpile such weapons in violation of the treaty, as it was sure that the US was doing the same[38]). In 1997 the Chemical Weapons Convention placed additional restrictions on chemical weapons and their precursors.[39]

---

[38]Stern, *The Ultimate Terrorist* (1999).

[39]The earliest regulation of chemical weapons came in 1675 with the Strasbourg Agreement's limitation on use of poison bullets. Hardesty, "Safety, Security and

Yet the risks of chemical, biological, and new agents made possible through synthetic biology are significant and both the understanding and development of these weapons could be accelerated through computer simulation. Such activities are easy to hide in plain sight: the difference between a vaccine and a bioweapon is whether or not the infectious agent is killed before it is put into the delivery system.

In fact, even conventional weapons could become more powerful with quantum simulation. Chapter 5 discusses the modeling of nitrogen fixation as a quantum computing application with tremendous human benefit. The flip side of that simulation is that nitrogen is a key ingredient in explosives. Governments will be intensely interested in developing more powerful explosives along with syntheses that are safe, cheap, and energy-efficient. And remember, unlike nuclear weapons, there is no taboo associated with using conventional weapons.[40]

As nations agree to forbear from nuclear testing or development of bio-warfare agents, inspection and monitoring efforts are necessary to ensure compliance. Nations must be able to demand access to facilities and to make sense of the equipment and materials found. Elaborate confidence-building measures have been developed to foster international trust in different areas of weapons control.

The 1992 Treaty on Open Skies (from which the US withdrew in the last days of the Trump administration on November 22, 2020) is an example of a confidence-building measure. Under that agreement, nations agree to a regime of aerial inspection of countries using limited sensors.[41] The idea is that these overflights allow political leaders to be confident about estimates of other nations' military capacity. The idea may seem antiquated in the era of the spy satel-

---

Dual-Use Chemicals" (2014).

[40]While nuclear weapons have retained a taboo, governments have been willing to use conventional weapons that have nuclear-like effects. In 2017, President Trump ordered the use of the Massive Ordnance Air Blast (MOAB), an enormous conventional bomb with a yield of approximately 10 tons of TNT, to destroy an ISIS base in Afghanistan – roughly a thousandth the yield of the US nuclear weapons that destroyed Hiroshima and Nagasaki in 1945. In 2019, Trump boasted that the US could kill 10 million people in Afghanistan, a quarter of the country's population, in a week relying only on *conventional* weapons. About 200 000 died in the Hiroshima and Nagasaki atomic attacks.

[41]The Treaty on Open Skies bans collection of electromagnetic signals in the radio band, and tops resolution of optical sensors at 30 centimeters, infrared at 50 centimeters, and side-looking radar at 3 meters.

lite, but aerial platforms generally have higher resolution, more flexible targeting, and lower cost of operation than platforms in space.
Also, over 30 nations have signed the treaty, and many of these nations do not have significant space programs. It is unclear if Open
Skies overflights could be supplemented by more precise quantum-
sensor-based position, navigation, timing (PNT) technologies (see
Section 2.3.2, "Sensing Location" (p. 51)). Even with low-resolution
images, a high frame-rate camera paired with quantum PNT and advanced post-processing could produce ultra-high-resolution images.[42]
These could be further enhanced with sophisticated spectral analysis. And this is before one even considers the possibilities of using
quantum-enhanced sensors.

Inspection and monitoring is where quantum computing could
address issues of strategic surprise for nuclear weapons, but not for
chemical or biological.

Nuclear weapons Even underground nuclear detonations are de-
     tectable remotely, through seismographic evidence and through
     atmospheric monitoring for ionizing radiation. Quantum sen-
     sors should make such detection efforts more accurate.

Chemical and biological These weapons are more difficult to de-
     tect, as they do not emit particles or radiation that are readily
     measured at distance.

Testing chemical and biological weapons requires large, secret
facilities to experiment with delivery mechanisms, especially those
involving aerosols. The testing itself must be carefully done, as accidents, such as the 1979 Sverdlovsk anthrax incident, signal cheating.

To detect such facilities, the Convention requires nations to identify vaccine manufacturing facilities, to share information about labs
that might have weapons capacity, and to release data on outbreaks
caused by toxins.

Cheating becomes easier when chemical and biological weapons
can be simulated in a computer. Barriers to development are lower if
compounds can be simulated, and if delivery methods could be modeled, and thus enhanced, without creating elaborate facilities that
have to both test agents and hide evidence of wrongdoing from others. Computer-aided research could bring a nation closer and closer
to a quicker, more effective development and stockpiling cycle.

---

[42]Note that the treaty requires disclosure of attributes such as frame rate frequency.

Here again, confidence-building measures can reduce the risk of these weapons. Such measures include records keeping, access to records, and on-site inspections. Indeed, the Biological Weapons Convention provides many layers of reporting and information-sharing requirements to surface illegal activity. However, it is vital that governments adopt and transition integrity mechanisms to digital signatures based on post-quantum algorithms as soon as they are available so that the records will continue to be regarded as authentic and unimpeachable.

8.2.7 *Strategic Surprise: Remote Sensing*

Quantum sensing will enable improvements in intelligence, surveillance, reconnaissance, positioning, navigation, and timing, and these improvements will have both strategic and tactical value.[43] Consider gravity. Using interferometry, we have created extraordinarily sensitive gravity wave detectors that ring when black holes collide. But much similar technology has been deployed into earth orbit to detect the location and movement of large masses on the Earth. (See p. 67 for details.) The small number of countries with space and quantum technology programs might be able to develop sensing platforms that combine gravity and electromagnetic sensing to detect not only other nations' underground natural resources, but also matériel. Quantum detection power exceeds classical abilities, because camouflage (tin-roofed airline hangars, concrete domes, or inflatable structures) and tactics such as operating at night can obscure heavy matériel from classical satellite observation, but camouflaged matériel will have signatures detectable using other sensing technologies.

We might imagine uses of satellite-based quantum sensors that would impose massive costs on a defender. Imagine that a nation maps out an adversary's entire critical infrastructure using quantum sensors from aircraft or satellite. This adversary cannot directly attack this infrastructure, because that would start a war. So the adversary nation does the next best thing: it anonymously publishes the map of every utility wire and natural gas pipe in a region. This kind of release could even be disguised as an "open data" effort. But such a data dump would elucidate dependencies in power infrastructure that could enable less sophisticated actors, say terrorists or even

---

[43]Gamberini and Rubin, "Quantum Sensing's Potential Impacts on Strategic Deterrence and Modern Warfare" (2021).

criminals, to attack and cause much larger outages than they could
without the information.

Quantum sonar and radar provide another area for strategic
surprise. The US invented and broadly deployed stealth technologies that absorb radar and other energy.[44] Stealth, known as low-
observable technologies, gave the US and its allies an advantage in
airpower. But the assumption that US stealth aircraft are practically undetectable by radar and that its submarines operate with
near-perfect acoustic stealth may be threatened by quantum sensing.
Low-observable technologies can still be seen with the kinds of lasers
described in Chapter 2. In addition, these quantum sensors themselves are "stealthy," meaning that detecting an adversary's sensing
may be impossible.

The implications for quantum technologies and submarine warfare cut both ways. On one hand, several kinds of quantum sensing could be deployed to detect submarines. On the other, submarines may gain additional stealth through quantum communications, which gives some advantages over existing methods (see Figure 7.4).

Turning to submarine detection, scientists have mapped out photonic, gravimetric, and electromagnetic sensing approaches,[45] as well
as proposals to use quantum computing to improve passive sonar.[46]
Because they are large, weighty vehicles full of electronics and heavy
metals, submarines have a geometry and composition unlikely to occur naturally. Sensitive quantum magnetometers or gravimeters (see
Figure 2.7) could be installed in the ocean to create a fence to detect matching geometries. Knowing more about where submarines
are has important implications for national security, because submarines are both part of a tenuous strategy to intercept first strikes
by ballistic missiles, but also their stealth and survivability help
make a "second strike" possible in a nuclear conflict. Upsetting assumptions surrounding submarine stealth with quantum radar and
sonar endangers key aspects of nuclear deterrence strategy.[47]

[44]Rich and Janos, *Skunk Works: a Personal Memoir of My Years at Lockheed*
(1994).

[45]Marco Lanzagorta, Jeffrey Uhlmann, and Salvador E. Venegas-Andraca, "Quantum Sensing in The Maritime Environment" (2015).

[46]S. E. Venegas-Andraca, M. Lanzagorta, and J. Uhlmann, "Maritime Applications
of Quantum Computation" (2015)

[47]Schelling, *The Strategy of Conflict: With a New Preface by The Author* (1980).

Unmanned aerial vehicles (UAVs), popularly known as "drones," have emerged as a key surveillance tool and offensive weapon as a result of technological, political, and cultural changes. Faced with the rise of Islamic militant violence and the failure of some states to police or exclude terrorists, President George W. Bush turned to drones to surveil militants with powerful sensors and then attack when the opportunity presented. Presidents Obama and Trump continued and expanded the program, in part because public support for fighting foreign wars, already weakening, further deflated after the second war in Iraq, but also perhaps because the growing documentation of the horrific impact of war on the war-fighter has made Western societies less tolerant of individual sacrifice in pursuit of geopolitical objectives.

UAVs have enabled successive presidents to use force in multiple theaters without committing troops, and to argue that their use of force is more proportionate and discriminant than traditional bombing campaigns. As we write this, it is publicly known that US drone strikes have been carried out in Afghanistan, Iraq, Libya, Pakistan, Somalia, Syria, and Yemen. Drones may also have been used to attack aircraft using missiles, and the US Air Force is developing a drone for aerial combat.[48]

Critics of the UAV program argue that UAV strikes are indiscriminate and disproportionate because of civilian casualties. These arguments find support in part because of the design of UAVs. Consider the "smart bombs" of the 1991 Persian Gulf War: these gave the military the chance to (very selectively) show footage of what appeared to be precise strikes against targets. This footage helpfully ended right at the moment of impact, leaving any human suffering off-screen and thus abstract. By contrast, the loitering capability of drones along with their more powerful sensors enables pilots to make final targeting adjustments as they see people running from Hellfire missiles and then carefully document the carnage, by attempting to count and even identify bodies and parts of bodies. One result of this is that UAV pilots, despite operating equipment far from the battlespace (often in Las Vegas, Nevada), frequently experience post-traumatic stress disorder (PTSD) symptoms similar to their forward deployed colleagues.[49]

---

[48]Pawlyk, "Air Force Will Pit a Drone Against a Fighter Jet in Aerial Combat Test" (2020).

[49]Wallace and J. Costello, "Eye in The Sky: Understanding The Mental Health of Unmanned Aerial Vehicle Operators" (2017).

Executives are unlikely to give up the UAV program since they
see strikes as necessary, and see civilian casualties as proportionate to
the gains of disrupting terrorist organizations. But could quantum
computing improve the targeting of UAVs, allowing them to find
flight behaviors that allow them to fly autonomously in contested
situations while being invulnerable to most countermeasures?

Berkeley professor Stuart Russell envisioned this scenario in a
popular video titled *Slaughterbots*, in which swarms of quadcopters
armed with tiny explosives pursue human targets using face recognition, setting off their charges that can "penetrate the skull" and "destroy the contents." A mysterious group obtains the technology and
uses it to selectively eliminate political opponents. Russell appears
at the end, urging viewers to support a ban on "killer machines,"
weapons that use computers to select targets and to make the decision to attack. In *Ghost Fleet*, P. W. Singer and August Cole describe
a near-future war with China, where UAVs play a major role. Singer
and Cole portray fighting UAVs that can perform maneuvers physically impossible for human pilots (because of gravity-induced loss
of consciousness) but also perfectly disciplined, such that the drones
can fly just above the ocean and obscure their presence by banking
into high waves. Clearly, as the offense gains advantages through
automation, defensive forces will also have to adopt automaticity.[50]

Two other military innovations point to quantum sensing as a
consequential technology. First, increasingly conflict can be waged at
great distances and with hypersonic vehicles. Nations have developed
hypersonic missiles (those that travel faster than five times the speed
of sound yet maintain the maneuverability of a cruise missile) and
even railguns capable of firing over 100 miles. These weapons have
created great worry both because of their speed and because their
use will occur with even fewer warning signs than ballistic missiles.
Quantum-enhanced sensing may provide earlier warning signs when
these weapons are used.

Second, developments in electronic warfare will change how conflict is waged, and these changes could make quantum technologies a
source of superiority. Consider that, in recent conflicts, the Russian
armed forces have been able to test out their electronic and cyber
warfare capabilities, showing them to be clever and capable.[51] Other

---

[50]Singer and Cole, *Ghost Fleet: a Novel of The Next World War* (2015).
[51]Creery, "The Russian Edge in Electronic Warfare" (2019).

evidence is mounting that nation states are using GNSS (Global Navigation Satellite System)/GPS (Global Positioning System) jamming and interference regularly.[52] A 2019 report by C4ADS found almost 10 000 suspected incidents of interference with GPS and other navigation systems, and estimated that "Russian forces now have the capability to create large GNSS denial-of-service spoofing environments, all without directly targeting a single GNSS satellite."[53]

Quantum sensing may be a possible solution to GPS jamming and other forms of electronic warfare. Companies and governments are developing "quantum positioning systems" to operate in GPS-denied environments.[54] Like the inertial and celestial guidance systems of the past, quantum positioning, navigation, and timing might perform a backup role to GPS.

8.2.8 Quantum Strategic Surprise: QKD and Quantum Internet

In quantum communications, advances may be so obvious as not to be surprises because there are already articulated concerns surrounding communications confidentiality and integrity. A nation that races ahead in quantum communications might not just deploy quantum key exchange technology, but may create entirely new communications systems and protocols to pursue confidentiality and integrity. However, it is not immediately clear to us why a nation would want to go beyond QKD and pursue a quantum internet. We believe that simply using QKD combined with AES-256, or even QRNG combined with post-quantum encryption protocols, would likely be sufficient to secure communications.

A quantum internet protocol, based on quantum effects, would not just provide randomness and thus strong encryption, but also reveal whether messages have been intercepted at all. This would be strategically relevant because currently, one can never know whether or where a copy of a communication has been made. Perhaps a nation that is skeptical of QKD or AES security might want this extra layer of assurance for confidentiality and integrity. Perhaps quantum

---

[52] The Coast Guard tracks and publishes incidents of GPS jamming, interference, and failure. Department of Homeland Security, US Coast Guard, "GPS Problem Reporting" (2018).

[53] C4ADS, "Above Us Only Stars: Exposing GPS Spoofing in Russia and Syria" (2019).

[54] Jones, "MoD's 'quantum Compass' Offers Potential to Replace GPS" (2014).

internet plans are products of a lack of trust in one's own network, or distrust of employees, who might be bribed or extorted to undermine the confidentiality and integrity of communications. Finally, knowing about interception means knowing whether adversaries have collected metadata about a communication. Metadata, even of encrypted transmissions, are surprisingly revealing. Nations have long sought clever methods to prevent metadata capture; perhaps excluding adversaries from access to metadata is worth the expense and challenges of developing a quantum internet. However, these technologies are sufficiently far in the future (decades?) that we do not consider them to be a credible policy issue in the near term.

A second implication of quantum internet is the ability to connect distant quantum computers through photonic entanglement. Consider IBM, which in 2020 claimed that it had 18 operational quantum systems. Presumably with quantum internet networking, it could link these systems to create more powerful ones. For instance, its Raleigh 28-qubit system combined with its 53-qubit Rochester device would be larger than any single device. Such a quantum network need only be a few feet from node to node.

Large implementations of quantum internet, however, would require infrastructure coordinated over a great distance, instead of just within IBM Research's lab. Practically speaking, and in the near term, quantum internet infrastructure is likely to depend on satellites, and this shapes the ability of governments to intercept information.

Experiments in dark fiber networks are promising, but quantum states degrade as photons travel through glass and this limits the distance over which fiber can be used to transmit information. Traditional networks use repeaters to cover great ranges. But until fully quantum repeaters are invented – ones that could hold the state in memory and still amplify it to traverse more fiber – each one of these repeaters offers an opportunity for classical interception and analysis.

It would seem that European nations would be poised to implement quantum communications, as relatively small countries could run optical links between cities. For instance, the Netherlands, where some of the most advanced achievements in quantum communications have occurred, might want to connect its seat of government (The Hague) with its capital (Amsterdam), which are only 32 miles apart.

Small nations and regions can use optical fiber to communicate, but larger ones will have to also use satellites to overcome the problem of repeating light signals. Satellite transmission is the only medium today that can distribute entangled photons over great distances. This is why China's Micius satellite is an important achievement. Recall that the satellite-linked base stations, combining both fiber optic and free-air transmission, create an entangled photonic channel. This means that the Chinese can beam quantum keys to two distant base stations simultaneously. However, nations that use satellites for quantum communication will need to focus attention on the security of these satellites similar to the ways that they must secure their physical, land-based fiber networks.

These developments in quantum communications are not a surprise we can foresee them and predict corresponding countermeasures. Intelligence and law enforcement agencies already have techniques to address strong encryption. With regard to what we might one day call the "classical internet," interception is easy and not detectable. Much of the Internet's traffic flows through the geographic borders of the United States, but, even for traffic that does not, "prepositioned devices" can quietly copy light from fiber optics at the bottom of the ocean. Because transport and content encryption is used to obscure these communications, and because content is so voluminous, intelligence and law enforcement agencies focus on metadata rather than content. After all, any major governmental or terrorist action requires coordination amongst many actors, activity that is revealed quite nicely by metadata in the form of link analysis. Even when content is at issue, adversaries can hack into devices and cloud services, often through the simple approach of password guessing. Thus, advances in quantum communications are likely to place even more emphasis on attacks using stolen passwords, hacked programs, metadata analysis, and human spies.

8.2.9 Quantum Strategic Surprise: Secrecy and Leakage

Secrecy will be important in a government-superior and -dominant landscape. Governments will seek to keep their quantum computing and sensing advances secret, because there are always countermeasures. The need for secrecy could limit the power that governments can exercise in a practical sense. Knowing a thing is helpful of course, but acting on knowledge can reveal sources and methods. Governments will have to generate cover stories and distractions from quantum programs, lest adversaries deploy countermeasures.

IARPA's Director articulated a series of questions for new proposals that help elucidate the risks of the government-dominant scenario. One asks: "If the technology is leaked, stolen, or copied, would we regret having developed it? What if any first mover advantage is likely to endure after a competitor follows?"[55] Indeed, whatever competitive advantage comes from the government-dominant approach is time-limited and could be perverse. It is time-limited, because the world is leaky and eventually the engineering secrets will diffuse to other nations and even companies. It is perverse because the government-dominant secrecy will hobble the broader market for quantum technologies. While government is dominant, secrecy excludes the private market from working its magic and training thousands of quantum computer programmers and engineers. Thus, the secrecy creates a short-term advantage that might be outweighed by a longer-term deficit in workforce and economic benefit. In fact, one could imagine quantum technologies diffusing in a copycatting country while the source of the innovation continues to treat it as a state secret, not allowing diffusion and growth of the technology in its own country. (This is largely what happened with electronic computing: the UK insisted on secrecy, and the ideas developed there took root in the US.)

To what extent will a government-dominant approach be leaky? In the US, our "five eyes" allies will probably learn, indirectly or not, about the nation's quantum technologies. Theft is a major risk as well. But one form of immediate technology dispersion comes from willingness to share with law enforcement. Law enforcement agencies would find much utility in quantum sensing. Sensitive magnetometers would allow detection of weapons and bombs, even at a distance, in public or even when concealed in a home or vehicle. Just as radiation detectors, X-ray technology, and sensitive microphones are used at the border, new quantum sensors might be used to detect contraband. Unlike physical searches, which focus on certain objects and occur at a discrete time, a quantum sensor "search" might happen remotely, passively, and continuously. A government-dominant scenario explored by the Center for Long Term Cybersecurity envisions

---

[55]The full list of questions developed by Jason Matheny is reproduced in Danzig, "Technology Roulette: Managing Loss of Control As Many Militaries Pursue Technological Superiority" (2018).

that quantum computers will put law enforcement ahead of every cartel and organized crime body.[56] But law enforcement agencies of less democratic countries might use the same capabilities to pin and skewer protest and opposition movements.

One obvious law enforcement use involves quantum sensors designed to detect contraband. A quantum sensor that could only recognize guns (perhaps it has been trained on a model of the most popular firearms), molecules of particular explosives, and of course, illegal drugs, would be useful with minimal privacy implications. Such a sensor's machine learning could be trained on every contraband item imaginable and be copied to other devices. The sensor would never tire, and be used continuously. Of course, there could be mission creep – why not detect counterfeit luxury handbags? Perhaps the sensor could even be mounted on aircraft and drones to detect weapons caches inside buildings through the roofs of private homes.

Finally, a government-dominant and -superior scenario has implications for the long-term success of quantum technologies. Technology sovereignty – the desire to have domestic champions – is needed to maintain both a strong and secret quantum technology industry. Thus, at the highest level, the secrecy and emphasis on government uses of the technology have long-term practical and public perception consequences. On a practical level, military and law enforcement uses might displace other pro-social uses of quantum technologies, such as drug discovery and materials optimization. The societal benefits of new classes of drugs could save many lives and improve the lived experiences of people. But a government-dominant approach might discount those benefits while seeking to retain its intelligence edge.

From a public perception perspective, it is important to reflect that attitudes towards computing are more positive today in the personal computer era than in the era of the mainframe. Before the personal computer revolution, only governments, militaries, and large businesses could computerize. Early computing empowered already powerful institutions. A government-dominant quantum computing landscape might feel like a replay of the mainframe era.

In recent years, some employees of Silicon Valley companies have renounced the Valley's defense department roots and have pledged not to work on the "business of war." This is a delicate position because many of the technologies developed by companies like Google

---

[56]Center for Long Term Cybersecurity, "Cybersecurity Scenarios 2025" (2019).

are dual use: computer vision projects for automated driving are easily repurposed for UAVs and autonomous weapons. Nevertheless, in a government-dominant quantum world, these employees might see quantum technologies as carrying the "taint" or "taboo" of the business of war. Military-first uses may make public perception of quantum technologies negative, even dangerous. Between the secrecy and quantum taboo, other humanitarian uses of quantum computing could be impeded, with consequences for medicine, materials science, and other scientific discovery.

8.2.10 Countermeasures in a Government-Dominant Scenario: Disruption, Denial, Degradation, Destruction, and Deception

Nations that could not compete in quantum technologies would likely prioritize development of quantum countermeasures. Indeed, all adversaries – quantum capable or not – would be likely to invest resources in some kinds of countermeasures. Such measures are typically classified as "D5" tactics: disruption, denial, degradation, destruction, and deception.

Experimental work suggests effective D5 tactics. For instance, the Chinese scientists discussed in Chapter 7 who achieved satellite-based quantum entanglement and communication had to generate millions of photons in order to overcome channel loss. The scientists had to manage beam diffraction, pointing error, and absorption and turbulence caused by clouds and the atmosphere generally. These challenges raise two vital points: first, interference similar to ordinary atmospheric events – even sunlight and rain, and in the case of underwater communication, water turgidity – can degrade quantum technologies based on photonics. Thus natural events might be simulated to stealthily interfere with the technology. We could imagine weather modification, such as cloud seeding, as a D5 countermeasure to some quantum technologies.[57]

Second, there is very little photonic loss in outer space; thus, there is incentive for operational systems to be placed in high orbit – much higher than the low earth path used in the experiment, and within the reach only of superpowers. One could imagine escalation and even a desire to develop space-based weapons in response.

---

[57]T. J. House et al., *Weather As a Force Multiplier: Owning The Weather in 2025* (1996).

> **Space Force**
>
> The elevation of the Space Force by President Trump has been met with some derision, perhaps because detractors imagine *Star-Trek*–like struggles with people in outer space.
>
> In reality, Space Force will work to manage threats to satellites, the targeting of which will be key in conflicts with the US, China, or Russia. Threats to satellites can be earth-based, but also come from other space vehicles. Although such efforts are veiled in secrecy, strategic opponents are reported to have developed space-borne anti-satellite weapons.[a]
>
> For example, an object that appeared to be space debris "made 11 close approaches to one of the rocket's discarded stages. Such an elaborate space dance would be possible only if the object had thrusters and enough fuel to maneuver very precisely." Sciutto also notes that China has "a satellite with a grappling arm capable of lifting other satellites out of orbit. China has now conducted multiple successful tests of this 'kidnapper satellite,' some of them at geostationary orbit, where America's most sensitive space assets reside, including satellites for communications, surveillance and early warning of a nuclear launch."
>
> ───────────
>
> [a]Sciutto, "A Vulnerable US Really Does Need a Space Force" (2019).

Each application of quantum technologies has different vulnerabilities. Still, several quantum technologies are uniquely resistant to existing D5 tactics and are being evaluated to operate in their presence. For instance, quantum clocks and location devices are seen as supplements to jamming-vulnerable GPS, and to guard against Digital Radio Frequency Memory (DRFM) jamming. A DARPA project focused on "micro-PNT" seeks to create chip-size quantum positioning systems (QPS) for UAVs, Unmanned Underwater Vehicles (UUVs), and navigators for missiles that do not rely on GPS.[58]

Quantum illumination enhances radar at a very low energy level, suggesting it will not be as susceptible to traditional jamming efforts. Recall that quantum radar involves sending entangled photons into the sky to detect things like missiles and jets, especially those that are cloaked with some kind of "stealth" technology. Thus like pho-

───────────

[58]Shkel, "Precision Navigation and Timing Enabled by Microtechnology: Are We There Yet?" (2010).

tonic communication, methods that interfere with the generation of
entangled photons and that scatter them in the atmosphere may be
effective to counter quantum illumination.

Quantum communications security is likely to be less consequential than metrology and sensing developments. This is because D5
tactics can be directed at other aspects of communications activities.
Modern encryption algorithms are (almost by definition) never the
weakest link in communications. Classical encryption affords such
great security that the only known attacks are on the ways that keys
are created or extremely clever "side channel attacks" that detect
information that leaks out of a presumably secure system. These
might include detecting subtle power or frequency variations when
a computer codes 0 or 1. Attackers also know that human deception
is relatively easy and simple phishing attacks frequently work, as do
attacks on cyber infrastructure.

The awareness of surveillance that quantum communications affords is a new factor that might prove more intriguing and useful
than communications confidentiality. Recall that because of the no-
cloning theorem, Alice and Bob can know something or someone
is interfering with their communication: there is no way for Eve to
eavesdrop on Alice and Bob, but an attempt to do so will alert Alice
and Bob that something is amiss! It is too early to say how nation
states will react to this signaling. One could imagine D5 strategies
that attempt to poison the channel by engaging in constant attempts
to intercept or block photons. Perhaps Alice and Bob can never generate a secure key if some foreign intelligence agency interferes with the
QKD. Another (more likely) D5 scenario would be to simply attack
Alice and Bob's devices before they communicate, so that one could
obtain information before it is encrypted or after it is decrypted.

On the other hand, if denial or degradation of terrestrial-based
fiber networks becomes routine, nation states could make their communications harder to reach through using point-to-point satellite
QKD.[59]

---

[59]Satellites could also use QKD for secure satellite-to-satellite communication. Another option for satellite-to-satellite communication is to use the 57 GHz to 64 GHz
band. Oxygen has significant radio absorption at 60 GHz, so any such signals will
not reach from space to the ground. For this reason, the 57 GHz to 64 GHz band
is available for use without license in the US, allowing gigabit wireless communications over distances of roughly 1 km, but only when it is not raining.

Finally, D5 tactics might be effective against quantum computers because the devices are so sensitive to environmental interference of all kinds. Simply creating a "noisy" environment with heat, wireless radio signals, and so on, might be sufficient to cause decoherence in quantum computers. Of course, they could also be targeted with conventional ordnance as well. For the foreseeable future, quantum computers will be large, intricate and delicate devices. They will be terrestrially based, in places where human expertise, a lot of electricity, and supercooling helium is available. As the next sections will make clear, these affordances make quantum computers subject to legal and policy interventions perhaps not possible against other quantum technologies, such as metrology and sensing devices that can be miniaturized and deployed in outer space.

Quantum interferometry and communications can be satellite-based, and thus the physical devices are out of reach of most nations' ability to physically destroy them. With powerful quantum intelligence, surveillance, and communications on satellite platforms, quantum technologies might in the coming years be another pressure encouraging the expansion of military force in space.[60] Thus, the handful of countries that both have space programs and quantum achievements might have incentives to invest in anti-satellite weapons. (The development and testing of anti-satellite technology does not appear to be illegal under the Outer Space Treaty, although the treaty does prohibit placing nuclear weapons in orbit, establishing military bases, or conducting military maneuvers on "celestial bodies."[61]) During times of crises, a nation with such capability might find it irresistible – or simply necessary – to destroy satellites in order to impair reconnaissance powers and communication routes of their adversaries.

8.3 Scenario 2: Public/Private Utopia

The government-superior and -dominant scenario naturally focuses on security-relevant developments, and thus government dominance takes on a certain patina. The government-dominant scenario helps elucidate how powerful, well-resourced actors might pursue a quan-

---

[60]Rabkin and John Yoo, *Striking Power: How Cyber, Robots, and Space Weapons Change The Rules for War* (2017); J. Yoo, "Rules for The Heavens: The Coming Revolution in Space and The Laws of War" (2020).

[61]Ortega, "Placement of Weapons in Outer Space: The Dichotomy between Word and Deed" (2021).

tum technology agenda. However, that scenario should not detract
from a scenario we think more likely: that the private sector makes
significant advances in quantum technologies and outperforms government labs, just as it did in electronic computing and cryptography.

In both electronic computing and in cryptography, the private
sector's emphasis on information sharing and commercialization eventually overcame government's first-mover advantage. From the 1950s
through the late 1970s the US government had the fastest computers
in the world at its disposal and the most mathematicians specializing in cryptography in its employ; neither was true by the end of the
twentieth century. Today the US government still has an impressive
array of systems at its disposal, but nearly its entire infrastructure is
assembled from commercial off-the-shelf systems. And while official
statistics are not available, it is widely assumed that there are more
cryptographers at universities and corporations than are directly employed by the government.

We believe that the same outcome is likely in quantum technologies as well. The benefits to individuals in terms of both prestige
and salary, combined with the commercial benefits that will accrue
to their employers, will be substantial in the coming years: this will
create incentives to further democratize quantum technologies. Governments will purchase off-the-shelf systems and will surely contract
with corporations to build secret devices. But the age-old pursuit of
profit drives actors in this scenario to apply quantum technologies to
solve all sorts of problems, all over the world. Quantum technology
won't be put back in the bottle.

We see a number of factors and incentives combining to make a
mixed government/commercial scenario the most likely one. Chapter 4 discussed the many efforts being made by cutting-edge technology companies in quantum research. This reflects the overall trend
of private-sector investment in research and development in the US.
In recent years, US research and development has continued to grow
and the most recent figure pegs it at $580 billion annually.[62] But
R&D characteristics have changed. The private sector is investing
more money than ever in R&D, with pharmaceutical development
being a leading contributor. The federal government's investment

---

[62]Congressional Research Service, "US Research and Development Funding and
Performance: Fact Sheet" (2020).

has largely flattened, although it is still primarily focused on basic research rather than applied research, technology development, or market creation.

Aside from a focus on development, private researchers operate with different incentives and constraints than those working in government labs or even universities. Private-sector researchers may have the advantages that make it possible to make breakthroughs in quantum computing. But private researchers *do* operate with constraints – they must have champions within the company willing to protect their funding for years. They must be able to show progress and results, and defend these goods against competing demands that directly contribute to the bottom line of a competitive firm.

The good news for these private researchers is that many of their companies are sitting upon huge amounts of cash. As of this writing, Amazon, Google, and Microsoft all have cash reserves in excess of $100 billion – meaning that these individual companies have more money in cash than the GDP of many Low or Middle Income Countries. Furthermore, private researchers have an advantage over academics in that they can devote their time to building devices instead of teaching, chasing funding grants, and earning tenure – although, even in corporate labs, there is still the pressure to publish in top-ranked journals.

Private researchers also have an advantage over government lab scientists because they are freed from the secrecy constraints imposed by security clearances. Although private companies can be very secretive, their researchers do not have to undergo the extensive background checks and hassles associated with maintaining a security clearance, which has implications for personal freedoms and for one's workforce in profound ways.[63] Private companies can also hire the best and brightest from all over the world, as citizenship and attendant concerns about loyalty will be less important than in government employment. Of course, hiring such individuals carries risks, but as we saw in Section 8.2.4 (p. 329), the government's background investigation process has not prevented the theft of secrets.

---

[63]Ben Rich laments that as Lockheed's Skunk Works took on sensitive projects, a huge portion of otherwise reliable employees had problems passing drug screens associated with the clearance process. Rich claims that 44 percent of applicants tested positive for drugs. Rich and Janos, *Skunk Works: a Personal Memoir of My Years at Lockheed* (1994).

Private sector researchers will not only be freed from many constraints that competing academic and government scientists face, their incentives will run towards non-national-security-related uses in the long term. This is because quantum technologies have so many commercial uses. Simply put, much more money can be made in commercial uses of quantum technologies because there are more buyers and a broader spectrum of uses outside national security. In the short term, companies may affect a national-security lilt, recruiting retired generals to their boards and emphasizing their DOD Projects. But this posture is likely temporary as companies use government projects for initial funding, and then sublimate company efforts to more broadly appealing commercial applications.

Quantum computing will have a host of non-security-related consequential uses. Competitors investing in quantum computing are focused on simulation of quantum mechanical events, in order to develop drugs, new synthetic materials, and engage in high-energy physics experiments. Some see quantum computing as a tool that will help us discover a room-temperature super-conductor or easier-to-control nuclear fusion. Others are focused on quantum computing's parallelism as a mechanism to build machine learning tools that can make sense of high-dimension datasets. The benefits could be legion. In any area where dimensionality is so high as to make analysis intractable or coarse, we can envision quantum computing making more sense of the world. Whether those applications are automobile traffic flow or logistics in the form of train or airplane scheduling, we can imagine a future with less waiting and more efficiencies.

8.3.1 How Quantum Technologies Could Change Governance and Law

As we explored the superior/dominant scenario above, we saw how nations might use quantum technologies to better understand other nations. In a world where private companies have quantum computers and sensing, their capabilities will similarly be trained on other companies and individuals, but this time in the search of profit. Thus, a threat discussion needs to contemplate how quantum technologies will contribute to power shifts between companies and individuals. Uses of quantum sensing and computing to govern human activity could displace democratic processes and become laws unto themselves.

Quantum sensing and computing will reinvigorate grand schemes to perfect society. Technological revolutions have long brought about utopian ideals for redesigning societies. These are "revolutions from the top," and they typically threaten individual autonomy in profound ways. In *Seeing Like a State*, Yale political scientist James C. Scott discusses several generations of social reformers who use new scientific insights to design putatively better systems – from more productive forests, heartier tomatoes, to more efficient cities. Scott terms these efforts "high modernism," an almost religious belief in technology to reorder natural and social systems. The most dangerous form is "*authoritarian* high modernism," where the coercive power of the state combines with scientism, creating a force that overrides markets and individual preferences in the pursuit of some ideal.[64]

Scott warns that high modernists, in their zeal, tend to discount complexity, local knowledge, and in particular *metis*, the ancient Greek word used to convey skills and learnings acquired by the skillful and clever. The concept of *metis* is best represented by Odysseus, the resourceful yet perhaps unprincipled[65] hero who solves problems pragmatically with little concern for ideological or moral purity or truthfulness. High modernist plans often fail to consider *metis*. After all, the point of *metis* is to act in a way that cannot be predicted by those who lack it.

Quantum computing could be enabling technology for several large-scale social experiments. High modernists will see it as the tool that can finally incorporate *metis* and other local knowledge, creating a kind of master system. We might imagine intrusions into the economy, our living circumstances, our bodies, and even our minds. As such, high modernist plans directly regulate people and become a form of law and governance through architecture and technology rather than through deliberative self-governance.

Friedrich Hayek and the Austrian School of Economics have definitively won the debate over the primacy of centrally planned or market-led economies. As Hayek recognized, there is just too much information in the forms of preferences, supply, and demand for a central

---

[64] J. C. Scott, *Seeing Like a State: How Certain Schemes to Improve The Human Condition Have Failed* (1998).

[65] "Tell me about a complicated man" begins Emily Wilson's translation. See Homer, *The Odyssey* (2018). Compare with Lattimore: "Tell me, Muse, of the man of many ways" and Fitzgerald: "[sing] of that man skilled in all ways of contending."

planner to sense and make sense of it. The twentieth century showed
planned economies to be slow adapting and both the Soviet Union
and China have shifted to more free-market economies, often with
aggressive state industrial policy or other economic action. But perhaps central planning will be revisited if a sufficiently large quantum
computer could make sense of the multifarious signals of an economy.

In such a scenario, the utility-maximizing individual loses its primacy and even its agency in favor of an economic oracle in the form
of a quantum computer.[66] One could imagine a long period of transition where data-heavy, sophisticated companies demonstrate winning strategies by ceding human instinct and control over marketing,
advertising, logistics, and other functions to a quantum computer.
Perhaps the first adoption will come from financial services firms
trading securities, as this is a field where computers already automatically analyze and conduct trades. Or perhaps it could be Ama-
zon.com, Inc., with its huge marketplace, computing power, and fantastic logistics system. If these first movers experience success, they
will pull away from competitors, offering lower prices while finding
savings and efficiencies identified by the quantum economic oracle.
Their successes could have a snowball effect that convinces other
sectors of the economy to trust more in automated analysis and execution. But if this happens, one of the most important bastions of
the liberal economic order – the notion that the emergent effects of
individual decisions make the best free market – could end in favor
of an increasingly centrally planned and coordinated economy.

The displacement of governance and law is most palpable in corporate efforts to reshape our lived environment. Efforts to perfect
our lives, such as "smart homes" and even "smart cities" require
tremendous sensing capabilities and computers for sensemaking. Efforts such as Google's "Sidewalk Labs" foresee a revolution in urban
planning, based primarily around redesigns and new thinking on mo-
bility. Among the ideas are to create an urban infrastructure that
can change as needs shift throughout the day. Traffic lanes might
change direction automatically and vehicular movement would be optimized to accommodate multiple modes of transportation, the need
for parking, and so on. Embedded sensors and mobile phone tracking are key for these endeavors, and instant sensemaking is necessary

---

[66]Evgeny Morozov explores attempts to perfect central planning with computers
in 1970s Chile, in Morozov, "The Planning Machine: Project Cybersyn and The
Origins of The Big Data Nation" (2014).

because the second-by-second decisions to control the environment could mean accidents or even death.

Like the quantum-computing planned economy, the smart city reflects the pathologies of high modernism, with its displacement of democratic governance and law. High modernists present these plans by showing only the benefits and omitting their less appealing implications. For instance, despite all its benefits, the smart city requires that individuals obey an arbitrary, unknowable authority – the algorithms that replace the laws and institutions and people that make up a government. Usually implicit in smart city schemes is that people would have to give up control over driving, a privilege thought to be a freedom for many Americans. And once that privilege or freedom is waived, the individual's needs can be subordinated to others. One's vehicle might stop to optimize overall traffic. One could imagine waiting for minutes as another flow of traffic is prioritized, perhaps to address a fomenting traffic problem elsewhere in the city. No longer would the car be the instrument of the individual's immediate self-interested needs.

There is no "opting out" of the system because the smart city is so interdependent. Even outside the car, individuals will have to submit to the system. A

Public/Private Utopia
Governments and the private sector advance state of the science, eventually commercializing sensing, computing, and communications.
Key Policy Characteristics
Industrial policy, need for liberalized export controls, relative openness in innovation and immigration.
Key Enabling Factors
Diverse set of competitors, market for components, availability of trained workforce.
Strategic Surprise
Entrepreneurs use quantum sensing and computing to shape society to their liking and increasingly to displace public governance with private decision-making systems.
Outlook
Because quantum technologies are in reach of even well-funded startups, a public/private outcome is the most likely scenario.

pedestrian might have to wait (or qo quickly) to ease traffic pressure
far from view. Already, in cities that are testing automated vehicles,
such as Las Vegas, Nevada, pedestrian barriers first erected to address drunken drivers plowing into sidewalks are being enhanced to
make it nearly impossible for pedestrians to jaywalk because automated vehicles are flummoxed by unpredictable pedestrians. Plan-
ners will have to design-in coercive architecture in order to ensure
that individual autonomy cedes to the oracle and to the vehicles that
could run over the individuals.

Both the planned economy and the planned city require individuals to sublimate their immediate self-interest for the goal of shared
efficiencies and gains. For instance in a 2019 blog post, Ford describes
how it used Microsoft "quantum-inspired" technology to simulate optimal traffic routes in Seattle. The team claimed it could achieve an
overall 8 percent reduction in traffic over a population of 5000 drivers,
but this reduction requires an alternative to what we are used to –
"selfish" routing.

Giving up on selfishness in favor of overall efficiency raises a series of practical, political, and even emotional challenges. Central
planning and control is a particularly difficult state to achieve because it asks individuals to pit their immediate, felt emotions and
needs against the abstract idea of collective benefits. These collective benefits are real. Minor efficiencies can indeed add together to
create significant savings for individuals, but these are far more subtle than the immediate rush of, say, putting the pedal to the metal.
And those most trusting of their inner instincts who are tempted to
ignore the commands of the smart city are probably the ones least
capable of self-reflection (and self-restraint).

For collective schemes to work, officials must also explain the
trust model carefully and convincingly and these models must be
subject to political scrutiny and consent. If some class of people, such
as the ultra-wealthy in Russia who put emergency lights on their cars
to evade traffic, get preferred treatment and quicker routes, this must
be explained and accepted in some way by the system's participants.
In modern cities, busses and high-occupancy vehicles enjoy reserved
car lanes, but we can both readily observe this compromise and agree
to it because of the social interests in efficiency. Google co-founder
Larry Page is known for his hatred of automobile traffic and has
invested in "flying cars" to solve the problem. As one sees Page's
car move swiftly through the smart city, will one think that like the

Russian oligarchs, the designers of the system get special treatment?

8.3.2 Implications for Human Primacy

How will a quantum-planned economy or society coexist with populist instincts to celebrate "independent" thinkers? Will *metis* be cherished, or be seen as sand in the gears of a fantastically efficient society?

On a deeper level, will the "intelligence" of these systems represent a turning point in the view of human intelligence and analysis as fundamentally special? The pendulum could swing back to a worldview where elites – the small number of people who operate and understand quantum technologies – have more command over ideas and the matters of what is correct and incorrect.

One could imagine a transition period where the veracity and benefits of quantum technology predictions make life better. Perhaps quantum computers could ease the transition by finding effective communication strategies to explain the sacrifices that individuals make for the broader efficiency of the system, or more directly, the benefits that the individual receives by forbearing from what appears to be the most self-serving, available option.

As the primacy of the individual recedes, how might humans seek to regain the status of being special? One could imagine genetic research and prediction would receive new attention in a world with quantum computers, leading to pressures to change both lifestyle and choices in reproduction.

Genetic prediction and personalized medicine (sometimes called precision medicine) was much hyped at the start of the Human Genome Project in the 1990s. Some scientists predicted a complete revolution in therapies flowing from the project, in which the US government invested billions. Heralds of the project conceived of discoveries of single genes that would predict morbidity, and thus relatively simple treatments and behavioral interventions. Yet decades later, the hype remains, but with little to show for it because so many diseases are not genetically determined and, among those that are, hundreds of genes may be involved in disease. In addition to the complexity of multiple genes, our health is a product of contingent environmental and behavioral variables, many of which are essentially unknowable. This is why, 20 years after the launch of the Human Genome Project, the leading business-to-consumer genetic testing company is still in essence an entertainment product, carry-

Embracing Probabilities

Several different theories have emerged to help explain the counterintuitiveness of quantum phenomena and the differences with classical physics. The Copenhagen interpretation, pilot-wave theory, and the theory of many worlds compete to account for quantum phenomena and provide some meaning for them in our lives.

In the soft sciences, experts are comfortable in conceiving of case outcomes, rules, and even facts probabilistically. Turning to law and policy, prediction of uncertain events, of court or regulatory decisions, is the stock-in-trade of lawyers. Law professors expect their students to predict that a court will "probably" come to a certain conclusion. They even teach that "facts" have some subjectivity. We do not know a jury's verdict and cannot observe a jury deliberation until it concludes. We could think of a verdict unsealing as a measurement of an uncertain process.

The law is rife with probabilistic standards to address the problem that there is imperfect knowledge of events, and what knowledge that does exist is colored by observer bias and misinterpretation.[a]

The law is satisfied establishing facts despite uncertainty, and does so by setting burdens of proof (e.g. preponderance of the evidence) and by assigning them (e.g. to be established by the plaintiff) so that matters can go forward and have resolution, even an imperfect one. As consequences become more grave, the law imposes higher burdens of proof and assigns them strategically, often to disadvantage the state.

The law lives with probabilistic standards because they embody a method that if applied systematically will produce justice, if not always a just outcome in each encounter with the law. That method must evolve with time, as society is shaped by new technologies, new norms, and new understandings of human behavior and expectations. In a systems-level sense, an embrace of a probabilistic universe does not threaten our basic methods and institutions.

---

[a]We allow police to check persons for weapons based on a "reasonable suspicion" that a suspect is armed. We allow the state to arrest people if officers reasonably have "probable cause" to believe that the suspect has committed a crime.

ing a lengthy "quack miranda warning" to disclaim the health claims the company implies with its marketing.[67]

But if the barrier to personalized health is the complexity of genes, behavior, and environment, might quantum computing's dimensionality be the answer? The promise of precision medicine is that knowledge about genes will create opportunities to act and prevent disease. As the knowledge puzzle begins to reveal a picture, a complementary development by Jennifer Doudna, CRISPR-Cas9,[68] provides a fast and low-cost way to manipulate genes. To take the decision now to alter a human is widely considered to be reckless and irresponsible. But might our attitudes change as quantum computers provide us with what we think is understanding of the relationships between genes and phenotype and the environment and disease? Combined, these developments could shift the risk–benefit calculus surrounding genetic manipulation.

What if personalized health still doesn't deliver the expected benefits? Advocates will say that the quantum computer needs more data, and there will likely be a movement to collect even more information about the inputs to a person's health: what you consume, where you walk and travel, the air you breathe, and details of physical activity. Only then will we learn the degree to which the messiness of health outcomes is determined by random chance out of control – which for many people, may be the most frightening insight of all.

Turning to our mental states, online advertising remains one of the chief reasons that companies surveil and make sense of ordinary people and their private activities in a quest to decipher their thoughts and preferences, termed *surveillance capitalism* by Shoshana Zuboff.[69] Despite the surveillance aperture of the online advertising model, online advertising itself is still quite coarse. Online

---

[67]In various places 23andMe describes its service as surfacing "health dispositions." At the bottom of several of its customer care pages is a disclaimer that includes the text: "The test is not intended to tell you anything about your current state of health, or to be used to make medical decisions, including whether or not you should take a medication, how much of a medication you should take, or determine any treatment." See 23andMe, "Choosing Which Reports to View" (n.d.).

[68]Emmanuelle Charpentier and Jennifer A. Doudna earned the 2020 Nobel Prize in Chemistry "for the development of a method for genome editing." See Doudna and Charpentier, "The New Frontier of Genome Engineering with CRISPR-Cas9" (2014).

[69]Zuboff, *The Age of Surveillance Capitalism: The Fight for a Human Future at The New Frontier of Power* (2019).

platforms have voluminous amounts of data on users. Some platforms not only know what websites people visit, where they go in the physical world, who their friends are, and how they spend money, they also know what people choose not to do (for instance, if one writes a message on a service, edits it, or decides not to send it). But advertising remains coarse, in part because of the size of the surveillance aperture.

Many people are familiar with the experience of being "retargeted" when considering an online purchase: search of "cheap mattresses" on Google, for example, or read a few reviews, and pretty soon mattress advertisements will show up on many websites that you visit. If you go to a website to actually make a purchase, then change your mind at the last minute, you'll start seeing advertisements for the specific mattress that you almost bought: this is "retargeting," and it appears to be effective in getting consumers to consummate their purchase.

The problem with today's information economy becomes clear *after* you click the "buy" button for the mattress. Despite the fact that a mattress is pretty much a once-a-decade purchase, you'll continue to see advertisements for mattresses. They won't go away for weeks. That's because the advertisers don't take into account that you've made that purchase decision and have stopped looking, even though the data should be relatively available.

Because of the data volume, no company can fully make sense of people, thus two strategies are taken: place users into an abstract category that captures their commercial characteristics (male versus female, high income versus low income household, etc.), and/or throw out old data.

As companies build larger quantum computers, advertisers – and other companies with surveillance incentives such as insurance firms – will take advantage of extra dimensionality to both create finer profiles and to analyze more historical data. What this means for people is that quantum computers will be yet another technology that makes individuals' desires, personalities, and lives more legible to powerful decisionmakers. The converse is likely not true – ordinary people will not train these same technologies to scrutinize powerful companies (other than to decide whether to invest in them).

Quantum sensing, in fact, might be the technology that fundamentally erodes what it means to be an individual. It is no accident that Google is a center for thinking about quantum technologies, but also about the concept of "technological singularity," a series of speculative technical advances that seek to create computers that can build faster, more intelligent computers, which would create more intelligent computers still. All of this seems pretty frightening, except that part of the singularity religion is that the computers we create will bring us along for the ride with advanced technologies that can unmoor humans' minds from physical bodies and allow them to merge with machines, creating some kind of advanced symbiotic "intelligence" – and achieving immortality in the process. To join the computers at this acme of intellectual accomplishment, we would need to make sense of and "copy" the structure and physical representations of memories and knowledge in the brain. This may be the ultimate use case for quantum teleportation.

For path-dependent reasons, these exciting and troubling applications of quantum computing are obscured in many accounts of the technology. The discovery of the Shor and Grover algorithms early in the history of quantum computing caused cryptanalysis to far overshadow other applications – even Feynman's existential quantum application of simulating physics. We think this is unfortunate. It is obvious that new and faster drug development and discoveries that lead to fusion energy are more consequential than code-breaking. In fact, it might be Grover's algorithm, so often presented by the media as a code-breaking tool, that delivers some of these breakthroughs, because Grover's underlying utility is that it speeds up mathematical search algorithms.

Quantum communications is promising but not as exciting as quantum computing in this scenario. Strong encryption has long been available to people, although it was awkward to use until recently.[70] In a short time however, a number of companies developed high-quality, widely adopted, usable communications tools with end-to-end encryption, such as Signal, software funded by former Facebook executives upset by the company's depredations of privacy.

If democratized, QKD could accelerate the trend of putting even stronger encryption into the pockets of ordinary people. But for most users, the difference between an encryption system that is computationally secure and one that is information-theoretic secure is not meaningful.

---

[70]Whitten and Tygar, "Why Johnny Can't Encrypt: A Usability Evaluation of PGP 5.0" (1999).

Quantum sensing, if we key this field's birth to NMR and MRI machines, has already contributed to the treatment and health of untold millions of people. As quantum sensors become smaller and can operate at ordinary temperatures, they can be moved closer to the patient, allowing for greater resolution.

In fact, medical uses for quantum sensors might be the "killer" application with a market for both in-facility and in-home devices that is vastly larger than military and intelligence ones. Consider how many people avoid diagnostic tests that we know are effective because of the indignities and fear associated with the test process itself. Imagine how many people would be delighted to replace an uncomfortable, invasive physical examination with a passive one performed by a quantum sensor. One's annual checkup might include a comprehensive body scan that could be compared to previous captures in order to detect unwanted changes in the body. Of course, full-body scans have been marketed to consumers for decades, but existing ones irradiate the body, produce false positives that result in dangerous procedures, and have not demonstrated a general medical benefit. The passive nature of quantum sensors with added resolution, paired with individuated analysis, offers a scenario with earlier diagnosis and, we hope, better health outcomes.

One could even envision an in-home device that provides a regular medical scan of individuals. Perhaps people with high genetic risk of cancer would be the first willing to pay for such a device. These individuals might have a daily scan for diseases of concern, and to be able to make other measurements about the body.

More broadly, a public/private scenario could include many forms of self-surveillance brought on by quantum sensors. Consumers have broadly bought into the "Internet of Things," internet-connected devices in the home, many of which make health and fitness claims. The demand for such devices is substantial, creating a virtuous cycle of new products with interesting new features, and stimulating competition among different vendors to provide operating systems for the home. But in practice, many of these devices are abandoned soon after they are bought, because their usability is poor and the services that they provide are trivial.

Internet of Things devices based on quantum sensing, because of sensitivity and passive information capture, could be a winning technology of the home. Consider a technology developed by MIT professor Dina Katabi that uses in-home radio waves to passively measure many kinds of physiological phenomena. Movement, breathing, heartbeat, and sleep patterns all subtly affect the low-power electromagnetic waves that are emitted by Katabi's device, reflected by water in the body, and then measured upon their return.

Katabi earned an Association for Computing Machinery prize for its development, and has expanded use cases for the technology into important areas such as fall detection, and contexts such as the hospital, where such passive monitoring would nicely replace the various devices to which patients are tethered. One can see why this technology might displace existing Internet of Things devices and be purchased for every hospital room: no one needs to wear anything or worry about finding the right charger for their tracker. There's no device to abandon, and so the sensor becomes more like a smoke detector that can be placed and function for years without user futzing. One can also imagine the quantum technology improvement on the approach: with even more precise timing and more resolution, more insight about the internals of the body can be had.

Industrial and commercial users may be the leaders in adoption, as well. For similar reasons of convenience, employers might want Professor Katabi's device to monitor worker efficiency and health. Perhaps with accurate and quick measurement of worker activity, one could train a robot to replace those workers, with their pesky breathing and heart rates and illnesses.

Oil services firms are among the biggest early investors in quantum sensing research and development. The industry clearly sees the potential for greater extraction activities brought on by quantum sensing. Absent more regulation on oil exploration and extraction, environmental threats will likely emerge as a problem in a private-sector-dominant quantum sensing world. Perhaps quantum sensing will drive a new wave of extraction activities, not only for oil and shale, but also for rare-earth materials and minerals. But one could also foresee a host of more complicated scenarios – more precise sensing might reduce exploratory drilling and prospecting activities, or it might make extraction more precise. For example, regulators could require detailed surveys of underground water flows before drilling or mining permits are granted.

8.4 Scenario 3: Public/Private, East/West Bloc

The previous section discussed a series of quantum technology successes brought about by enthusiasm and cooperation among govern-

ments and the private sector. In a way it described a technology
utopia, a mythical, perhaps perfect place. Yet it should be remembered that *utopia* is a combination of the Greek words for "no" and
"place" with a Latin -ia ending. A more realist version of the scenario
takes on a Cold War patina, one where East races West in its pursuit
of quantum technology dominance.

Technology development is a focus of national competition, with
economists increasingly elucidating the links between government incubation of basic research and private-sector payouts.[71] Historians
too are making the connections between Silicon Valley's rise and
generations of government investment in infrastructure and military
research efforts.[72] Technology research occurs on a canvas with increasing nation-state divisions. After decades of public policy that
sought Westernization of China through empowering its middle class,
the US changed direction under Presidents Trump and Biden. Europe's cohesion strains under economic pressure and from immigration tensions that contributed to the 2016 "Brexit" referendum on
the United Kingdom's membership in the European Union.

Technology competition is now a major topic of international
relations.[73] Consider that after Brexit, the European Union excluded
UK companies from participating in its Galileo satellite navigation
program. The UK is struggling to establish its own "sovereign" space
program. The US, UK, and EU face a common challenge in China.
China's Belt and Road Initiative proposes a major reinvestment in
infrastructure across Asia, Africa, the Middle East, and even Europe
itself. Participants will not only receive funding for massive capital
projects, but also new strategic partnerships with China. In 2019, the
Italian government signed a memorandum of understanding to join
the Belt and Road Initiative. Liberal observers are concerned that as
China's infrastructure and investment spreads, a new Silk Road will
speed China's sphere of influence, bringing authoritarianism, China's
breed of state capitalism, and the spread of China's military presence
elsewhere in the world.

---

[71]Mazzucato, *The Entrepreneurial State: Debunking Public Vs. Private Sector
Myths* (2015).

[72]Nash, *The Federal Landscape: an Economic History of The Twentieth-Century
West* (1999); O'Mara, *The Code: Silicon Valley and The Remaking of America*
(2019).

[73]Farrell and Newman, "Weaponized Interdependence: How Global Economic Networks Shape State Coercion" (2019).

Under President Trump, the US took increasingly aggressive measures to cabin China's technical might. These have included a new focus on export controls; strategic deterrence of China's most competitive companies, such as Huawei; imposing restrictions on suppliers to Chinese firms in order to harm the country's competitive posture; the threat to allies to withhold intelligence support unless they remove Chinese components from their networks; and even the criminal prosecution of faculty members alleged to have received funding from China that was improperly disclosed.

These trends could produce a scenario where two factions, one including China, Russia, and perhaps even some Westernized nations enticed by Belt and Road, and a second representing the US, Japan, and Europe, compete to reach quantum technology superiority.

The East/West bloc scenario is not necessarily a *dystopia*. Viewed through a practical lens, a quantum technology national competition – on sensing, computing, and communications – is a less risky one than tussles focused on weapons systems. It is more akin to the outer space race than an armaments race, as is the competition between the UK and the EU for sovereign space programs. Such national competitions are also likely to prompt huge amounts of public investment in research.

East/West Bloc Scenario

Governments and private sector collaborate, but in sharp competition divided between China and the US and EU.

Key Policy Characteristics

Secrecy, limits on immigration, and industrial policy in pursuit of technological sovereignty.

Key Enabling Factors

Bloc scenarios are more likely if quantum technologies are more difficult to create than currently thought, and if countries choose technology stacks of differing effectiveness.

Strategic Surprise

A nation achieves superiority by pursuing a successful quantum computing substrate that others cannot.

Outlook

More dependent on international relations than any single technology. Decoupling, technology/data sovereignty make a bloc scenario more likely.

Governments won't be able to complete that research alone; much funding will flow into universities and the private sector. Indeed, to take the UK's post-Brexit space race as an example: instead of building its own program at the cost of billions, the UK is investing in domestic aerospace company OneWeb.

Secrecy and export controls would be one cost of the competition scenario. These controls could slow down innovation and the democratization of quantum technologies. They might also posture development towards military and intelligence uses of quantum sensing and computing rather than to ones that will directly benefit people in their lived experiences. For instance, development in a market economy might naturally flow to healthcare applications of quantum sensing. But in a scenario where worries surrounding technology leaks abound, the government will not want powerful and potentially portable sensing technology in every hospital.

Indeed, some early entrants to the quantum computing race, such as D-Wave Systems, sold devices to clients. But as covered in Chapter 4, quantum computing is likely to evolve into a cloud model. The East/West bloc scenario might cement the cloud approach in fact. This is because the cloud model provides companies a thick veil of secrecy for the devices themselves. The secrets of engineering, the hard-won tradecraft learned in assembling and maintaining a quantum device, all stay in a secure room available only to company technicians. The cloud model allows companies to secretly implement enhancements and keep them proprietary in a physical sense. Of course militaries will demand to have on-premises devices, and these will be guarded like their cloud-based siblings. But it won't be possible for a company to simply buy a device and reverse engineer it in order to learn the easy way.[74]

Experts from these different blocs may be unwilling to participate in knowledge exchange opportunities and even employment at international firms. In fact, East–West competition could bring about the same sort of lifetime employment and loyalty that was seen during the Cold War research boom.

In the long term, the competitive scenario presents a mixed picture for technology development. Many innovations are path-dependent, a result of initial development success that leads to waves of

---

[74]Some speculate that Google's purchase of a D-Wave Systems machine in 2013 was for the purpose of reverse engineering the device.

greater investment and lock-in to certain assumptions. For instance, in classical computers, silicon is the medium that dominates architecture, and hardly anyone considers alternative media. In quantum computing, everything from hardware to software is up for grabs. The medium for mastering quantum effects could be based on several competing alternatives, from topological approaches touted by Microsoft to the superconducting circuits used by Google and IBM. No matter what physical medium is chosen, control systems and software matters must be settled.

With so much so uncertain, East and West may choose different quantum computing paradigms, different technology stacks, and different software approaches. The divergence could be dramatic and the differences important. The divergences could identify the best hardware and software and possibly undo the path dependence that might happen without competition.

For instance, if the West pulls ahead in quantum technologies and establishes a software stack written in English, language alone will provide the kind of advantage that makes it easier for English speakers to enter the field, as it did in both the first computer revolution and the first two decades of the Internet.

At the same time, secrecy could result in siloed approaches, or even the identification of a certain approach as virtuous or lacking virtue.[75] One need only look to the history of steam and electricity to see an example where a dominant technology (steam) was romanticized as honorable and superior in attempts to resist electrification. We might see similar values attributed to hardware and software approaches; some might be called "red" instead of merely different and possibly better.

One would hope that after current hostilities and suspicions de-escalate, a period of cooperation would follow, and this period would benefit from the experimentation and different paths chosen by East and West. We could imagine a new period where globalism trumps nationalism, and an opportunity presents itself to identify the best of approaches explored by different factions.

But during the period of conflict, what we are willing to do to win might surprise us. Take intellectual property theft. It is safe to say that American norms towards intellectual property are relatively pious. A large group of innovative American companies have saber-

---

[75]Juma, *Innovation and Its Enemies: Why People Resist New Technologies* (2016).

rattled for years about China, complaining of dramatic losses of trade secrets, lost revenue from pirated movies, and about eerily similar copies of domestic inventions. Intellectual property theft became an executive-level concern during the Obama administration, resulting in a complaint to President Xi.

The desire to win may also change our attitudes toward stealing innovations. These attitudes are malleable, if one takes an historical perspective. When the US was an upstart nation struggling to develop an industrial base of its own, our forefathers were impious towards intellectual property and unrestrained in their appropriation of others' inventions.[76] In pursuit of technological superiority or sovereignty, might we adopt the tactics of using spycraft to steal and copy others' innovations?

8.5 Scenario 4: Quantum winter

Consider the shade cast on quantum computing by quantum computing skeptic Mikhail Dyakonov:

> "In riding a bike, after some training, we learn to successfully control 3 degrees of freedom: the velocity, the direction, and the angle that our body makes with respect to the pavement. A circus artist manages to ride a one-wheel bike with 4 degrees of freedom. Now, imagine a bike having 1000 (or rather 2^{1000}) joints that allow free rotations of their parts with respect to each other. Will anybody be capable of riding this machine? ...
>
> "**No, we will never have a quantum computer**. Instead, we might have some special-task (and outrageously expensive) quantum devices operating at millikelvin temperatures."[77]

---

[76]Ben-Atar, *Trade Secrets: Intellectual Piracy and The Origins of American Industrial Power* (2004).

[77]Dyakonov, *Will We Ever Have a Quantum Computer?* (2020).

What if, as some critics like Dyakonov argue, quantum computing is just too complicated and too hard a problem to solve – at least for the next few decades?[78] What if, as happened in artificial intelligence in the 1970s, and in cold fusion, quantum technologies experience a "winter," a period where enthusiasm and funding for the entire class of technologies lags?

In the quantum winter scenario, quantum computing devices remain noisy and never scale to a meaningful quantum advantage. Perhaps research on quantum computers and machine learning leads to optimizations for classical algorithms, but classical computers remain faster, more manageable, and more affordable. In this scenario, "quantum" might remain a serviceable marketing term, but companies will soon figure out that classical supercomputers, simulators, and optimizers outperform them. After a tremendous amount of public and private monies are spent pursuing quantum technologies, businesses in the field are limited to research applications or simply fail, and career paths wither.

Quantum Winter

Large-scale quantum computers do not emerge within a decade.

Key Policy Characteristics

Policymakers recognize failure, reallocate funding. Need mechanism to reassess, recognize thaw.

Key Enabling Factors

Scaling strategies unsuccessful, as mid-size quantum computers don't trigger virtuous cycle of device growth.

Strategic Surprise

Nations reorganize educational systems, spend billions in quantum computing that never produces new innovations; nations that invested in other technologies pull ahead and prosper through automation and advanced services.

Outlook

While quantum computing flounders, quantum sensing still flourishes. Quantum communications loses steam as the cryptanalysis threat diminishes.

---

[78]Dyakonov, "When Will Useful Quantum Computers Be Constructed? Not in The Foreseeable Future, This Physicist Argues. Here's Why: The Case against: Quantum Computing" (2019).

In this scenario, funding *eventually* dries up for quantum computing. Academics and scientists in the field either retool and shift, or simply appear irrelevant, even embarrassing. As the winter proceeds, hiring priorities shift to other disciplines, further sidelining quantum technologies as a field. Even where important developments are made, they are given short shrift, viewed with skepticism, or simply seen as irrelevant to computing praxis.

One of the greatest risks of a short-term failure scenario is whether we are willing to recognize it. One sign that quantum winter is approaching would be for quantum technology advocates to continually "move the goalposts," and insist that grand discoveries are around the corner if we just keep funding the dream. The politicians, military leaders, scientists, and CEOs who invest in quantum technologies will become diehard defenders of them – until they stop or are replaced. If we do not recognize failure, investment in quantum computing will continue to be at the expense of other, more promising fields. To take a current example mentioned above, the billions of dollars invested in precision medicine have not delivered on promises of revolutions in therapy or life extension. Its advocates, perhaps because their professional reputations are tied to its promise, keep the faith.[79] Meanwhile, public funding for precision medicine has appeared to come to the detriment of tried-and-true investments, such as public health interventions.[80]

The primary danger of a quantum winter isn't the wasted resources and careers – it's that research abruptly stops, resulting in economic dislocation and delaying discoveries that aren't around the corner, but may be just over the horizon. The AI winters (there were two) stunted some research efforts that eventually proved successful, and killed others outright. The failure of modern AI systems to incorporate systematic approaches for knowledge representation and explainability – two hallmarks of the earlier AI waves – may be a lasting negative impact.

A quantum winter would be in keeping with the boom/bust cycle of many technologies in the West. Before the bust, there is general technology optimism, boosterism from news media and investors, emphasis on growth over sustainable operations, and inability to criti-

---

[79]Marcus, "Covid-19 Raises Questions About The Value of Personalized Medicine" (2020).
[80]Bayer and Galea, "Public Health in The Precision-Medicine Era" (2015).

cally judge innovations – all could contribute to a refusal to recognize failure. Then comes the bust.

Quantum technologies, because of their complexity and the secrecy surrounding their research and development, are well poised to fall victim to these dynamics. Consider the relatively recent failures among firms that have presented themselves as "technology companies" such as office-space-leasing firm WeWork and German payments company Wirecard AG. Sometimes investors give traditional companies a pass by placing them in special categories with less oversight, because the firm is seen as a "technology" company instead of an ordinary one that uses technology. This regulatory misclassification, with looser scrutiny because of "technology," appears to have helped Wirecard AG evade earlier detection.[81] Private companies also enjoy less transparency, and in some cases, loose norms that enable inventive accounting. Ordinary investors might be confused by these norms, because publicly traded companies have more defined benchmarks and different scrutiny from regulators.

Modern, privately traded "technology companies" can manipulate key benchmarks surrounding sales and use them to make it appear that they are much more promising than in reality. For instance, the recent craze over home-delivered "meal kits" and claims surrounding booming subscriber statistics omit the key problem that firms pay huge amounts of money to acquire new customers, most of whom cancel soon thereafter.[82] Or take the enthusiasm surrounding electric kick scooters. To the public, these companies appear to be enormously successful because scooters appeared on every corner, seemingly overnight. The technology press fanned the optimism, but a few outlets, such as *The Information*, reported on the underlying economics of scooter business models, which reveals them to be unsustainable.[83]

---

[81] Storbeck and Chazan, "Germany to Overhaul Accounting Regulation after Wirecard Collapse" (2020).

[82] "[M]eal kit subscription services are plagued with an incredibly high churn rate – 19 percent of US adults have tried a meal kit service, but of that 19 percent, only 38 percent are still subscribing." See PYMNTS, "The Meal Kits Crowding Problem" (2018). Transparency into these pathologies tends to come from third parties, such as payment companies, that have incentives to accurately report how people are using their accounts.

[83] These scooters cost about $500, on average only receive a few rides a day, these rides generate just a few dollars, and the scooters only last a few months. Vandalism, operator injuries, confiscation by authorities, and simple theft also create

Throughout history, publics have fallen victim to secretive, cult-like profitmaking claims. From Charles Ponzi's international postal stamp arbitrage scheme to Elizabeth Holmes' drop-of-blood-testing Theranos to Wirecard AG's illusory successes in payments, these schemes work because of the same elements currently present in technology generally – optimism, boosterism, secrecy, and a network of people invested who could make a fortune if the company succeeds in the short term. In-the-know insiders often cannot whistle-blow because companies pressure them with non-disparagement agreements and threats from lawyers (and sometimes even the government). When attacked, company loyalists defend the firm, and markets tend to ignore claims of impropriety until the charade plays itself out. Ponzi, Theranos, and Wirecard all had leaks pointing to the truth of their operations, but the promise of profit kept investors optimistic.[84] And such schemes are not restricted to the West, as the Russian company MMM demonstrated in the 1990s.

When the state is invested in the technology enterprise, the technology could itself become part of national identity. Consider the Soviet campaign of Lysenkoism. Lysenko proposed an alternative to Mendelian genetics that aligned with Marxist theory and was embraced by Stalin. For decades, Lysenko's view reigned in the Soviet Union, with adherents to mainstream genetic theory ejected from academia and some even executed.

If a nation bets big on quantum information science, will it be able to admit failure? Or is it more likely that big bets will come with a kind of psychological investment in the technology?

Many of the elements that obscured the dead-end truths about other technologies are present in quantum technologies, and the stakes are growing. Quantum technologies' complexity, the elite na-

---

losses overlooked by many. In October 2018, authorities removed over 60 scooters dumped in Oakland's Lake Merritt.

[84]Going back to Ponzi, he enjoyed a chorus of support from individuals who were indeed paid early in Ponzi's schemes and thus had made demonstrable gains from the fraud. It was difficult to counter these first investors' successes (Zuckoff, *Ponzi's Scheme: The True Story of a Financial Legend* (2005)). Theranos used elaborate efforts to hide shortcomings of the firm, ranging from Secret Service-like security and seclusion for Elizabeth Holmes to a high-powered law firm (Carreyrou, *Bad Blood: Secrets and Lies in a Silicon Valley Startup* (2018)). Wirecard AG hired a former special forces soldier and the former head of intelligence of Libya to investigate its critics in what it called operation "Palladium Phase 2" (Murph, "Wirecard Critics Targeted in London Spy Operation" (2019)).

ture of its scientists, secrecy mandates, incentives to maintain funding, incentives to appear innovative and profitable, and lack of third parties in a natural position to inspect and report on performance, all could combine to obscure the prospects of quantum technologies. Quantum information science itself could also become a form of nationalistic Lysenkoism, because the concepts of indeterminacy and entanglement provide endless fodder for philosophical exploration and even breathing room for strained religious doctrines, such as mind–body dualism.[85]

The failure scenario has different implications for quantum communications and sensing. In communications, many of the underlying technical achievements have been made to support deployment of commercial technologies. QKD-based hardware is commercially available today for militaries and companies interested in it. If quantum communications fails, it won't be because the technology doesn't work: it will be because the technology isn't needed, or because its use is limited due to network effects, or other market conditions, or prohibitions on its use that cause firms not to adopt the technology.

In sensing too, the failure scenario does not mean that quantum technology is a complete bust. Quantum sensors have worked for decades in the form of medical imaging devices, and sophisticated, well-heeled entities will continue to invest in them. For instance, the oil and gas industries, early patrons of the supercomputing industry, are already poised to take advantage of quantum sensing. Governments will continue to create demand for satellite-based sensing, and for sensing to counter electronic warfare capabilities as discussed in Chapter 2. They just might avoid using the word *quantum*.

This means that even in a quantum computing failure scenario, quantum sensing technologies would still likely create national winners and losers. From a military and intelligence perspective, quantum sensing, when paired with a satellite network, will give nations a different aperture. It will be difficult to hide heavy matériel from these nations, and low-observable stealth technologies will become more detectable.

Yet the public might be a loser in the failure scenario. The failure scenario will lack the virtuous cycle of competition, research,

---

[85]Deepak Chopra has written several books tying quantum physics to healing, and specifically the remission of cancer. Professor Chopra was awarded the *Ig* Nobel prize in 1998 "for his unique interpretation of quantum physics as it applies to life, liberty, and the pursuit of economic happiness."

and price reduction that gave rise to the personal computer. Instead, we are likely to see a much slower growth cycle of quantum sensors and communications – just as we saw with AI from the early 1990s through the mid 2010s. Cutting-edge industries will be willing to invest and experiment because the payoff could be high. But the advantages of quantum encryption and quantum sensing will more slowly diffuse to other players. Industries that depend deeply on sensing, such as healthcare, will be willing to invest in quantum sensors. But without a virtuous cycle, these sensors will never enter the consumer marketplace and will only remain in reach of businesses.

Other losers include big-ticket government investments. The billions spent on quantum technologies and artificial intelligence – priorities voiced in the Trump administration budgets – come at a cost to the budgets of the National Institutes of Health and the National Science Foundation. As such, the quantum science and artificial intelligence priorities displace the priorities that would have been identified by expert program officers at those agencies. The commandeering of such a large amount of money also assumes that American research universities and companies have the capacity to perform so much research in quantum information science. As paylines at agencies become more constrained, principal investigators will be tempted to jam "quantum" into their proposals to support their ordinary work.

Governments and companies are pouring billions into quantum technologies. Where does a quantum failure scenario leave the people and institutions who have invested their money and careers into quantum technologies? Alas, the outlook for these people will remain bright even in the failure scenario. The skills and training required, and the multidisciplinarity of the quantum technology enterprise will be adaptable to other fields.

8.6 Conclusion

Exploring technology scenarios helps us envision how governments, companies, and people will use quantum technologies. Governments will prefer to be both technologically superior and dominant in quantum technologies, and use this advantage to supplement military power. But we are no longer living in the Cold War military/industrial research era. The private sector competes with governments in development, and there is good reason to believe that the private sector could build a quantum computer before or soon after a government does. Unlike stealth jets and bombs, development in quantum technologies is likely to have many potential buyers and many unforeseen uses, much like the modern personal computer. Private companies seeking economic return will broadly democratize access to quantum computing services. Yet we must also contemplate the possibility that it is simply too soon for the quantum age, that investments won't pay off in the near term but possibly decades in the future.

In this chapter we have presented four visions of the future: three that imagine different ways that successful quantum information technologies might be employed by nations and corporations, and a fourth in which quantum sensing and communications are widely used but quantum computing is a bust. These scenarios painted many problematic futures that are brought about by or accelerated by quantum technologies. The next chapter turns to policy options to advance the good while mitigating the negative effects of this innovation.

A Policy Landscape

I N this chapter we present our recommendations for how the policy landscape in the US and other liberal democracies should respond to the opportunities and challenges brought on by quantum information science. These recommendations are informed by the four scenarios of quantum futures we presented in Chapter 8, combined with the understanding of technology capabilities we discussed in Part I.

The most important social and political changes resulting from quantum technologies will not be felt uniformly: there will be winners and losers. But this is not a zero-sum game: with good policy choices, there can be *dramatically more* winners than losers, and we can use other mechanisms to mitigate the negative impacts.

Policymakers have already decided to make large, but not historically unprecedented, investments in quantum technologies. Such investments are known as *industrial policy*, because they are intended to stoke a nation's prowess in science and technology. As these political *bets* reach maturity and begin to pay off, some quantum technologies will diffuse into society. How can we manage the policy challenges raised by those technologies?

We begin this chapter by putting our cards on the table and presenting our policy goals. We then explore how to achieve these goals using traditional policy levers: direct investments, education, and law. We conclude with a discussion of national security issues.

9.1 Quantum Technology's Policy Impact

To ground our policy discussion, we start by articulating our high-level policy goals that we hope will be shared by most readers:

1. Quantum technologies have the potential to profoundly benefit human society, particularly if non-military, non-intelligence uses predominate. To take just one example, there are clear paths to improved detection, diagnosis, and treatment of disease from quantum sensing and quantum simulation. A public/private sector approach that enables commercialization of quantum sensing and computing is likely to produce a market for medical and other pro-social uses of quantum technologies.

2. We think there is an important contextual difference between intelligence and military technology uses on one hand, and law enforcement uses on the other. While we understand the need to use quantum sensing for the first, these technologies would allow unprecedented surveillance and intrusion into private spheres. Therefore we seek to avoid having quantum sensing devolve to law enforcement and proliferate to private actors in advance of significant public discussion and approval, lest we become inured to the privacy invasions that these technologies would likely enable.

3. The capabilities brought about by quantum sensing and quantum computing could result in devastating destabilization of civilian infrastructures and undermining societal trust and integrity mechanisms, public and private law, and even the historical record. As such, civic society needs to embark now on a fact-based, science-based discussion of these capabilities and appropriate mechanisms for controlling them, similar to the discussions in the 1950s and 1960s regarding the control of nuclear weapons and nuclear energy.

Next, we surface two of our assumptions regarding quantum technology, the first regarding technological determinism, the second regarding technological novelty of quantum information science:

Moderate technological determinism We view QIS technologies as political artifacts, in the tradition identified by Langdon Winner (see Section 8.1, p. 305). We do not view this technology as policy-neutral. Quantum technologies are powerful and will tend to push policy discussions in a specific direction, absent political will to redirect. We may be in the driver's seat, but the car is in motion and it is proceeding down a highway with limited offramps and forks in the road.

The invention and growth of the Internet provides a good example of the power and limits of technological determinism. It also shows how *predictions* of where the car will travel depend strongly on each forecaster's beliefs, principles, and hopes. In the initial adoption of the computer networks, visionaries like Ithiel de Sola Pool and John Perry Barlow predicted that the technology would promote democratization, individual empowerment, and exclusion of government power and action.[1] They may have been excellent forecasters, or they may have been merely expressing their hopes as prediction: both were self-described libertarians.

History has shown the Internet's impact is more complex, but also dependent on *implementation specifics*, the social contexts in which the technology was deployed. In liberal democracies cyberspace largely erased restrictions on speech, commerce, and intellectual property. In nations such as China, the government spent significant effort to transform the Internet from a technology of freedom into a technology of control – and it was largely successful. The effect is that the Internet has strengthened China's political institutions.

We embrace the idea that quantum technologies are inherently political, while rejecting the notion that our future is determined by them. We can anticipate the effects of quantum technologies and work so their deployment supports liberal values, but the longer we wait, the harder it will be to do so.

Novelty that's limited but nevertheless game-changing In some cases, quantum technologies offer fundamentally new capabilities, but in other cases they offer merely enhancements for capabilities that we have long had at our disposal. In part this is because many quantum technologies, particularly those of quantum sensing, date back to the 1950s.

We believe that casting quantum technologies as entirely novel is itself a political act, because the claim of novelty is frequently noth-

---

[1]Sola Pool, *Technologies of Freedom* (1983); Barlow, "A Declaration of The Independence of Cyberspace" (1996).

ing more than an ideological appeal against government regulation of the marketplace.

That is, while some might argue that quantum technologies are "novel" and that regulating them now might kill the goose before it lays its first golden egg, we argue that making this argument is itself a wolfish, anti-regulatory political argument against regulation, wrapped in the sheep's clothing of technological exceptionalism that only partially applies. It is an argument designed to limit the ability of policymakers to make sense of what are in reality predictable futures.

* * *

In this chapter, we emphasize strategically and legally relevant differences between classical and quantum technologies. Because the landscape of implications is so large, leading to complex, contingent policy conflicts, and because this quantum age as we conceive of it is so new, we strive to remain at the options level rather than solve specific policy issues.

9.1.1 Game-Changers: Code-Breaking and Possibly Machine Learning

Based on our analysis in the preceding chapters, we believe that the two key areas where quantum's impact will be the greatest are code-breaking and machine learning. We discuss code-breaking extensively in Chapter 5, but we mention machine learning only in passing. This is because far more is known about quantum computing's impact on the first than the second.

We *know* that a sufficiently large quantum computer will be able to crack nearly all of today's encrypted messages, because we have mathematical proofs that show a sufficiently large quantum computer will be able to factor large numbers and compute discrete logarithms in polynomial time. If we can build a large enough machine, today's encryption algorithms are toast.

Quantum-assisted machine learning is at a much earlier point in its development. There is no scientific consensus on whether or not quantum-assisted machine learning will offer fundamental speedups in training machine learning algorithms. For example, many algorithms require that training data itself be stored in some kind of quantum memory – something we don't know how to build. Even if quantum computing dramatically reduces the time and power requirements for training machine learning algorithms, there is no mathematical proof that perfectly training statistical classifiers will offer breakthrough capabilities not enjoyed by today's systems. Therefore, for the remainder of this chapter, we explore the policy implications of instantaneous, perfect, and all-powerful realized machine learning applications, without addressing the question of whether or not quantum computing will ever get us there.

We believe that the most likely near-term quantum technologies to be realized, the quantum-simulators, are unlikely to have game-changing, breakthrough policy implications. However, as we argued in Chapter 5, the process of creating teams to realize quantum simulators, and access to the simulators themselves, will make it more likely for an organization to realize the other game-changing benefits of quantum computing that we mentioned above.

9.1.2 Quantum Technology Dominance

Accepting that there is a role for policymaking in promoting the goals we articulate above, an important question to answer is, *What is the appropriate governmental level to engage in that policymaking?* Should there be QIS treaties among governments, similar to the way that the Treaty on the Non-Proliferation of Nuclear Weapons was designed to promote the peaceful use of nuclear power while preventing the spread of nuclear weapons? Is quantum education something that should be promoted at the community level, with school boards advocating for the establishment of science-based courses in "quantum thinking" for children in secondary school aged 12 through 14, and quantum physics being taught alongside mechanics for students destined for college?

To put it in the language of defense doctrine, is it possible for a nation to achieve *quantum dominance*? By "dominance" we mean, is it possible for a nation to take unilateral actions on matters of quantum technology research, development and deployment, while simultaneously denying state-of-the-art quantum technology to others?

Achieving and maintaining quantum dominance would require a unification of industrial policy, education policy, significant support for research, and strong export controls. We discuss these options in this chapter.

At the same time, the race to build working quantum systems lays bare the fiction of other national attempts to achieve and maintain various forms of technological sovereignty. At the end of World War II, Operation Paperclip successfully scooped up Germany's rocket scientists, giving the US a brief head start in the space race, but the Soviet Union quickly pulled ahead in both rocketry and space exploration. Likewise, the Soviet Union was able to eliminate US nuclear dominance through a combination of espionage and scientific ingenuity.

9.2   Industrial Policy

Whether governments should invest in quantum technologies is a settled policy issue: they are doing so, generously, but not at levels that are historically unprecedented, such as the Manhattan Project ($28 billion in adjusted dollars) or the Apollo Space Program ($190 billion). The pursuit of quantum technologies is now a significant *industrial policy* priority in the US and abroad. Industrial policy is "a strategy that includes a range of implicit or explicit policy instruments selectively focused on specific industrial sectors for the purpose of shaping structural change in line with a broader national vision and strategy."[2] Industrial policy can be general, in the sense that tax breaks or incentives for investment are shaped to broadly advantage domestic business interests. Industrial policy can also be specific, in that the government can organize policies to aid a particular vertical industry, such as price supports for corn farming, tax-subsidized grazing fees for cattle ranchers, and requirements to add ethanol to gasoline.

9.2.1   National Quantum Investments outside The US

The embrace of quantum technologies by national governments clearly flows from lessons learned by observing the US technology miracle. The US has enjoyed a decades-long period of technological superiority, culminating with the internet boom and the vast production and concentration of wealth, thanks to strategic investments in computing, microelectronics, packet networking, and aerospace between 1940 and 1980.

Quantum technologies provide an opportunity for a reordering of technical might that should concern US policymakers whose goal is to maintain the nation's technological superiority. The EU and

---

[2]Oqubay, "Climbing without Ladders: Industrial Policy and Development" (2015).

China are desperately seeking opportunities to overcome the asy metric advantages that the US has enjoyed from incubating Silicon Valley. For example, the Internet, as a global communications system, is still largely seen by other nations as America's playing field. Political scientists now recognize how American power is exercised through control of others' access to and use of networked systems like the US-dominated Internet.[3] Many nations have acknowledged the continuing disadvantage of having their domestic communications structured by the Internet and often delivered by US dominant companies. This is another lens for understanding the ongoing antagonism between US policymakers and Chinese communications firms such as Huawei.

Both the EU and China have established significant quantum information science efforts that include basic research funding. This funding often goes beyond the development of specific quantum technologies, and supports basic, theoretical research, workforce preparation, educational outreach, and even funds inquiry into the philosophy of quantum mechanics.

In 2018 the EU funded a €1B ($1.2B) quantum initiative, supporting both multiple corporate and academic research groups and funding specific projects. Europe's investment also builds upon a number of domestic competitors in quantum computing, communications, and precursor technologies, such as high-end cooling devices and precision-machined equipment.

China appears to have invested about $3B in quantum technology, according to a report warning of the country's muscularity and devotion to surpassing American innovation in the space.[4] But there are many popular reports claiming many billions more are invested in China's quantum technology, and in infrastructure for massive technology integration centers. For instance, it is reported that China invested $10B in support for quantum internet science based at the University of Science and Technology of China in Hefei. As detailed in Part I, China has implemented the longest publicly known fiber quantum network, distributed quantum keys by satellite intercontinentally, created the most powerful (albeit single-purpose) quantum computer, and appears to be developing game-changing quantum

---

[3]Farrell and Newman, "Weaponized Interdependence: How Global Economic Networks Shape State Coercion" (2019).

[4]Kania and John Costello, *Quantum Hegemony? China's Ambitions and The Challenge to US Innovation Leadership* (2018).

sonar technology that could one day be deployed to hotbeds of conflict, such as the South China Sea. Many of these accomplishments are not heralded by state media, but rather by peer-reviewed articles in *Science* and *Nature*.

Press accounts of national quantum policies frequently focus on pan-EU projects and overlook individual national initiatives. As early as 2014, the UK embarked on an academic/industry program investing £270M ($375M) to establish hubs focusing on sensing, communications, and quantum technology development. These UK national quantum technologies (UKNQT) hubs involve many universities and scores of private partners. A related initiative is pouring over £167M into graduate training in QIS – Brexit is giving the UK additional incentives to compete technologically with Europe. Germany announced an additional €650M in funding in early 2020, but after the COVID pandemic's effects were realized, Germany introduced a €50 *billion* ($60B) stimulus package in "future technologies," which explicitly earmarks €2B ($2.4B) for quantum technologies, as well as €300M ($360M) for development of a Munich Quantum Valley.[5] France has committed over €1B to QIS as well.

Nations in Europe with their own quantum industrial policies are engaged in a two-sided strategy. These nations want to be part of the EU funding compact, which is characterized by regional sovereignty and technology superiority goals. Such sovereignty carries with it the East/West bloc downsides we discuss in Chapter 8. But by investing in their own national quantum portfolios, EU nations straddle the divide between closed sovereign strategies and the open collaboration typical of scientific inquiry. The two-sided approach enables nations to attain more independence from the EU and have more opportunities to engage the US and foreign companies that might end up developing breakthrough insights.

Russia appears to be late to the competition and is absent from state-of-the-science developments in quantum technology. Not until December 2019 did the country announce a major initiative to fund quantum research, and when it did, the amount specified – $790M over five years – was underwhelming given the country's population, ambition, and early contributions to the field.[6]

---

[5] Bundesministerium für Bildung und Forschung, "Die Zweite Quantenrevolution Maßgeblich Mitgestalten" (2020).

[6] Schiermeier, "Russia Joins Race to Make Quantum Dreams a Reality" (2019).

India too has recently announced a major initiative in QIS research, with a $1B commitment made in its 2020 budget.[7] India's investment should be seen in context with the nation's outer space program, which it funds in the billions, and that has launched vehicles to the Moon and Mars.

9.2.2 US Quantum Technology Industrial Policy

The US government quickly changed its posture in response to EU and Chinese investment. Previously, the US had spent hundreds of millions pursuing various QIS projects, many of which were funded through the Department of Defense, making them difficult to track. Responding to the foreign interest and investment, Congress quickly introduced and enacted the National Quantum Initiative Act.[8] Signed by President Trump in December 2018, the NQIA authorized $1.2 billion in research and education, to be coordinated by the White House's Office of Science and Technology. The NQIA's National Quantum Initiative (NQI), led by NIST, NSF, and the Department of Energy, in turn coordinated government/industry/academic relations to promote the development of QIS and quantum technologies.[9] NQIA also formally established the Subcommittee on Quantum Information Science (SCQIS) of the National Science and Technology Council. Congress specified that this new body will be chaired jointly by the Director of the National Institute of Standards and Technology (NIST), the Director of the National Science Foundation (NSF), and the Secretary of Energy, and has participation by the Office of Science and Technology Policy (OSTP), Office of the Director of National Intelligence (ODNI), Department of Defense (DOD), Department of Energy (DOE), National Institutes of Health (NIH), and the National Aeronautics and Space Administration (NASA).

In 2020, the Trump administration named appointees to the National Quantum Initiative Advisory Committee (NQIAC), which was established by the NQIA to advise the new subcommittee. Advisory committees are typically constituted of experts from outside government; initial appointees are prominent academics and participants from startup, defense industrial base, and established technology

---

[7]Padma, "India Bets Big on Quantum Technology" (2020).

[8]US Congress, *National Quantum Initiative Act* (2018).

[9]Christopher Monroe, Raymer, and J. Taylor, "The US National Quantum Initiative: From Act to Action" (2019).

firms in the space.[10] The body is charged with regularly making reports to the President and Congress, and to give advice on progress made in implementing the quantum initiative, management and implementation issues, American leadership strategy in QIS, potential for international cooperation in QIS, and whether "national security, societal, economic, legal, and workforce concerns are adequately addressed by the Program." The first meeting took place on October 27, 2020.

Following the NQIA, President Trump proposed doubling research funding for QIS by fiscal year 2022. In August 2020, the administration announced the creation of five quantum information science centers coordinated by Department of Energy Labs (the Argonne, Brookhaven, Fermi, Lawrence Berkeley, and Oak Ridge National Laboratories). In addition to a $625 million commitment of federal government funds, the project is complemented with over

---

[10]The body was chaired by Dr. Charles Tahan, OSTP Assistant Director for Quantum Information Science and Director of the National Quantum Coordination Office, and by Dr. Kathryn Ann Moler, Dean of Research at Stanford University. The initial appointees were: Professor Timothy A. Akers, Assistant Vice President for Research Innovation and Advocacy, Morgan State University; Professor Frederic T. Chong, Seymour Goodman Professor, University of Chicago; Dr. James S. Clarke, Director, Quantum Hardware, Intel Corporation; Professor Kai-Mei C. Fu, Associate Professor of Physics and Electrical and Computer Engineering, University of Washington; Dr. Marissa Giustina, Senior Research Scientist, Google, LLC; Gilbert V. Herrera, Laboratory Fellow, Sandia National Laboratories; Professor Evelyn L. Hu, Tarr-Coyne Professor of Electrical Engineering and Applied Science, Harvard University; Professor Jungsang Kim, Co-Founder, IonQ and Professor of ECE, Physics and Computer Science, Duke University; Dr. Joseph (Joe) Lykken, Deputy Director for Research, Fermi National Accelerator Lab; Luke Mauritsen, Founder/CEO, Montana Instruments; Professor Christopher R. Monroe, University of Maryland; Professor William D. Oliver, Associate Professor EECE, Professor of Practice Physics, and MIT-Lincoln Laboratory Fellow, Massachusetts Institute of Technology and MIT-Lincoln Laboratory; Stephen S. Pawlowski, Vice President of Advanced Computing Solutions, Micron; Professor John P. Preskill, Director of the Institute for Quantum and Matter, California Institute of Technology; Dr. Kristen L. Pudenz, Lead for Quantum Information Science, Lockheed Martin; Dr. Chad T. Rigetti, Founder and CEO, Rigetti Computing; Dr. Mark B. Ritter, Chair, Physical Sciences Council, IBM T.J. Watson Research Center; Professor Robert J. Schoelkopf, Sterling Professor of Applied Physics and Physics, Yale University; Dr. Krysta M. Svore, General Manager of Quantum Systems, Microsoft Research; Professor Jinliu Wang, Senior Vice Chancellor for Research and Economic Development, The State University of New York; Dr. Jun Ye, JILA Fellow, Professor of Physics, National Institute of Standards and Technology.

$300 million in commitments from academic institutions and companies.

It is important to recognize that research funding has many paths in the US. In addition to NQIA funds, quantum technology projects receive support directly from the Department of Defense, under its Research, Development, Test, and Evaluation (RDT&E) budget. This budget now exceeds $100 billion annually; the DOD 2021 budget estimates for RDT&E mention the word "quantum" on 27 pages of the 1094-page document.[11] As this manuscript goes to publication, President Biden and other policymakers proposed an extra $250 billion in funding for general high-technology research. With this level of money flowing into the field, the question becomes one of talent: are there enough people with the rarefied, specialized forms of training that quantum technologies require? Below, Section 9.3 (p. 401) focuses on the challenge of workforce training.

9.2.3 Industrial Policy: Options and Risks

With billions being spent by many nations, quantum technologies are clearly part of many nations' industrial policy. We note, however, that the spending is not at the levels of previous big technology feats, such as when Russia and Europe each found the need to replicate the US GPS constellations (see Figure 9.1).

Quantum technologies make a good case for vertical industrial policy interventions under a framework applied by Vinod Aggarwal and Andrew W. Reddie. Writing in the cybersecurity context, one that shares strategic characteristics common with quantum technologies, the authors explain that governments pursue industrial policy to create markets (market creation), to facilitate markets, to modify markets, to substitute for market failures (market substitution), and to set rules to control technologies created by markets (market proscription).[12]

In this section, we consider the risk of market substitution for quantum key distribution, quantum networking, and quantum computing in general. In all three categories of quantum technologies, market substitution appears to be necessary to support continued

---

[11] Office of the Secretary of Defense, "Department of Defense Fiscal Year (FY) 2021 Budget Estimates" (2020).

[12] Aggarwal and Reddie, "Comparative Industrial Policy and Cybersecurity: a Framework for Analysis" (2018).

Market Substitution

In the literature of industrial policy, the phrase *market substitution* occurs where "instruments of political authority are used to allocate or distribute resources or control conduct of individuals or organizations..."[a] Aggarwal and Reddie point to several examples in the cybersecurity context. For instance, In-Q-Tel is a privately-held not-for-profit venture capital firm that is funded by the US Intelligence Community and other federal agencies to help the government stay atop cutting edge technology developments. Governments also substitute for cybersecurity market failures by promoting educational and workforce training efforts.[b] Such moves can "prime the pump" by supporting a new market until there is sufficient demand. Market substitution is a more controlling approach than market *facilitation*, where incentives are shaped to spur the private sector into useful action – for example, by eliminating the liability shield for cybersecurity vulnerabilities that many software and service providers currently enjoy. The control inherent in substitution means that choosing properly, and choosing in the public interest – instead of the interest of the choosers – is a challenge in industrial policy.

[a]R. G. Harris and Carman, "Public Regulation of Marketing Activity: Part II: Regulatory Responses to Market Failures" (1984).
[b]Aggarwal and Reddie, "Comparative Industrial Policy and Cybersecurity: a Framework for Analysis" (2018).

development of these technologies for an indeterminate amount of time.

QKD Market Substitution

While there are obvious commercial uses for quantum metrology and sensing among the most sophisticated and well-resourced companies (such as oil services firms, mining firms, and medical imaging), the National Academies report estimated that there are only limited short- to medium-term commercial uses for quantum communications such as QKD.[13] One of those limited uses of quantum communications is to secure point-to-point links used by banks and trading houses – organizations that have both the resources to procure

---

[13]Grumbling and Horowitz, *Quantum Computing: Progress and Prospects* (2019).

private fiber connections, and the risk of loss necessary to justify investments in QKD.

Otherwise, despite the excitement surrounding QKD, commercial justifications for it are thin. To date, most public deployments of QKD are better regarded as technology demonstrations, rather than the first step in creating significant new markets. For example, in 2007 the Swiss government allowed a domestic company to use quantum encryption to transmit election information to a central government repository, with the justification provided by Geneva state chancellor Robert Hensler, that QKD would "verify that data has not been corrupted in transit between entry and storage."[14] The irony here is that QKD does not provide data integrity, it provides secrecy against some future attacker with a code-breaking quantum computer *who also captured and made a permanent recording of the encrypted transmission.* But the use of QKD by the Geneva government did result in having *New Scientist* note that "three companies [are] pioneering the field – BBN Technologies of Boston, US; MagiQ of New York, US; and ID Quantique of Geneva, Switzerland."

Today's commercial QKD systems send their flying qubits down a single strand of fiber-optic cable that's typically 10 km to 100 km in length. This is ideal for exchanging encryption keys between a data center in lower Manhattan and a data center in Hoboken, NJ. A near-future satellite-based QKD system might send pairs of entangled photons simultaneously to an embassy in Moscow and a government office in London, assuring that no future Russian government might be able to crack RSA encryption keys that are used today (although another way to address this threat would be to use a human courier to deliver a year's worth of AES-256 keys in a secured briefcase). However, it is inconceivable that businesses or consumers would opt for QKD technology to encrypt the packets that they send over today's Internet: there is no way that the pairs of photons could be routed to the correct destination to be used for decryption. Quantum encryption for the masses will need to wait for a quantum internet, and that might be a very long wait.

Where QKD might play a role in the commercial Internet would be ISPs using it to encrypt specific internal, high-risk long-haul links. The distance from Moscow, Russia, to Kyiv, Ukraine, is 865 km; in a few years this might be within the service range of a QKD system.

---

[14]Marks, "Quantum Cryptography to Protect Swiss Election" (2007).

Western businesses with offices in Moscow might be willing to pay a premium for an internet connection from the Ukraine that is encrypted using QKD. However, if they do, it is our opinion that they will be wasting their money unless they also have 24-hour guards to protect against having their laptops stolen, perform detailed background investigations of all their employees, and undertake similar measures to protect themselves from a wide range of electronic surveillance.

Another possible customer of QKD is backbone providers and others that have private ("dark fiber") networks. Such providers typically have more control over elements of the network and their protocols, and are interested in protecting point-to-point connections. Some of these network owners may also have particular concerns about nation-state spying, either by adversaries digging up their private fiber and tapping it, or by bribing or extorting company engineers to provide access. For instance, as discussed in Chapter 7, in 2017 South Korea's SK Telecom claimed that it had secured its network backhaul with a QKD system, offering additional protection to a wireless network serving over 350 000 mobile users in Sejong City. Given that the cost of QKD network encryption devices is similar to the cost of a few full-page advertisements in a leading newspaper, this may be money well-spent, even if it is just for bragging rights.[15] That's because QKD protects today's encryption tomorrow: any possible fallout that would be protected by a QKD-based system won't take place for years, or even decades.

We thus believe that the commercial prospects for QKD are poor, because of a lack of incentives, coordination problems, and primarily the sufficiency of classical encryption alternatives. Furthermore, although the QKD protocols are information-theoretic secure, the actual QKD *devices* can still be hacked.[16] Market substitution will be required to create a viable QKD industry.

Quantum Networking Market Substitution

The near-term case for quantum internet is even poorer than the case for QKD for one simple reason: although commercial QKD systems can be purchased and used today, working quantum network-

---

[15]Kwak, "The Coming Quantum Revolution: Security and Policy Implications, Hudson Institute" (2017).

[16]Anqi et al., "Implementation Vulnerabilities in General Quantum Cryptography" (2018).

ing systems appear to be even further in the future than large-scale quantum computers.

Consistent with the market substitution approach, in 2020, the Department of Energy and University of Chicago announced plans to build a national quantum internet framework.[17] Such a fully quantum internet would use entangled photons for communication, thus giving communicants security against quantum computing attacks, the ability to detect interception or blockage of the signal, and the ability to connect quantum computers over distances. Nevertheless, quantum internet is still an experimental concept. Most designs call for a fiber optic network passing entangled photons between quantum computing elements to maintain and communicate quantum states. Many fundamental engineering problems need to be addressed. And even if some kind of quantum network is created, such a network would be a para-internet, for specific use cases, and not a general communications infrastructure.

The power of the Internet that we have today is that it is a general network. Although the Internet started as a slow-speed network capable of sending email and allowing users to log on to remote computers, by the 2000s the Internet was being used to transmit all manner of broadcast and interactive content. Slowly legacy networks such as telephone systems were reworked so that they traveled over the Internet. But this was not a surprise: even in the 1970s, it was clear that the Internet would one day encapsulate all other communications networks. (Xerox's Palo Alto Research Center demonstrated the first packet network voice system, called the "Etherpone," in 1982, before the Internet adopted TCP/IP.) No such technology roadmap is envisioned for quantum networks.

No similar claim can be made for a quantum internet. Although some authors claim that quantum networks will be able to transmit vast amounts of data faster than the speed of light, such claims are inconsistent with both our vision of quantum networks and the laws of physics as we currently understand them (see the sidebar "Alas, Faster-than-light Communication Is Not Possible" on page 301). Instead, it appears that the advantage of quantum networks is they would allow quantum computers to engage in quantum communications algorithms that would decrease the number of required steps

---

[17]Dam, *From Long-Distance Entanglement to Building a Nationwide Quantum Internet: Report of The DOE Quantum Internet Blueprint Workshop* (2020).

for certain operations. Such a network would also allow for a quantum computer to connect to a remote quantum database (if one existed) to search that database using Grover's algorithm, without the database operator learning what had been searched and what had been retrieved (blind quantum computing). But such fantastic applications seem decades in the future, if they are even physically possible.

For these reasons, as governments promote development of the quantum internet, the best-case scenario is a para-internet for certain applications, and of course, the learning-by-doing inherent in research and development. After all, quantum communications devices are merely small quantum computers that compute with flying qubits. Governments investing in quantum communications are also preparing their scientific and technical workforce for the eventual emergence of large-scale quantum computers, although there may be more efficient ways to do so.

Quantum Computing Market Substitution

Turning to the industrial policy case for computing, some companies are beginning to experiment with quantum computing, but there is no broader market for quantum computing services. Classical computers still outperform quantum ones in all practical applications. Although there is a growing commercial market for quantum computing, this use is limited to experimentation and training. That is, at the present time, researchers are focused on researching quantum computing, rather than on using quantum computers to do research. Simply put, there is no market to facilitate with ordinary incentives. Thus, market substitution, in the US case, through massive funding of research, is in order for the time being.

Consider that a wide range of companies are testing a variety of applications for quantum optimization using cloud-based quantum computers and annealers. One promotional video by a quantum computing company summarized projects at:

- BMW (robotic manufacturing)

- Booz Allen Hamilton (satellite placement)

- British Telecom (placement of antennae)

- Denso (ride sharing)

- DLR (aircraft gate assignment at an airport)

- Los Alamos National Laboratory (face recognition, social networks of terrorist groups, and attack prediction)

- NASA/Ames (cybersecurity of aircraft traffic management systems)

- Ocado (robot product picking in a warehouse)

- QBranch (election modeling)

- Recruit Communications (real-time bidding in online advertising)

- Volkswagen (vehicular traffic analysis),

- ... and a former academic researcher focused on prediction of health outcomes even where relevant data are missing.

This same promotional video explained that four institutions had installed its systems, perhaps for secrecy reasons, and these systems were mostly focusing on aspects of optimization:

- Google/NASA Ames/USRA,

- Lockheed Martin Corporation/USC ISI,

- Los Alamos National Laboratory, and

- Oak Ridge National Laboratory[18]

But to date, the aspects of these projects that have been shared publicly are aimed entirely at simply getting model problems to work, rather than developing cost-effective solutions to problems that the companies are currently facing.

For companies outside quantum technologies – that is, most companies – buying quantum computing services is still not worth the investment. The National Academies lamented in 2019 that broadly appealing commercial uses of quantum computers have not been developed, and that investment in applications is necessary to kickstart a "virtuous cycle" of innovation in quantum computing. One of

---

[18]D-Wave Systems Inc., "Quantum Experiences: Applications and User Projects on D-Wave" (2019).

the group's main findings was that "There is no publicly known application of commercial interest based upon quantum algorithms that could be run on a near-term analog or digital NISQ computer that would provide an advantage over classical approaches."[19] By *commercial*, the Academies essentially means quantum-enhanced computation or service that would give a company a competitive advantage sufficient to justify its cost.

9.2.4 Innovation and The Taxpayer

Until commercial and consumer applications take root, quantum technologies will need some kind of research sponsor to substitute for a market. In the US, the government, major technology firms, and private foundations have been patrons for QIS. These efforts are matched by the EU and China's government-patronage approach. The EU and China seem to be trying to replicate the US success with the Internet in their funding of QIS.

Indeed, there is compelling proof that sustained federal investment over decades in an industry or region can yield ample rewards. Consider California. Prior to the commercialization of the Internet as a tool for connecting consumer and business devices, "the military-industrial complex was the West's biggest business in the cold war years," writes Gerald D. Nash in his economic history of the West. "The size and scale of the new federal [military] establishments were unprecedented. Congress poured more than $100 billion into western installations between 1945 and 1973."[20] Margaret O'Mara observes that Lockheed, which minted billions creating cutting-edge military hardware, including the P-80, the Polaris missile, the U-2, the SR71, GPS satellites, and the stealth attack aircraft (see Figure 2.12), was the largest high-technology employer in Silicon Valley until the Internet boom.[21] Joan Didion elucidates nineteenth-century forms of federal largess, such as waterworks, dams, irrigation subsidies, railroads and other infrastructure that set the stage for development of the region, again complicating the California narrative of self-reliance and self-made fortunes.[22]

---

[19]Grumbling and Horowitz, *Quantum Computing: Progress and Prospects* (2019).

[20]Nash, *The Federal Landscape: an Economic History of The Twentieth-Century West* (1999).

[21]O'Mara, *The Code: Silicon Valley and The Remaking of America* (2019); O'Mara, *Cities of Knowledge: Cold War Science and The Search for The Next Silicon Valley* (2015).

[22]Didion, *Where I Was From* (2003).

Today's internet companies emerged from a region where an educated middle class with a focus on engineering was groomed over generations, thanks to the largess of the federal government and the American taxpayer. Companies like Apple built revolutionary products and services but in context these products can be seen as remixes and masterful re-implementations of technologies developed for the military at taxpayer expense.[23] Other Silicon Valley darlings might flounder if they lacked the ability to freely depend on taxpayer-provided infrastructure such as GPS or even the nation's highway system.

Consider the story of Konrad Zuse (Chapter 4). Zuse built a cutting-edge, switch-based computing device in 1936, four years before the British Bombe and eight before a similar project at Harvard University. However, the German government did not embrace computing in the ways the British and the US did. After World War II, the British failed to capitalize on their lead, in the interest of preserving the secrecy of Bletchley Park. (Tommy Flowers, who designed and built the code-breaking Colossus computer, was blocked from re-implementing or commercializing the technology and spent the rest of his professional career working on telephone switching systems.)

The absence of credible competition from overseas allowed the US to dominate the nascent field of electronic computing. In the US, the military, scientific, and defense communities aggressively adopted computing, giving the US a lead that it held for decades. Visionary scientists such as J. C. R. Licklider anticipated the importance of computers and invested in them long before their uses were fully apparent. Licklider convinced legendary defense industrial base company Bolt Beranek Newman Inc. (BBN)[24] to buy not one but two early computers, the most expensive laboratory devices that BBN had ever purchased, before the firm even had uses for them. Of course, such uses quickly became clear. The need for ever-intensive machine analysis during the Cold War funded computer and com-

---

[23]Mazzucato, *The Entrepreneurial State: Debunking Public Vs. Private Sector Myths* (2015).

[24]Discussed earlier in Section 4.4.1 (p. 146). BBN Inc. eventually became BBN Technologies, and was acquired by Raytheon in 2009. In 2012, President Barack Obama awarded Raytheon BBN Technologies the National Medal of Technology and Innovation, the highest award given by the nation to technologists, recognizing "those who have made lasting contributions to America's competitiveness and quality of life and helped strengthen the Nation's technological workforce."

Figure 9.1. Major science, technology, and military projects (2021 inflation-adjusted dollars, not to scale). Precise figures for quantum technology investment are elusive because funding flows through both specific authorizations and separately through the Department of Defense.

ponent manufacturers and drove employment of untold number of programmers. With the advent of the personal computer, computing was democratized, resulting in a cycle where computers became both less expensive and faster. And the US was at the center of that virtuous cycle.

At the dawn of internet commerce, it was not clear at all that the web would even succeed as a medium. Other similar systems had failed: France's "Minitel" was widely used, but it had not spurred an economic revolution. Likewise, the US online service Compuserve had 1.5 million subscribers in 1993, but it was not a vibrant marketplace. Today's most profitable companies, such as Amazon.com, spent years trying to perfect a web platform for commerce. In the process, the company developed its web services platform, which today is responsible for the bulk of the company's operating profits.

Despite these facts on the ground, it is European thinkers and policymakers who primarily promote the belief that governments can be effective market creators in technology,[25] and that these new fields need government incubation to eventually become successful. But Europe suffers because it lacks both Silicon Valley's affluent and gamblesome venture market, and the Valley's highly efficient labor market – the highly educated high-tech workers who, because of state law, can leave an employer when a better deal or more promising technology comes along and go work for a startup or even a competitor.[26]

Turning to the development of quantum technologies, US government funding and technical achievements abound. Scientists at NIST developed the first quantum circuit. That agency's scientists have been in the vanguard of quantum technologies, with three Nobel Prize recipients in this field alone. This book recounts many examples of scientific achievements realized by Department of Defense research institutions, the Department of Energy National Laboratories, and the federal government's medical research gem, the National Institutes of Health. US government agencies were critical for both convening events to develop the theory of quantum computing, and for developing a vision and strategy for funding investment in the field. The state of the science in quantum technologies has advanced

---

[25]Mazzucato, *The Entrepreneurial State: Debunking Public Vs. Private Sector Myths* (2015).

[26]Saxenian, *Regional Advantage: Culture and Competition in Silicon Valley and Route 128* (1996).

because of US taxpayers' dollars supporting a strong science and technology industrial policy.

In the larger political conversation, there is rhetoric rising to the level of reaction formation against government involvement in new technology in Silicon Valley. Many technology advocates parrot libertarian ideas from John Perry Barlow's ahistorical statement on internet freedom:

> Governments of the Industrial World, you weary giants of flesh and steel, I come from Cyberspace, the new home of Mind. On behalf of the future, I ask you of the past to leave us alone. You are not welcome among us. You have no sovereignty where we gather ...You have not engaged in our great and gathering conversation, nor did you create the wealth of our marketplaces. You do not know our culture, our ethics, or the unwritten codes that already provide our society more order than could be obtained by any of your impositions.[27]

Barlow's essay and others like it argue that governments lacked the competence to understand and to act on the Internet. We find such arguments disingenuous, given the US government's widely known and dramatic investments in science and technology that occurred during his lifetime. More broadly, we argue that this brand of libertarianism is bad policy, dangerous, and smacks of hypocrisy. It's bad policy because if the US taxpayer had not supported basic science research, the twentieth century might have been defined by innovation in Japan or Europe. It's dangerous because libertarianism animates extremist anti-government actors, such as Oklahoma City bomber Timothy McVeigh,[28] and because the ideology shares overlapping space with nationalist movements. And it's hypocritical because many of the greatest advocates of libertarianism have themselves been the beneficiaries of significant public largess: we note that between 1971 and 1988, when he ran his family's cattle ranch with his mother,[29] Barlow's business was heavily subsidized by the US government and favored by US tax policy.

---

[27]Barlow, "A Declaration of The Independence of Cyberspace" (1996).

[28]Ayn Rand's hero, Howard Roark, blows up a public housing complex in response to slights from government bureaucrats.

[29]Schofield, "John Perry Barlow Obituary" (2018).

Many of today's US policy debates flow from a libertarian frame, and the idea that government impedes innovation is widely shared. Perhaps this is why Aggarwal and Reddie observe that there is a "puzzling gap in the [industrial policy] literature with regard to the role the state has played in driving investment in the high-tech industry."[30] Such patronage is an explicit goal in Europe and China's quantum initiatives. Other nations seem to be learning from what the US has done, rather than what various influential opinion leaders have said about industrial policy.

9.2.5 The Risk of Choosing Poorly

One risk of industrial policy is that of choosing poorly: choosing the wrong technology, or investing just enough money to crowd out private investments without sufficient funds to kick-start an industry, or investing more money than can be spent by the available talent, leading to waste and making it more difficult for valuable contributions to stand out.

Governments around the world are trying to position their industrial centers for the future, and quantum technologies are but one possible focus. Governments are also focusing on the promise of automation and machine learning; big bets are being placed on battery and photovoltaic technology development.[31] Innovation is also shaped by other policy concerns, such as environmental impact, that have intersections with quantum optimization. For instance, the European Union is seeking to arrange the economy "circularly," so that technologies used in the future are serviceable and repairable, resulting in less waste.[32]

Consider what happens if governments excessively fund quantum technologies for a decade and the technologies do not create self-sustaining markets: at that point, governments might significantly curtail funding, leaving companies, faculty, and graduate programs fighting amongst themselves for the few remaining scraps. Many people who had spent years mastering difficult quantum technologies would suddenly find themselves without jobs: some would success-

---

[30] Aggarwal and Reddie, "Comparative Industrial Policy and Cybersecurity: a Framework for Analysis" (2018).

[31] The German government is in the midst of an ambitious plan called *Industrie 4.0*, designed to leapfrog ahead with a focus on the Internet of Things and automation.

[32] European Commission, "A New Circular Economy Action Plan" (2020).

fully transition elsewhere, others not.[33] It might take quantum information science 10 or 20 years to be taken seriously again, and when it came back, it might be in a very different form. This is the *quantum winter* scenario, based on the "AI winters" of the mid-1970s and the late 1980s.

We think that this is a real risk. Quantum sensing is already paying off, so there are clear reasons to believe that some investments in quantum technologies are a good bet. But while quantum sensors have similar physics requirements to quantum computers in terms of controlling noise and managing materials, quantum sensors do not run algorithms the way quantum computers do. Some skills from quantum computing are transferable, others not.

There are also strategies governments can pursue to lessen the consequences of a bad technology choice:

1. Governments can invest in basic quantum research, rather than applied research, development, or marketization. This is because the basic challenges in quantum technologies are so great and we are so early into their development. In classical computing, the transistor is the basic technology used to create bits, and that technology scaled dramatically from the 1960s, with transistors getting smaller, chips getting larger, and the number of transistors per chip increasing geometrically (not exponentially!) over time. But the basic idea of silicon-based transistors has not changed. Contrast that with quantum computing, where no consensus has emerged for the fundamental qubit technology, in part because scaling is so much more difficult when scale requires control over quantum-level phenomena. Basic research to find the transistor-like invention for quantum states does not bet on any single technology, and if successful, will revolutionize the field.

2. Governments can pursue diverse research and development efforts. Because the fundamentals of quantum computing are so uncertain, government money is better spent funding smaller,

---

[33]Consider the Japanese Fifth-Generation computing project, one that started in 1989 to develop artificial intelligence and that sought to make breakthrough gains in natural language processing. The Japanese project is considered a failure; even mid-project stream reviews of the project were disappointing. The one main benefit of the project seems to be the training of Japanese people in computer programming, a field that the nation was considered to be behind in at the time.

more innovative projects that are high-risk, high-reward, and ultimately less likely to produce workable systems. Placing many bets on different breakthrough approaches might result in winning the quantum computing technology lottery. If the lottery is lost, it still provides training opportunities for multi-disciplinary researchers who could bring diverse insights to the winning technology.

Market-leading companies such as Google, IBM, and Microsoft have immense amounts of cash on hand, and incentives to develop quantum technologies as quickly as they become financially viable. These companies can decide to spend their treasure to pursue quantum computing, and they can pull back if they believe that the market is premature. (Nathan Rochester, an IBM research scientist, was one of the organizers of the 1956 conference on artificial intelligence.[34] But after IBM received negative publicity for its research into AI, Rochester was directed to other tasks.)

We believe that it is too early to bet on a specific physical medium for quantum computing. At present, the risk of locking in to a specific quantum technology seems low, and none of the current technologies may be the one that ultimately carries the day. Indeed, as the National Academies report states, no technological approach currently demonstrated can scale to a fault-tolerant quantum computer.[35]

3. Governments are better positioned to evaluate the implications of international collaboration for their national security and overall global stability than are multinational corporations. Government regulators and policymakers have access to information obtained from many non-public sources, are able to plan using longer timescales, and have a wide range of tools available to realize their policy goals.

Current industrial policy is tilting towards the East/West bloc scenario we present in Section 8.4 (p. 361), where nations choose sides and pursue research efforts independent of each other. This stands in opposition to other grand-scale science projects, such as the

---

[34]McCarthy et al., "A Proposal for The Dartmouth Summer Research Project on Artificial Intelligence" (1955).

[35]Grumbling and Horowitz, *Quantum Computing: Progress and Prospects* (2019).

Large Hadron Collider (LHC) built by the European Organization for Nuclear Research (CERN), or the ongoing attempt to create a workable fusion reactor at ITER, the International Thermonuclear Experimental Reactor (a collaboration that includes China, India, Japan, Russia, South Korea, and the US).

One compelling reason to continue an individual nation approach is that unlike the LHC and ITER, quantum technologies do not require massive engineering efforts, the retraining of significant numbers of workers, or thousands of workers with hard hats. Both the LHC and ITER are projects that only rich nations can afford. In quantum computing, startup companies relying only on private funding are able to assemble NISQs.[36]

Another compelling reason is that, unlike the LHC and ITER, a successfully realized quantum computer would immediately have implications for national security and intelligence gathering efforts.

Perhaps the deeper industrial policy concern surrounds betting on QIS at all, instead of putting more money into artificial intelligence powered by classical computers or some kind of new approach for organizing electronic computation, such as the Fujitsu "quantum-inspired" digital annealer.[37] Much like the first 60 years of nuclear fusion research, quantum computing is a field where its advocates predict that fundamental advances are at hand, yet these advances remain, like the Chimera, on the horizon but out of reach.

In addition to funding, an industrial policy could make technical mandates, and this is an area where the government could pick winners and losers. To achieve a fully quantum internet, communications must be both generated and relayed by fully quantum devices. This would seem to require that networks not only be quantum, but also fully optical, as the technology works most robustly with photons. Thus, laying fiber optic, a major priority in Europe and China, should also be a focus in the US. Satellite networks also enable quantum communications, and a number of competitors are attempting to make worldwide broadband systems through low-earth-orbit mini-

---

[36]The startup company Rigetti required less than $100 million in funding to develop its 19-qubit superconducting "Acorn" system in 2017. By 2020, Rigetti offered "Aspen-8," a 31-qubit superconducting system, connected through Amazon's cloud. As of this writing, Rigetti accomplished all of this with only $174 million in funding, just $8 million of which came from a US government source (DARPA).

[37]Aramon et al., "Physics-Inspired Optimization for Quadratic Unconstrained Problems Using a Digital Annealer" (2019).

satellites. The choice of physical infrastructures for communications will lead to long-term policy consequences surrounding access to and control over communications.[38]

9.3 Education Policy

Public policy can be shaped to realize quantum goals, but no matter the goal, human capital is necessary.

National governments can increase the availability of human capital through education policy, training programs, tax credits, and even immigration policy. Of these, education is among the slowest but potentially the most effective in the long term.

9.3.1 Graduate Training in QIS

Most academic research in Western nations is performed by graduate students pursuing doctorates under the guidance of a faculty advisor. Thus, the number of graduate students pursuing doctorates in QIS is as critical as the availability of funding: without the supply of students who can work at all hours of the day and night, explore new ideas, and immerse themselves in new possibilities, money spent on basic research is frequently money wasted. One of the best ways to measure productivity of graduate students as a group is to count the number of dissertations and theses published each year.

We searched ProQuest Dissertation and Theses Global seeking QIS-related graduate research output[39] and found 10 242 results in March 2021.[40]

In examining graduate output over time, there is clearly a steadily increasing number of students training in QIS-related areas (Figure 9.2).

---

[38]Musiani et al., *The Turn to Infrastructure in Internet Governance* (2016).

[39]The search expression used was: (noft(quantum) AND (noft(compu*) OR noft(communic*) OR noft(sensor OR sensing) OR noft(entangle*) OR noft(superposition) OR noft(``cloning theorem'') OR noft(wave AND particle))). That is, the search was limited to the term quantum plus a technology or quantum effect, such as superposition appearing in the title, abstract, or keywords (full text was excluded).

[40]"ProQuest Dissertation and Theses Global is the world's most comprehensive curated collection of dissertations and theses from around the world, offering 5 million citations and 2.5 million full-text works from thousands of universities all over the world." ProQuest claims, "PQDT Global includes content from more than 3000 institutions all over the world." See "ProQuest Dissertations and Theses Global" (n.d.).

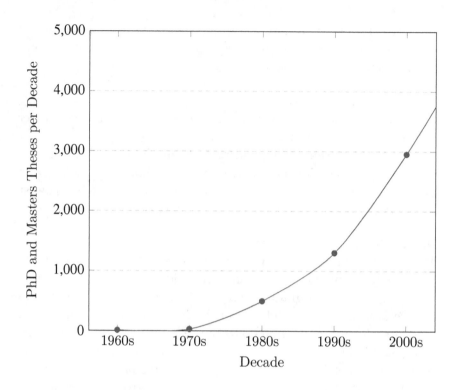

Figure 9.2. Graduate research output in QIS

ProQuest also produces subjects related to the graduate work. Here are the subjects associated with the corpus of quantum-related graduate output, as shown in Table 9.1. The disciplines represented also signal how difficult it would be to form a credible quantum information science academic department. Such a department would have to unify and ensure rigor amongst chemists, computer scientists, electrical engineers, and physicists just to cover the most popular disciplines in QIS represented with more than 150 works. Below that threshold, many other disciplines emerge, from astrophysics to information theory to music theory.

The ProQuest data also help us understand where graduate students are training. As suggested by Table 9.2, US institutions have a strong lead in QIS. Even work being performed outside the US is largely being written in English (Table 9.3). And while academic institutions broadly collaborate, they also compete fiercely; Table 9.4 indicates who is currently on top in the race for academic quantum superiority.

Table 9.1. Subjects associated with QIS graduate theses and dissertations (limited to subjects with more than 100 works)

Subject	Number of Works
Electrical Engineering	1591
Optics	1214
Quantum Physics	940
Physics	894
Condensed Matter Physics	836
Theoretical Physics	742
Atoms and Atomic Particles	720
Computer Science	682
Condensation	662
Chemistry	652
Materials Science	632
Particle Physics	568
Mathematics	463
Physical Chemistry	441
Nanotechnology	294
Inorganic Chemistry	202
Nanoscience	179
Molecules	176
Organic Chemistry	175
Analytical Chemistry	160
Nuclear Physics	152
Chemical Engineering	148
Biochemistry	137
Mechanical Engineering	137
Computer Engineering	136
Biophysics	132
Astronomy	128
Electromagnetics	124
Applied Mathematics	114
Astrophysics	111
Engineering	108
Total	13 650

Table 9.2. Nations and number of QIS theses and dissertations

Nation	Number of Works
United States	6494
England	1249
People's Republic of China	1053
Canada	536
Scotland	201
Sweden	88
Hong Kong	66
Northern Ireland	55
Germany	47
Finland	34
Wales	34
Ireland	29
Netherlands	26
Republic of Singapore	24
Switzerland	23
Total	9959

Table 9.3. Nations and number of QIS theses and dissertations

Language	Number of Works
English	8963
Chinese	1039
French	29
German	14
Spanish	4
Dutch	3
Polish	3
Afrikaans	1
Catalan	1
Finnish	1
Total	10058

Table 9.4. Institutions with more than 100 dissertations and theses published on QIS

Institution Name	Number of Works
Massachusetts Institute of Technology	253
University of California, Berkeley	225
University of Oxford	198
University of Illinois at Urbana-Champaign	176
Purdue University	165
University of California, Santa Barbara	159
Princeton University	156
University of Maryland, College Park	156
Harvard University	148
University of Cambridge	144
University of Toronto	138
Stanford University	121
Northwestern University	118
University of Michigan	117
Cornell University	111
California Institute of Technology	110
Tsinghua University	110
Imperial College London	109
The University of Texas at Austin	108
University of Rochester	105
University of Colorado at Boulder	103
The University of Wisconsin - Madison	101
Total	3131

We can derive several observations from these tables. First, research in quantum technologies is attracting attention in many nations and regions. Second, despite the strategic advantages made possible by quantum technologies, a healthy amount of research is being openly published. Indeed, nations and individual scientists are competing for prestige with their quantum research portfolios. Finally, while quantum publications are emerging from many nations, most graduate training in the field is in US institutions. All three of these observations should inform the policy discussion on industrial policy, immigration, and secrecy.

Education policy interacts with immigration policy. Many US graduate students in science and engineering fields hold temporary "student" visas. These students do not automatically qualify for permanent residence upon graduation under current US policy. Instead, the graduating students must return to their home country unless they can find an employer to sponsor the graduate for one of the limited number of H-1B visas. Such a policy might make sense for disciplines in which there is a surplus of graduates, such as PhDs in English or Art History, but seems short-sighted in science and technology – unless the purpose of the policy is to train students in the US and then send them home to seed high-tech hubs in China and India.

According to the National Center for Science and Engineering Statistics at the National Science Foundation, between 1999 and 2019 the number of doctorates granted in science and engineering fields rose from 25 997 in 1999 to 41 519 in 2019. At the same time, the number granted to temporary visa holders rose from 7500 (28.8 percent) to 15 801 (38.1 percent).[41]

In Computer Science, Computer Engineering, and Information Technology the numbers are even more lopsided. According to the 2019 Taulbee survey, 60.4 percent of the PhDs awarded in 2019 went to "nonresident alien students."[42] (For comparison, the survey found that only 13.2 percent of bachelor degrees were awarded to nonresident aliens.) Sadly, the Taulbee survey does not separately recognize quantum computing as a computer science specialization.

The Taulbee surveys tell us how many of these newly minted nonresident PhDs manage to stay in the US, or return to the US at some

---

[41]National Center for Science and Engineering Statistics, *Doctorate Recipients From US Universities* (2019).
[42]Zweben and Bizot, *2019 Taulbee Survey* (2019), p. 10.

later point, but it does give us an upper bound. The Taulbee survey asks the fields, economic sectors, and geographical areas where graduates get their first job, but only has data for 1362 of the 1860 graduates. Of those, 7.5 percent find their first job "outside North America." But given that employment type and location is unknown for 26.8 percent of the cohort, it is likely that many of these graduates couldn't be reached because they had already left the country. So as many as 34.3 percent may find their first job after graduating with a US doctorate in the service of the country's economic competitors.

9.3.2 The Human Capital Challenge

In 2015, the European Commission estimated that only 7000 people were working on QIS *worldwide*.[43] Presumably, if a quantum technology virtuous cycle takes hold, many more people will be needed to invent, research, design, program, test, market, and deploy quantum technologies.

The US can stay ahead on quantum technologies by investing in research, by preventing other, hostile countries from getting the technology through theft, sale, or rental (as in commercial cloud or satellite offerings), and by attracting the brightest minds from the world to work on quantum technologies for team USA. That is, solving the human capital challenge requires integration between education policy, export controls, and immigration policy.

Immigration is an important part of the human capital equation because the skills are in short supply, the time to create a quantum PhD, postdoc or assistant professor is long, and these people are highly sought after. Absent restrictive emigration policies, some human capital will flow between nations – both for research fellowships lasting a few years, and permanently.

One need only look at the biographies of those working on quantum projects to see that quantum information science is staffed with experts from around the world. The esoteric, multidisciplinary skillset and focus on difficult-to-grasp quantum mechanics concepts is a rare fit for job applicants.

In the US, uncharitable immigration laws combined with government policies that are increasingly hostile to aliens and immigrants have the potential to create a "brain drain"[44] that might push quantum scientists and engineers to countries such as Canada, Germany,

---

[43]Omar, "Workshop on Quantum Technologies and Industry" (2015).
[44]Moller, "How Anti-Immigrant Policies Thwart Scientific Discovery" (2019).

or the Netherlands. These countries heavily support quantum technology research and offer high-quality of living.

When we interviewed him about IBM's quantum computing research within the US, Dr. Robert Sutor, who was then the vice president for Q Strategy and Ecosystem at IBM Research, made it clear to us that there is no US strategy: there is a single IBM strategy, and it is international. "All we can really say there is that we have teams working on Quantum. If you look at the papers, you can follow the addresses. It's primarily in the US, in New York, in California at our Alamaden Lab, in Japan, in Switzerland, in Zurich. We do have a couple of people here and there, but everybody in the countries that I mentioned are working together," Sutor said.

Indeed, even within the US, he said, IBM's team is an international one. "More than half the people at IBM, at last count, are from outside the US We get people from all other countries."

One might think that China has the raw population numbers to find domestic talent that checks all the boxes. But even scientists in China rely on international collaborators. China's "father of quantum," Jian-Wei Pan, wrote to us that "Over the past decade, my laboratory in China has received more than 20 international students and visiting scholars from the United States, Canada, the United Kingdom, Germany and other countries...As a physicist who has been devoted to quantum information research for 20 years, I would like to emphasize that quantum information technology has a long way to go before it can be widely used. Active international cooperation and open exchanges are imperative."

We believe that nations that wish to succeed in quantum technology will be pushed to adopting liberal immigration policies that ease administrative burdens when it comes to short-term visits for conferences and other scientific and technical exchanges, medium-term visits lasting up to two years for extended bouts of collaboration, and easily obtainable residency for an indefinite period – what the US calls a "green card." The human capital market will select against countries with more restrictive policies.

9.3.3 Faculty Research Incentives

The intricate engineering and resource intensity of building a quantum device is significant. Some scientists we spoke with signaled that their full ambitions were difficult to realize because the need to spend time building a device competed with teaching, service, and even publication expectations.

In fact, part of the requirements for building quantum devices seems to be the creation of intermediate steps that provide publication opportunities. In the National Science Foundation's 2019 workshop on quantum simulation, for instance, a consensus statement valorized the approach of creating experimental simulators that in themselves were worthy of study.[45] The timeline suggested would keep faculty publications coming as expected.

Universities are in competition with private companies and research labs to make discoveries in QIS. In fact, universities are in competition with their own faculty, in a way, because so many faculty form private companies to supplement their basic science work free from institutional red tape, to spend money while avoiding rules and competitive bidding requirements, to hire and keep their brightest students, and of course to make more money. Universities might benefit from creating more research professorships to give faculty time to develop quantum devices free from other responsibilities. Universities should also have policies that discourage or prohibit faculty from hiring students prior to the student's graduation, as such business relations between faculty and their students present many opportunities for conflicts of interest. (For example, MIT's Policies and Procedures generally prohibit faculty from hiring their students at the faculty's startup, for example.[46])

A separate question concerns whether educational institutions should create quantum information science departments. Table 9.5 demonstrates why department creation is a challenge: quantum technologies draw from so many different, well-established disciplines that unifying them in a single department presents quality and rigor-control challenges. Theoretical physicists, for instance, might not feel prepared to evaluate colleagues from materials sciences or applied science fields and vice versa. This disciplinary diversity explains why so many institutions have pursued academic "center" models that leave faculty in their home departments while providing support for collaboration across relevant fields.

---

[45]Altman et al., "Quantum Simulators: Architectures and Opportunities" (2019).
[46]MIT, "Outside Professional Activities" (2018).

Table 9.5. Fields associated with quantum technology

Field	Number of Papers
Optics	3780
Physics Multidisciplinary	3737
Physics Applied	2297
Physics Atomic Molecular Chemical	2182
Engineering Electrical Electronic	1873
Computer Science Theory Methods	1527
Physics Mathematical	1314
Quantum Science Technology	1261
Materials Science Multidisciplinary	1202
Physics Condensed Matter	1168
Multidisciplinary Sciences	1079
Computer Science Information Systems	597
Nanoscience Nanotechnology	585
Telecommunications	476
Physics Particles Fields	446
Chemistry Physical	429
Computer Science Artificial Intelligence	412
Chemistry Multidisciplinary	377
Computer Science Hardware Architecture	360
Computer Science Interdisciplinary Applications	269
Computer Science Software Engineering	244
Mathematics Applied	242
Automation Control Systems	158
Mathematics	135
Engineering Multidisciplinary	100
Total	26 250

Education Pipelines

Over the longer term, the US and other nations would be wise to build in quantum physics to grade-school curricula. Such an approach could both grow the number of students exposed to quantum physics and help diversify potential candidate pools for the workforce. In 2020, the National Science Foundation and the White House Office of Science and Technology Policy created a partnership anchored at University of Illinois Urbana-Champaign and University of Chicago to promote K–12 education (see the sidebar "Key QIS Concepts for K–12 Students" on page 412). Called Q2Work, the group will develop online educational material and modules for in-person learning, presumably so that these will diffuse to school systems. The partnership includes participation from big players in quantum computing, including Google, IBM, Microsoft; DIB companies Boeing and Lockheed Martin; and startups Rigetti and Zapata.

Q2Work builds upon a NSF workshop that defined key quantum information science concepts to be taught in schools. The workshop output, a high-level, five-page summary, "Key Concepts for Future Quantum Information Science Learners," reflected input from leading QIS researchers, and teachers and officials from public and private schools. We note in Appendix A that without training, people may be familiar with how everyday objects behave, but will have little intuition about how angstrom-sized objects behave. Education in the K–12 years could start developing that intuition. Yet, basic questions about QIS education in schools are still unanswered. For instance, what learning goals are appropriate for grade and secondary school students? What do we expect the average student to be able to do with the knowledge? What advantages and risks come from reforming education so that it is QIS-first, for instance, by teaching quantum mechanics before classical mechanics?

9.4 National Security and Quantum Technologies

Quantum technologies can give nations strategic advantages. This section focuses on how nations might consider the advantages and disadvantages of export control and other tools to hinder adversaries' development of quantum technology. The section then turns to other limits and dynamics implicated by quantum technologies: the effect on nation-state competition in space and in cyberspace.

Key QIS Concepts for K–12 Students

A March 2020 NSTC/NSF workshop produced the following high-level concepts for teaching QIS in K–12 schools.[a]

1. Quantum information science (QIS) exploits quantum principles to transform how information is acquired, encoded, manipulated, and applied.

2. A quantum state is a mathematical representation of a physical system, such as an atom, and provides the basis for processing quantum information.

3. Quantum applications are designed to carefully manipulate fragile quantum systems without observation to increase the probability that the final measurement will provide the intended result.

4. The quantum bit, or qubit, is the fundamental unit of quantum information.

5. Entanglement, an inseparable relationship between multiple qubits, is a key property of quantum systems necessary for obtaining a quantum advantage in most QIS applications.

6. For quantum information applications to be successfully completed, fragile quantum states must be preserved, or kept coherent.

7. Quantum computers, which use qubits and quantum operations, will solve certain complex computational problems more efficiently than classical computers.

8. Quantum communication uses entanglement or a transmission channel. to transfer quantum information between different locations.

9. Quantum sensing uses quantum states to detect and measure physical properties with the highest precision allowed by quantum mechanics.

---

[a]Alpert, Edwards, and Freericks, *Key Concepts for Future QIS Learners* (2020).

9.4.1 Export Controls

According to the US International Trade Administration, "The United States imposes export controls to protect national security interests and promote foreign policy objectives. The US also participates in various multilateral export control regimes to prevent the proliferation of weapons of mass destruction and prevent destabilizing accumulations of conventional weapons and related material."[47] US export controls are administered by the Bureau of Industry and Security (BIS) within the US Department of Commerce.

Export controls and other approaches for preventing the spread of advanced technology can be effective in the short term, but in the long term they can inadvertently create independent foreign tech ecosystems that are resistant to any controls. Three illustrative cases are the US Global Positioning System (GPS), the US attempts to regulate the export of cryptographic technology, and the proliferation of nuclear weapons.

GPS

Originally developed by the US military, for military purposes, at an inflation-adjusted cost of $14 billion, the Global Positioning System (GPS) is now available to the public freely.[48] Over the course of two decades, the US launched the GPS constellation, with Europe following with the Galileo network, Russia with GLONASS, the Japanese with the Quasi-Zenith Satellite System, which enhances the resolution of the US system, and India with the Indian Regional Navigation Satellite System (IRNSS).

Reflecting US concern that a high-precision location service might be used by its enemies, the original GPS system had two tiers of service. The US military received an encrypted, highly accurate service. The unencrypted service had noise intentionally added, a practice that the US called "selective availability." Industry found ways around selective availability, and the lower quality helped spur interest in the Russian and European alternatives. In response, President Clinton ended selective availability in 1990, meaning that civilians can reliably obtain a signal accurate within 4 m, with the military and other users obtaining greater accuracy through capturing more signals or by augmenting the GPS data. Unencumbered civilian use

---

[47]International Trade Administration, "US Export Controls" (2021).
[48]Posen, "Command of The Commons: The Military Foundation of US Hegemony" (2003).

of GPS has contributed to unimaginable benefits and exciting innovations.

The Internet has had a similar founding although a more complex path to commercialization that nonetheless has transformed our
economy.[49] American companies dominate the Internet in important
sectors, even overseas, where usage rates of Google Search exceed
those of domestic competitors created to fend off the American company. The situation is different in China, where direct blocks on
American internet services combined with more significant language
differences allowed the country to develop its own domestic internet
ecosystem.

Quantum Technologies and Export Control

Should quantum technologies, to the extent it is possible, be open
for similar public use and extension? This question relates to the
above-discussed industrial policy issues. Industrial policy often seeks
to benefit domestic companies, in an attempt to reach technological
sovereignty. If quantum technologies are sufficiently open, no one
country is likely to dominate the field.

In the US, several quantum technologies, particularly quantum
sensors, and their precursors are already subject to export controls.[50]
Under the Trump administration, the US retained a market proscription posture, and funding models that make it easier for the
government to restrict openness of research outputs. In November
2018, the Department of Commerce's Bureau of Industry and Security released an advance notice of proposed rulemaking seeking comment on whether a broad series of technologies should be considered
for export control under the Export Control Reform Act of 2018.[51]
This initial regulatory exploration suggested that quantum sensing,

---

[49]Clark, *Designing an Internet* (2019).

[50]The US has traditionally followed a policy making applied research subject to
more restrictions than basic research. "It is the policy of this Administration
that, to the maximum extent possible, the products of fundamental research remain unrestricted." "'Fundamental research' means basic and applied research in
science and engineering, the results of which ordinarily are published and shared
broadly within the scientific community, as distinguished from proprietary research and from industrial development, design, production, and product utiliza-
tion, the results of which ordinarily are restricted for proprietary or national
security reasons." National Security Decision Directive 189 (1985).

[51]Department of Commerce, Bureau of Industry and Security, "Review of Controls
for Certain Emerging Technologies" (2018).

computing and encryption are "foundational technologies," indicating that they are "emerging technologies that are essential to US national security, for example because they have potential conventional weapons, intelligence collection, weapons of mass destruction, or terrorist applications or could provide the United States with a qualitative military or intelligence advantage."

The Department of Commerce sought how to define and thus bound the definition of quantum technologies so that identifiable products could be included on an export control list. Initial reporting suggested a narrow set of restrictions, yet one technology identified as possibly controlled is the "quantum diluted refrigerator," a device used to supercool some quantum devices with helium (see the sidebar "The Helium Challenge" on page 251).[52] For this reason, national competitors may be dependent on foreign makers of low-temperature devices. Companies such as Cryomech (New York based), Sumitomo (Japan), Oxford Instruments (UK), and Bluefors Oy (Finland) all offer helium coolers, while some competitors offer low-kelvin devices that do not use a cryogen (a cooling agent such as liquid helium or liquid nitrogen). Presumably export control of dilution refrigerator devices will hinder China and Russia in their efforts. Yet, competitor nations can build their own domestic cryogenic industries, or rely on devices already circulating in the market. As early as 2012, the Cryogenic Society of America claimed on its website that "Dilution refrigerators are a common technique for reaching temperatures below 1 K ... reliable dilution refrigerators are in fact a commercial product and can be purchased as turnkey systems from vendors." IBM is creating its own custom supercooling device in anticipation of building a large superconducting machine. If a single private company can build a cooler, it would seem not to be much of a challenge for other nations.

European governments generally approach quantum technologies as something that should be relatively open. The €1 billion European initiative to promote quantum technologies explicitly embraces openness, calling for "end-user-inspired applications" in quantum networks and inclusion of quantum random-number-generation-based encryption in even "cheap devices."[53] The European posture sug-

---

[52] Alper, "US Finalizing Rules to Limit Sensitive Tech Exports to China, Others" (2019).

[53] European Commission, High Level Steering Committee, DG Connect, "Quantum Technologies Flagship Final Report" (2017a).

gests support for an end-to-end quantum internet for the average person to use. This anti-surveillance interest also aligns with a series of high court opinions in Europe that object to intelligence gathering on European citizens by American agencies.

It is unclear what posture China will take toward openness of quantum technologies. Chinese scientists are publishing their work in top journals and are genuinely interested in engagement. However, national competition between the US and China has led both companies to discuss and implement economic *decoupling* policies, that is, deliberate strategies to separate technology supply chains from other nations. For instance, US policymakers have made a priority of removing China-made Huawei equipment from domestic and even foreign telecommunications networks. At the same time, China is creating domestic industries, such as helium capture plants, to address gaps left from decoupling.

At the moment, it would seem that both the US and China would lose in a decoupling scenario. US domestic manufacturers of quantum components and optics sell their wares to a large foreign market. For instance, examining Jian-Wei Pan's Jiuzhang device reveals it to have an astonishing number of components from US-based Thor-Labs and from Israel-based Raicol Crystals (see Section 6.7, p. 250)). America will lose out on those high-precision manufacturing sales as China in-sources technology manufacturing. Conversely, as decoupling intensifies, we should expect more explicit export control to prevent Chinese-developed and -manufactured technologies from diffusing into the US and Europe.

This discussion makes it clear that rather than asking whether governments should export-control innovations in quantum technologies, one should begin by considering whether it is even possible. Imposing export controls will have different implications for our categories of quantum technologies. In metrology, interferometry is already widely dispersed, indeed many of its applications were demonstrated by European investigators. Jian-Wei Pan's Jiuzhang quantum computer is a masterful implementation of interferometry (see Section 6.6, p. 243). Some sensing technologies can be miniaturized in part because they lack supercooling requirements, thus making controls practically more difficult. Quantum computing technologies, on the other hand, rely upon expensive, complex and sensitive hardware/software ensembles that are more readily controlled. Miniaturization is unlikely in quantum computing in the near future.

Adding to the market proscription complexity is that private companies play lead roles in quantum communication and computing development. Yet, there are ways to bring private companies into the fold and make it difficult for them to diffuse discoveries to potential adversaries. The Trump administration strategy was to encourage private sector participation, including financial outlays from the private sector, with government research money vested in Department of Energy Labs. In August 2020, the Trump administration allocated over $600 million in funding to five national labs, with over $300 million in commitments from academic and industry companies. These private-sector partners include many of the recognized leaders, including IBM, Microsoft, Intel, Lockheed Martin, and Rigetti. Notably absent is Google, and its absence is not for a lack of merit. Google and other companies may be avoiding government entanglement so as to keep its inventions in the public sphere.

The Energy–labs centered approach signaled that the Trump administration was taking a market prescription strategy, by funding companies lavishly and aligning incentives to keep the technology restricted to domestic actors. This has elements of the longtime domestic defense firm practice of "paternalistic socialism."[54] Interestingly however, this strategy is limited in efficacy. Despite efforts to keep domestic aerospace firms well sated, these same firms often pay large fines for export violations.

The capture of industry through the military embrace approach is becoming more complex with the rise of the power of the private sector. Most quantum technology companies are located in liberal, Western democracies, and many already have military funding in the form of leased computer time or purchases of devices, or they are angling for it (for instance, by having former high-level military officials on their boards).[55] Many technology companies are depen-

---

[54]Paternalistic socialism is where the government spreads money around several competitors to ensure that America has multiple options for companies to hire for projects. Rich and Janos, *Skunk Works: a Personal Memoir of My Years at Lockheed* (1994). Particularly in aerospace, the need for government patronage of the private sector is explicit: "the development in the United States of a dynamic and innovative private-sector space industry will be indispensable to future US space leadership." Independent Working Group on Missile Defense, *Missile Defense, The Space Relationship, and The Twenty-First Century: 2009 Report* (2009).

[55]Rigetti Computing's board features three PhDs, the obligatory representative from a venture capital funder, and a former chair of the Joint Chiefs. ColdQuanta

dent on military investment; some seem to abhor this investment. For instance, in 2018, Google employees objected to "Project Maven," an effort to improve the object recognition capabilities of the Department of Defense.[56] Google is widely agreed to be among the leading companies in the quantum computer research space. Will its employees forgo military markets for quantum technologies, many of which have no other obvious buyer than governments? Google's closest rivals in the quantum technology space, IBM and Microsoft,[57] both have extensive government consulting practices and are unlikely to turn away from military and intelligence services.

Theft is an additional complexifier. Nations that follow others in technical might can develop their own quantum programs, but it is probably easier to copy the leader. Cybersecurity vulnerabilities are among the newest ways that competing nations have lifted secrets from American companies, and in some instances, companies have lost huge portions of their intellectual property portfolios to attackers. There is no reason to believe this will not continue. In academia as well, thefts of secrets occur, but also bribery which is masked as scholarly accolades. The Chinese government in particular has bought access to American scientists through its Thousand Talents programs, where faculty members receive what appear to be prestigious honors (often accompanied by money) for collaboration with Chinese institutions. In recent years, faculty members have been targets of criminal prosecutions for pursuing these relationships while not disclosing "honoraria" to their own institutions and the US government.

Tools for Controlling Quantum Technology Proliferation

The US and other nations have several tools to block diffusion of technology. For inventors seeking a patent, the government has a broad power to impose secrecy on the invention, even if the inventor

---

has a strategic board with former officials from several intelligence agencies.

[56]Unnamed Google Employees, n.d. Project Maven had clear implications for the unmanned air vehicle program and for weaponry that needs to make target distinction decisions in situations where humans cannot. But a deeper problem with the employee objections is that all of Google's commercially focused computer vision and artificial intelligence research can contribute to military objectives; the technologies are inherently dual use. It is unclear how Google will ever comply with these employees' demand to never "build warfare technology" when the root of so much of Google's discoveries is easily deployed for ISR or offensive purposes.

[57]B. Smith, "Technology and The US Military" (2018).

is a private person. Outside the patent system, government can use export controls to bar sales and services.

Patent secrecy may be an attractive option to prevent diffusion of quantum technologies, Under the Invention Secrecy Act, the federal government has broad powers to force secrecy of an invention if its publication is "detrimental to the national security."[58] The Federation of American Scientists tracks secrecy orders under the law, and finds that almost 6000 patents are subject to secrecy orders. Most of these pertain to government-funded inventions, but in any given year, a few dozen "John Doe" secrecy orders are imposed on private citizens or companies that independently sought patent rights in a sensitive technology. The Act provides for criminal and civil penalties, and those who disclose the secret patent "abandon" it under the statute, thereby losing any economic benefits of the invention.

One might think that patent secrecy orders primarily deal with nuclear bomb-making plans and the like,[59] but the scope of inventions that could be detrimental to national security is seen as much broader. The Federation of American Scientists' Steven Aftergood has obtained summary statistics and identifiers of formerly secret patents. Conventional weapons building and targeting systems appear in many formerly secret patents. Patent secrecy orders concern stealth aircraft countermeasures, radar resilience, anti-radar technologies, and encryption. Quantum technologies will likely contribute to these same fields, making quantum technologies likely targets of secrecy orders.[60]

But what about sensitive, non-nuclear technologies that are sold directly as goods or as services? The government has three primary controls for such technologies. These controls can be focused on technologies, individual firms, and nation states.

The Department of Commerce's Bureau of Industry and Security owns the Export Administration Regulations (EAR), which focus on

---

[58]Secrecy of certain inventions and withholding of patent, 35 USC § 181. Consulted agencies include the Department of Defense, Department of Justice, NASA, Department of Energy, and the Department of Homeland Security.

[59]A separate provision of the US Code creates criminal penalties for disclosure of atomic weapons design-and-manufacture information if the person has "reason to believe such data will be utilized to injure the United States." This is the "born secret" provision of US law, 42 USC § 2274.

[60]If a secrecy order is rescinded, a patent does not explicitly state that it was subject to an order. However, secret patents sometimes have a filing date that precedes an issuing date by decades, hinting that it was subject to suppression.

control over export of dual-use technologies. Dual-use technologies are those that have both commercial and military uses, and these are broadly defined to include commodities but also software. Thus, allowing a download of software, even in the US, to a foreign person could be an "export." The Department of Commerce's Commerce Control List (CCL) identifies a lengthy list of controlled technologies; those listed cannot be transferred to or through certain countries without a license.

Several quantum technologies are explicitly identified in the CCL, including superconducting quantum interference devices (SQUIDs) of a certain resolution, gravimeters, quantum wells, quantum cryptography, and post-quantum cryptography. The CCL also identifies precursors to quantum computing, encryption, and sensing technologies to stop their spread to designated nations.

The Department of State oversees the International Traffic in Arms Regulations (ITAR), which blocks the transfer of military-related technologies, and information about their design, to non-US persons. The transfer carries civil and criminal penalties, on a strict liability basis (many violations of the regime do not require ill intent). Almost all the dominant US defense firms have paid fines or settlements for ITAR violations, and these are large, often in the eight-figure range.

Keying a violation on transfer to non-US persons means that sharing technical data, even inside the country, can be a violation if the recipient is a foreigner. This means that foreign (defined as people lacking permanent residence) graduate students and employees have to be excluded from ITAR-regulated projects (absent special permission). ITAR does not apply to public domain information, which includes research performed at universities that is intended for publication. This would seem to be a large loophole that gives researchers significant freedom. However, as explained above, only a small amount of research in quantum technologies is funded by private foundations. Most flows through the NSF, Department of Energy Laboratories, and a panoply of Department of Defense agencies that can condition work on these sponsored projects to be in compliance with ITAR.

A wide set of technologies related to quantum sensing and communication fit under ITAR's "United States Munitions List," an enumeration of technologies that is now over 33 000 words in length. Many quantum technologies fall under the current munitions enumeration because the broad categories include sonar and radar technologies, quantum clocks, gravimeters, communications systems that are difficult to intercept, cryptographic and cryptanalytic systems, and computer systems for modeling weapons.

Companies need to carefully monitor ITAR restrictions to understand the rules for technologies that really can only be made in America. Policymakers too need to monitor the commercial landscape because if foreign firms can create quantum technologies and are willing to sell them to designated nations, ITAR restrictions make the US less competitive. The most recent example of this phenomenon came in satellite technologies, where ITAR restrictions on US firms enabled foreign companies to capture a significant share of the space market.[61]

Finally, under federal law, the President has a sweeping power to declare emergencies in peacetime that, in turn, enable declaration of sanctions and other interventions to shape economic activity.[62] Over two dozen such emergencies are currently declared, with some identifying broadly scoped, potentially worldwide emergencies, such as weapons proliferation, transnational criminal activity, and the scourge of cyber-related intrusions and influence. The Department of the Treasury's Office of Foreign Assets Control (OFAC) oversees the primary mechanism used to block economic transactions under the declared emergency. This agency is charged with enforcing trade sanctions and other international relations policy positions through economic deterrence.

OFAC does so through the Specially Designated Nationals and Blocked Persons List (SDN). US persons, companies, and, perhaps most importantly, banks, are prohibited from engaging in transactions with any entity in the database. Because of the network effects and surveillance power in international banking,[63] being designated effectively locks sanctioned entities out of mainstream value transfer mechanisms and other businesses.[64] The SDN database is now

---

[61]Zelnio, "The Effects of Export Control on The Space Industry" (2006).

[62]50 USC. §§ 1701 et seq.

[63]Farrell and Newman, "Weaponized Interdependence: How Global Economic Networks Shape State Coercion" (2019).

[64]Some wily actors find ways of buying goods despite being designated. For a fantastic case study of SDN evasion focusing on North Korea and Kim Jong-un's acquisition of an armored Mercedes-Maybach S600 Guard, see Kuo and Arterburn, *Lux and Loaded: Exposing North Korea's Strategic Procurement Networks* (2009).

sprawling. It is used to enforce over 60 trade sanction or policy regimes, including to punish Russians involved in hacking the US elections. The SDN is over 1400 pages long and contains the name Muhammad over 3800 times. Suffice it to say that as a general matter, no quantum technology can be sold to any entity on the list. But more broadly, if quantum technologies are associated with weapons proliferation, for instance the use of quantum computing to simulate more effective biological and chemical agents, the SDN is another tool the government can use to block relevant entities, nations, and people from transactions.

9.4.2 Quantum Technology and Space Law

The seminal Outer Space Treaty of 1967 declares that the use of space will be "for the benefit and in the interests of all countries..." and "exclusively for peaceful purposes." The Treaty further prohibits stationing any weapon of mass destruction in space. But despite that proscription and affirmative obligation for peaceful purposes, nation states have many options for using force in space.

The US military sees the U.N. Charter's inherent right to self-defense language as limiting the exclusively peaceful purposes language. And once the door to self-defense is opened, many "defensive" preparations resemble offensive ones.[65]

There are other loopholes allowing weaponization as well. As Jeremy Rabkin and John Yoo explain in their book analyzing next-generation weaponry and conflict, the treaty does not prohibit ICBMs, as they are not installed in space but rather pass through it.[66] Nor does the treaty explicitly ban intelligence and surveillance activities,[67] even those that support or enhance force in conflict. The treaty

---

[65] A fascinating 2002 study by RAND signals the US government's interest in and options for space weapons. Celestial weapons are attractive in part because they give nations the ability to attack anywhere on Earth without pesky complications of weather and troop deployment and supply chain concerns. See RAND, *Space Weapons: Earth Wars* (2002).

[66] Rabkin and John Yoo, *Striking Power: How Cyber, Robots, and Space Weapons Change The Rules for War* (2017).

[67] United Nations, *Principles Relating to Remote Sensing of The Earth From Outer Space* (1986). The affirmative command of "peaceful purposes" creates ambiguity. A subsequently enacted UN statement broadly allows remote sensing in space, but does not mention surveillance and defines remote sensing as observation performed for environmental purposes. Consider that a launch-monitoring satellite is key to waging war, but at the same time provides monitoring essential for nuclear peace.

has not stopped the advance of anti-satellite weapons, including by China[68] and India.[69]

Quantum technologies' utility in outer space is evident. Companies angling for government contracts have often appointed board members and advisors with former leadership roles in Department of Defense agencies with a space focus, such as the National Geospatial-Intelligence Agency (NGA) and the National Reconnaissance Office (NRO). As MASINT becomes more important, NGA and NRO will be key agencies for deployment of quantum technologies.

Quantum technologies also appear to have even more leeway than other military-related activities in space. Even when used in a force-enhancing role, quantum technologies in no way trigger the traditional concerns of weapons regulation, which are indiscriminate or superfluous injury, or of widespread, permanent environmental damage.[70] In fact, these technologies might be de-escalatory, in that they help nations understand adversaries through better intelligence, and in conflict, they may enable more discriminate applications of force.

Quantum technologies may be lawful in space, but they still could change adversaries' strategies. Nations may find it compelling, even necessary, to make first strikes at space-based vessels to silence or blind the handful of superpowers that have both a space program and quantum technology. If *jus ad bellum* requirements (the rules for initiating armed conflict) or rules for engaging in self-defense, are met, it would seem that *jus in bello* considerations (the rules for the actual waging of war) might mitigate in favor of striking at space-faring platforms. This is because targeting satellites could be justified as a discriminate attack on military infrastructure and that does not directly harm people, thus minimizing human suffering, in the sense of injury and death.

Nevertheless, the psychological harm from a satellite attack could be substantial. People, particularly in developed nations so dependent on communications, may panic as uncertainty deepens with normally chatty devices going mute. Another side effect, analogous to the long-term, serious destruction of habitat, may be discounted: attacking space vessels can create clouds of space junk that remain in orbit for years, endangering all space programs.[71]

---

[68]Kan, *China's Anti-Satellite Weapon Test* (2007).
[69]Brumfiel, "India Claims Successful Test Of Anti-Satellite Weapon" (2019).
[70]Boothby, "Space Weapons and The Law" (2017).
[71]Zissis, "China's Anti-Satellite Test" (2007).

Quantum sensing could be so powerful that a national policy of parallel contingent restraint is appropriate. That is, nations may find it expedient to voluntarily limit where and when quantum sensing is deployed so long as others do so as well. In some cases, superpowers have refrained from developing technologies and in militarizing spaces because of the inherent destabilizing or weapons-race effects they can have. For instance, at times, superpowers have refrained from creating anti-ICBM defenses, for fear that their very presence could change the game theory of nuclear strikes and be escalatory. Turning to terrestrial forbearance, the Antarctic Treaty System prohibits militarization (both offensive and defensive uses) in Antarctica, making it more strictly regulated than outer space.

Generally speaking, intelligence systems are seen by policymakers as providing more context and information to adversaries, and thus, traditionally, espionage has been a tolerated activity of statecraft.[72] As uncomfortable as intelligence systems may make us feel, we have to contemplate that they can make us safer.

9.4.3 Quantum Technology and Cybersecurity

In his discussion of designing a next-generation internet, David Clark recounted how early internet designers relied upon contacts within the intelligence community to model security threats. According to Clark, two salient principles emerged: that endpoints should be the focus of security (because it was hopeless to provide security for the voluminous infrastructure between endpoints), and that endpoint security had to resist nation-state-level determination and ingenuity. The result of these emphases is that there is little trust for confidentiality and integrity "in the network."[73] As a result of this architecture, one does not know whether internet intermediaries are trustworthy, whether they relay information faithfully, or whether they copy or alter data for their own purposes. We use encryption to reduce the risks of intermediary opportunism. Yet, intermediaries

---

[72]German Chancellor Angela Merkel provides an example of this ambivalence. After documents were released purporting that the US NSA had intercepted her wireless phone conversations, Merkel allowed herself to be photographed holding her phone aloft, in a kind of protest. Less well known is that behind the scenes, Germany has been clamoring to join the US "Five Eyes" partnership with Australia, Canada, New Zealand, and the UK. See Spiegel, "Angela Merkel Eyes Place for Germany in US Intelligence Club" (2013). A follow-up investigation found no evidence that the NSA had targeted her phone.

[73]Clark, *Designing an Internet* (2019).

can still infer the meaning of messages from monitoring metadata. One might address these problems by routing information differently, but the classical internet makes this difficult. Could quantum networks change the game theory of surveillance?[74] Recall that quantum technologies change communications in two ways: first, quantum key distribution makes it possible to enjoy communications confidentiality and integrity that is invulnerable even to a quantum computing attack. But that is not so different than the situation today, with proper AES or post-quantum encryption. Content is protected, while metadata can be observed.

The second quantum communications change is more consequential: a quantum-entangled communication network would enjoy full end-to-end quantum encryption, meaning that interception (whether by spies or by natural events that interfere with the transmission) will be apparent. In essence, a quantum internet gives its users no need to rely on fraught network trust. How might governments react to that?

One could imagine that governments will double-down on interception, perhaps in the form of creating noisome interference that blocks photonic communication. Having an eavesdropper present could deny communicants the ability to establish a secure session because "listening" would interfere with the quantum states. Eavesdropping might also have a signaling function that has utility in a "defend forward" security posture, one characterized by penetration into third party networks.[75] Currently, such eavesdropping on networks is easy because internet traffic is both copied multiple times and is routed circuitously, sometimes leaving national boundaries, which has legal consequences for its protection.[76] Unless the current infrastructure of the Internet changes, nation states will have many opportunities to physically access fiber optic cables and "listen," even if they cannot understand what is being sent.

On the other hand, QIS could also make the very design of the Internet change, such that the network is more resilient against interception. One could imagine an investment in quantum entangled networks coming with careful planning surrounding the routing of the fiber, and security measures for it. Rather than implement the

---

[74]Tambe, *Security and Game Theory: Algorithms, Deployed Systems, Lessons Learned* (2012).

[75]Springer, *Cyber Warfare: A Documentary and Reference Guide* (2020).

[76]Kerr, "The Fourth Amendment and The Global Internet" (2015).

system in existing fiber used by others, one could foresee a factionalization of networks, with nation-state controlled, central trunks, much like China's Beijing to Shanghai fiber network.[77] For regions such as the EU and countries like Russia and China, the promise of an interception-resistant channel might make it worthwhile to reroute the physical layer so that it is more controlled and so that one might choose the paths that important data take to avoid likely interception points. Still, if these routes are not defended, nation states might dig up fiber lines and place devices that interfere with quantum states.

Another, likely approach to the hardening of network privacy is to erode endpoint security.[78] That is, to discover ways to degrade the security of end users' devices. As discussed in Chapter 8, even if communications links are perfected and users adopt quantum encryption for their local data, data has to be unscrambled for people to use it. Intelligence and law enforcement agencies that gain control of endpoints through faked software upgrades or other exploits will be able to see all data stored on them. Another network-hardened scenario is that the future of cyberattacks becomes physical, in the sense that spies or crooks simply steal devices from targets at gunpoint. They will ask you to unlock your phone before leaving.

9.5 Quantum Technology and Privacy

Privacy rules, which take the form of constitutional rights, statutory limitations (from the many different sections of the US Code from the criminal law to evidence rules), administrative regulations, to social and business norms, might blunt the kinds of transparency that quantum technologies will provide. This section discusses how we might arrange privacy rules to prevent a quantum technology privacy meltdown.

Military and intelligence technologies tend to devolve to law enforcement and proliferate to nongovernmental actors.[79] Law and custom provide few limits on the kinds of technologies that even local law enforcement can obtain. Recent examples include "eye in the sky" monitoring that can provide moment-by-moment surveillance

---

[77]Liao et al., "Satellite-Relayed Intercontinental Quantum Network" (2018).

[78]Kadrich, *Endpoint Security* (2007).

[79]Consider the scenario of the "GEOINT Singularity," conceived as "the convergence, and interrelated use, of capabilities in artificial intelligence, satellite-based imagery, and global connectivity, where the general population would have real-time access to ubiquitous intelligence analysis." Koller, *The Future of Ubiquitous, Realtime Intelligence: A GEOINT Singularity* (2019).

of entire cities, cell-phone-hijacking "Stingray" devices, encryption-circumventing device forensics platforms, and malware that collects secret information from users.

Over time, invasive monitoring equipment finds its way into the private sector as well. A 2017 Rand Corporation market analysis of surveillance systems relying only on unclassified sources found "examples of SIGINT capabilities outside of government that are available to anyone [with applications in] maritime domain awareness; radio frequency (RF) spectrum mapping; eavesdropping, jamming, and hijacking of satellite communications; and cyber surveillance."[80] Such technologies are used by private investigators, stalkers, and employers that tend to see themselves as having a kind of dominion over workers similar to that of parents over children.

We should be prepared for a similar devolution of quantum technology. Military and intelligence agencies are likely to lead the deployment of these technologies. But with time, the same technicians that build, operate, and provide service for military and intelligence actors will naturally cross over to federal law enforcement agencies. Joint federal–local activities will further diffuse quantum technologies. Incentives to grow the marketplace will naturally cause quantum technology companies to find commercial and employment-related uses. Before long, we will have to ask what is to stop the average person from looking into the home of their neighbor. In most people's minds, technical might makes actions right. How can we create norms now to prevent a new era of forced transparency?

9.5.1 Secrets and Their Time Value

All individuals and institutions have secrets. Most of these secrets are only valuable for a limited time. For instance, business strategies might be relevant for a few years, the secret sauce of an invention may only be valuable until competitors figure out how to copy it, and the encryption on entertainment media might only need to be strong enough to protect the movie or music as long as people are willing to pay to enjoy it. Immutable personal facts, such as one's Social Security Number, might need protection for a lifetime.

Turning to secrets of the United States, policy dictates periods of protection for government materials. The Obama administration set

---

[80]Weinbaum et al., *SIGINT for Anyone: The Growing Availability of Signals Intelligence in The Public Domain* (2017).

a policy of automatic declassification of agency documents.[81] Many records will be declassified after 25 years, but the policy also envisions longer periods of protection for certain sensitive documents, keyed at classification lengths of more than 50 and more than 75 years. Outside the intelligence field, other secrets are time-limited. Most notably, the US Census keeps individually-identifying information secret for 72 years, meaning that in 2022 the 1950 Census records will be released.[82]

These dates give us some guidance for how we might think about the protections for encrypted data and when the things we write or keep today will lose their sensitivity. Again, if a large quantum computer is built, economics dictates that most owners of the device will make more money synthesizing chemicals and materials than cracking old messages. But cryptanalysis is a real risk among governments, which will carefully task the highest-value keys in their attacks. Owners of sensitive information must consider the time value of data, along with the proposition that the first quantum computers will be large machines owned by large companies and governments, but over time, the technology will shrink, become less expensive, and be democratized. These risks are unlikely to be realized in the next decade, but 20 to 50 years from now, quantum cryptanalysis could be a much larger risk.

9.5.2 Regulation of Decryption

On first blush, it might sound preposterous, but policymakers could weigh a simple prohibition on decryption of others' data. Such a prohibition would not be futile because of the affordances of quantum technologies. To start with, practically speaking, because quantum computers are so expensive to build and maintain, the technology will not be democratically distributed for some time. This gives regulators the opportunity to police a few big players, some of which will want to avoid the negative reputational taint of being linked to decryption efforts. There are economic constraints too. Companies will want to capture profits from the devices, and there will be

---

[81]President Barack Obama, "Classified National Security Information, E.O. 13526" (CFR2010).

[82]In the meantime, to maintain its confidentiality duties, the US Census releases datasets processed in some way to prevent reidentification of individuals in the enumeration. Similarly, many governments release datasets under the assumption that the data cannot be tied to particular individuals.

more money to be made in drug discovery and similar efforts than cybercrime or descrambling decades-old prescription records.

Of course, this argument will not be true with respect to all government agencies and their contractors. Public sector quantum computing users will have to be policed in other ways – through constitutional tort and political oversight.

Protections for Encryption

Avoiding a new era of eroding lines between personal and public space requires revisiting the capabilities of quantum technologies. Two broad areas of concern are present: attacks on widely used encryption and the different ways quantum sensing will give institutions powers to perceive phenomena in new ways.

In the encryption threat scenario, recall that quantum computing will degrade (but not render useless) the most widely used encryption for stored files – AES, the Advanced Encryption Standard. Confidentiality of stored information is critical because so many of our communications and other interactions in the world are now recorded and retained somewhere. Even if one has "nothing to hide"[83] – but we all do – these stored files might contain commercial secrets, passwords, financial information that might be exploited by swindlers, information about third parties, such as clients or children, who have not agreed to publicity, and so on.

Recall from Chapter 8 that passwords are essential to security but that their crypographic hashes could be reversed more quickly with quantum computers. We think this an unlikely use of quantum computers. Classical computing techniques, and simple trickery such as phishing, offer inexpensive, and too frequently, effective ways of getting into accounts.

Policymakers should focus on situations where, over time, information aggregates about people, creating particularly valuable attacks. One example is email. With the advent of limitless-storage email services, it is now easier to keep all emails than to segregate out material that should be deleted. The result is that if one can guess an email password, years of embarrassing, or simply valuable, data (think about yet-to-be-used gift card numbers and the like) are easily exfiltrated, mined, and sometimes made public. Increasingly multi-factor authentication is available for high-value accounts (and

---

[83]For a comprehensive critique of the "nothing to hide" argument, see Solove, " 'I've Got Nothing to Hide' and Other Misunderstandings of Privacy" (2007).

patient users), but the reality remains that once access is obtained, all this data can be quickly exfiltrated.

Recall that protections for stored data, notably AES, are resilient to quantum cryptanalysis. It would seem sensible to start storing email archives with such encryption. Such archiving is what Professors John Koh and Steven Bellovin have proposed in Easy Email Encryption (E3), an approach that focuses on encrypting the stored emails that many people use as a kind of backup method for information.[84] Currently this information is protected while it is sent by the user, and by login authentication. But once an email password is guessed, all bets are off. The E3 approach downloads email, encrypts it, and throws away the original message. Breaches of such a system only expose the most-recently received messages. An attacker who used a quantum computer to break the password would then have to break an AES-protected archive.

Several classical computing techniques could frustrate mass decryption by a hypothetical quantum computer.[85] A simple way of countering Grover algorithm attacks (typically against stored data), which in effect cuts symmetric key sizes in half, is to lengthen key sizes, thus re-imposing fantastic levels of computational costs.[86] With respect to asymmetric encryption systems widely used for payments and communications, "forward secrecy" is an option. In forward secrecy, each session key is unique, thus a compromise of one does not degrade the confidentiality of all messages.[87] Forward secrecy is available in the free Signal voice, text, and file encryption app. Shor's, Grover's, and yet to be discovered quantum algorithms have caused the updating of security standards,[88] and even experiments to determine whether new technologies are readily deployable.

Those working on "post-quantum" cryptography seek to enhance existing encryption or create new systems that will withstand a hypothetical, general purpose, powerful quantum computer.[89] Certain problems are uniquely tractable by a quantum computer; post-

[84]Koh, Bellovin, and Nieh, "Why Joanie Can Encrypt: Easy Email Encryption with Easy Key Management" (2019).

[85]Bernstein and Lange, "Post-Quantum Cryptography" (2017).

[86]Grumbling and Horowitz, *Quantum Computing: Progress and Prospects* (2019).

[87]Goldberg, D. Wagner, and Brewer, "Privacy-Enhancing Technologies for The Internet" (1997).

[88]National Security Agency and Central Security Service, "Commercial National Security Algorithm Suite and Quantum Computing FAQ" (2016).

[89]Bernstein, "Introduction to Post-Quantum Cryptography" (2009).

quantum researchers test measures that are intractable for quantum computers. For instance, company PQ Solutions developed a technique that involves injecting random noise into each message. In 2016, the Open Quantum Safe Project was formed to create open source versions of quantum-resilient encryption. Already other companies, such as ID Quantique SA, offer quantum encryption featuring QKD and QRNG.

Getting Rid of Data

Until recently, the modus operandi of technology companies was to keep information forever. But now even Google, the standard-bearer for information hoarding, has started efforts to randomize identifiers associated with searches and to delete them. This came in response to both FTC guidance and European regulation that encourage or require companies to limit how long identifiable information is maintained to "reasonable" business necessity. To do otherwise risks the creation of what Paul Ohm has called the "database of ruin," aggregations of even pedestrian facts that could haunt us.[90] One can imagine that behavior considered perfectly acceptable at one time could mar one's reputation in the future. But even documentation of perfectly legal behavior has been weaponized to degrade individuals' reputation, resulting in a drip-drip-drip of revelations about public officials, exposing what appear to be inconsistencies between their public and private lives. UK political theorist William Davies speculates that such banal revelations are triggering a crisis for liberal governance.[91]

Establishing ceilings for how long data is kept, even if those data are pseudonymous,[92] would seem to be a worthwhile intervention in the face of quantum computing. But once regulators limit data retention to reasonable business necessity time periods, one must consider *how* to delete information. Of course, data are encoded on disks and other physical media; however, when erased, most businesses de-

---

[90]Ohm, "Broken Promises of Privacy: Responding to The Surprising Failure of Anonymization" (2009a).

[91]Davies, *This Is Not Normal: The Collapse of Liberal Britain* (2020). The idea is that large-scale transgressions now matter less than minor revelations that impugn the authenticity of a political actor. When authenticity becomes the coin of leadership, the result is the rise of uncompromising, yet authentic, political personalities on both the left and right.

[92]Because of the advent of machine learning-enabled reidentification techniques.

stroy the data logically rather than physically.[93] A physical layer deletion approach requires data collectors to actually destroy media with equipment such as disintegrators, which grind hard drives into a mash of metal bits. When one's business is "in the cloud," physical destruction is impossible because the data reside on another company's physical media. Thus, logical approaches, including formatting and simple encryption of the data, are common practice. Weak encryption – anything less than AES-128 – used for deletion purposes will fail in the presence of quantum computing.

Several quantum computing innovators have created cloud-based devices for the public to use.[94] This is an ingenious strategy because it allows the company to study how users manipulate the device and to identify the most talented programmers. It also allows the quantum computer owner to keep its engineering secrets private, locked away in some secure cloud facility that makes reverse engineering impossible.

The cloud strategy is likely to be a winning one because few companies will be able to afford their own quantum computers. Providers thus become a chokepoint that can monitor their cloud for signs of decryption, just as one can look for signs of child pornography trading or spam transmission today. Importantly, a cloud monitoring strategy fails if *blind quantum computing* is achieved, because its functions will be encrypted end-to-end and obscured even from the cloud quantum computer operator (see Section 7.5, p. 293).

Finally, regulating decryption may seem futile, but US law already regulates many forms of information manipulation that are technologically easy to perform. These are attempts to set norms, and they are sometimes effective. US copyright law prohibits the circumvention of digital rights management technologies (often a form of encryption) that protect copyrighted works.[95] The Fourth Amendment and the wiretapping laws prohibit warrantless interception of communications content,[96] even though such activity is technologically simple for private investigators, law enforcement, and the intelligence community. Just as it is creepy to wiretap others, and dishonest to watch movies without paying, we might be able to create

---

[93]Reardon, *Secure Data Deletion* (2016).
[94]M. Harris, "D-Wave Launches Free Quantum Cloud Service" (2018).
[95]17 USC § 1201.
[96]18 USC § 2511.

norms that prevent most people from using quantum technologies to spy on each other.

9.5.3 Challenges of Government Power

Constitutional law precedent will likely apply to some kinds of privacy invasions brought about by quantum technologies. Chapter 8 describes capabilities that law enforcement agencies would pursue, such as UAV-mounted quantum sensors that search for firearms, explosives, and contraband drugs. One could imagine a city (but do not discount the privacy invasions of well-resourced advocacy groups) scanning entire neighborhoods for the presence of guns in the homes of people who are disqualified to own them: for instance, convicted domestic abusers or those on supervised release (probation, parole, or house arrest).

Investigatory Power

Yet, as private spaces and conduct become more vulnerable to sensing at a distance, courts have adapted and expanded Fourth Amendment protections for the home. For instance, in *Kyllo*, the Court interpreted the use of infrared cameras to detect heat emanating from homes as a Fourth Amendment search.[97] *Kyllo* would be strong precedent for the proposition that home-directed quantum sensing is exceptional and requires a warrant.

In recent years, the Fourth Amendment has had a kind of renaissance, embraced by both liberal and conservative justices. For instance, the Supreme Court has expanded privacy protections concerning information outside the home. As wireless phones have proliferated and made it possible to track individuals continuously, the Court has increasingly brought such devices and even the data they generate held by third parties under the ambit of Fourth Amendment protection.[98] As the Court contemplates how modern privacy protection requires government restraints on data held by the private sector, there could increasingly be warrant preference and other limits on data held by third parties.

As exciting as the Fourth Amendment renaissance is, the Court's actions merely establish a warrant requirement or "preference." The warrant preference, upon inspection, is a limited protection. Many

---

[97] *Kyllo v. US*, 533 US 27 (2001).
[98] *Carpenter v. United States*, 585 US ____ (2018); *Riley v. California*, 573 US 373 (2014).

people simply waive their right to privacy when the government asks to do a search, so no warrant is needed. And where the government does obtain a warrant, the exercise is more paperwork-intensive than substantive. That is, a lot of paperwork and procedure is involved, but as a substantive matter, all the government must show is "probable cause" that the place to be searched has evidence of a crime. The word "probable" leads many to think the government has to have more than 50 percent proof – that it is more likely than not that the suspect's private space has evidence of a crime. But that is not the standard. Courts interpret "probable" to mean a "fair" probability, something less than a 50 percent chance that evidence is present.

Thus, the question that civil libertarians should be considering is: is a warrant a sufficient safeguard against quantum-enhanced remote sensing? Traditional searches of homes occur a single time and are performed by people who may overlook contraband or forbear from an exhaustive search. But a quantum sensor, perhaps with millimeter resolution, would not just see more finely but also enable continuous searches. Just as we use quantum sensors at the borders to detect radioactive material (see Section 2.1, p. 36), we could foresee a day where searches are comprehensive and easy. Daily quantum searches might be in store for certain populations, for instance those with reduced expectations of privacy because they are on supervised release.

The wiretapping "superwarrant" standard may be apt for quantum sensing searches. In wiretapping, an activity that now includes the monitoring of many kinds of communications, even with wireless phones, the government has to comply with extra safeguards. These "superwarrant" limitations include the requirement that wiretapping only be used to police serious crimes, that irrelevant conversations be purged, and that surveillance occur only for a time-limited period. Importantly, the government must also explain why wiretapping is necessary, that is, why the investigation cannot proceed using other investigatory methods. These substantive and procedural safeguards could be adapted to quantum sensing searches to make such searches exceptional, time-limited and to exclude them from routine police procedure.

Sensemaking Power

The above discussion primarily deals with situations where the government is seeking to collect information. Indeed, civil libertarians have long sought to limit government power by keeping the government in the dark and stopping it from collecting data. That strategy's efficacy erodes as the government is involved in more aspects of our lives, giving it opportunities to collect data, and as the government gains greater power to make sense of the data it possesses than other actors have.

A further conceptual step is necessary to impose limits when the government lawfully obtains information and subjects it to some quantum-enhanced scrutiny. As Orin Kerr observes, Fourth Amendment analysis focuses on the government's acquisition of data, not on the depth and cleverness of the subsequent analysis of such data.[99] Thus the government is free to attempt to make sense of ciphertext, in the same way it is free to decode puzzling mysteries associated with a crime.

A series of parallel developments in machine learning may cause us to rethink whether the government's power of analysis requires additional regulation to protect existing civil liberties.

Today we have so much data about the world that many academics and policymakers think that the world is comprehensible to the average person. However, data have no meaning until they are given context. Increasingly it is clear that access to data is not enough: the process of sensemaking, the ability to evaluate data and convert it to information and knowledge, is critical. Yet, there is a dramatic sensemaking gulf between the ordinary person and governments and companies.

Already, sophisticated actors can examine evidence more deeply, and for a longer period of time, than can individuals or small organizations. This ability to interrogate data may itself become an independent basis for concern and rationale for limiting future government activity. For instance, sophisticated computer vision algorithms combined with a massive archive of imagery allowed an agent at the Department of Homeland Security to identify a child sex offender living in Las Vegas because their face appeared in two different photos. One was a grainy, oblique photo of his face that appeared in

---

[99] Kerr, "The Fourth Amendment in Cyberspace: Can Encryption Create a Reasonable Expectation of Privacy?" (2001).

a Syrian Yahoo user's account showing a young girl being sexually abused; the other was a thumbnail-sized image of him standing in the background of a family vacation photo.[100]

In 2009, a student project at MIT called "Gaydar" discovered that it was possible to reliably infer the sexual orientation of many MIT students by analyzing their online social networks.[101] Thus, some scientists claim that merely viewing a photograph that a person chooses to display on their social media profile can reveal that person's sexual orientation.[102] Since then, scientists have shown that it is possible to infer a person's sexual orientation using "minimal cues"[103] – and if such cues can be inferred by humans, then surely they can be inferred by machines as well (although rigorously controlled scientific experiments to answer this question have yet to be conducted). Another study showed that the photos that a person posted to their Instagram feed could be analyzed for depression, and that the results were just as accurate as diagnostic tests currently in use.[104]

An entire industry now sees emotion as fair game for manipulation by computer, with applications ranging from voting to buying to workplace conduct.[105] Presumably, higher-dimensional analyses only possible with quantum computers will accelerate these trends, making it difficult in practice to avoid revealing facts that, for whatever reason, we would rather not reveal.

Sensemaking is powerful, and the power to make sense is becoming concentrated. As quantum computing and sensing enhance machine learning, there will be even more troubling advances in com-

---

[100]The match was made possible by Clearview AI, a company that later came under attack for the way in which it has quietly downloaded over a billion such photos from social network websites and made the tool available to law enforcement and others. See Hill, "Your Face Is Not Your Own" (2021).

[101]Jernigan and Mistree, "Gaydar: Facebook Friendships Expose Sexual Orientation" (2009).

[102]Y. Wang and Kosinski, "Deep Neural Networks Are More Accurate Than Humans at Detecting Sexual Orientation From Facial Images" (2018). For a critique of Wang and Kosinski, see Katyal, "Why You Should Be Suspicious of That Study Claiming A.I. Can Detect a Person's Sexual Orientation" (2017).

[103]Rule, "Perceptions of Sexual Orientation From Minimal Cues" (2017).

[104]Reece and Danforth, "Instagram Photos Reveal Predictive Markers of Depression" (2017).

[105]Zuboff, *The Age of Surveillance Capitalism: The Fight for a Human Future at The New Frontier of Power* (2019).

puter vision and sensemaking that contribute to government investigatory power in public spaces. Consider these scenarios:

- Quantum illumination might make darkness no longer a barrier to observation with cameras. Private actors or the government might use low-light cameras to film people in darkened areas.

- Perhaps through quantum sensing, dense objects such as firearms will be remotely detectable through clothing.

- Single-quanta sensors and machine learning might contribute to a technique known as blind signal separation, tying individual voices to specific speakers even in a chaotic, loud environment. Such a world would change from the "masquerade ball" conception of identity in public[106] spaces to one with perfect identity and speaker attribution.

- Finally, these sensing techniques could be augmented with a range of machine-learning-based analytics claiming to predict personality, predisposition to crime, and so on. Quantum computing could enhance such analyses through optimizing machine learning, or at least add a patina of credibility to underlying pseudoscience.

Police departments might find such applications attractive because they would effectively allow officers to conduct a "Terry Stop" or "Stop and Frisk" of everyone on a public street. In court, the defenders of the practice would say that analysis of lawfully acquired data observed in public is fundamentally no different than observing a bulge in a person's pocket from a handgun.

9.5.4   The European Approach to Privacy Rights

European human rights and rules provide one attractive approach that is technology-neutral in order to prevent new techniques from evading legal controls. Article 8 of the European Convention on Human Rights (ECHR) establishes privacy as a human right, and specifies that the right to privacy shall not be interfered with unless the interference "is in accordance with the law and is necessary in a democratic society in the interests of national security, public safety or the

---

[106]Bailey, *The Open Society Paradox: Why The 21st Century Calls for More Openness – Not Less* (2004).

economic well-being of the country, for the prevention of disorder or crime, for the protection of health or morals, or for the protection of the rights and freedoms of others." This framework requires states to put people on notice of special investigative measures with specific, enabling legislation. A broad range of police conduct is considered an "interference" including the mere collection of data about individuals in police files, but also special investigative techniques, such as the use of phone-number collecting pen registers, and even the recording of suspects while in a jail cell.

Interferences with privacy must be lawful, necessary, and proportionate. Lawfulness is satisfied by enacting a domestic law authorizing the special measure in question; the law must be specific enough to put the individual on notice of the consequences of the investigative measure. That is, the law must impart guidance to the individual, so that the individual can foresee what the government technique might lead to.

Necessity and proportionality are judgement calls relating to the power of the state, and the kinds of interests that the state seeks to satisfy. European courts are more likely to authorize special measures in response to specific security threats, but to reject them when applied to general criminal deterrence. As part of the analysis, European courts consider whether the technique is effective in addressing articulated state interests, and whether there are alternative techniques that are less invasive of privacy. In this respect, the European approach is different from the Fourth Amendment to the US Constitution. Courts have interpreted the Fourth Amendment to be *transsubstantive*, that is, privacy protections apply with equal weight regardless of the crime suspected. US persons' privacy is the same whether the substance of the crime is murder or mere vandalism. There are many advantages to transsubstantive approaches, but one serious downside from a civil liberties perspective is the ability to scale up police powers to address serious crimes while disallowing high-power approaches from being unleashed in investigation for petty crimes.

Under the European framework, many investigative techniques are indeed lawful, because of the need to provide national security or security against crimes. But in some cases, particularly when government interests pursue general deterrence, even in anti-terrorism matters, courts have curtailed government power. As this book goes to press, a United Kingdom appellate court ruled that a face recognition system used *in public* – a widespread practice in the US – violated ECHR's Article 8. Despite having implementing legislation, the court found the law too vague in that it failed to specify who might be targeted by the system or where it would be physically implemented. On separate grounds, the court found the government violated an anti-discrimination law for failing to test whether the facial recognition system produced biased results based on race and gender.[107] In a separate case, the European Court of Justice held that broad mandates for data retention among communications providers are illegal for general crime fighting and even national security purposes. Only specific, serious national security threats justify mandates that providers keep data about users, and only for a "strictly necessary" time.[108] Meanwhile in the US, police are free to deploy face recognition even to deter petty crime, and police need not consider bias; they are also free to order providers to retain users' data for almost any crime and without having to ask a judge.

In addition to substantive checks on government power, the procedural aspects of the ECHR framework have real value. The requirement of enacting a law forces a public debate about government power. In regard to quantum technologies, this debate, and the law flowing from it, would have to be sufficiently specific to warn the public about the kinds of powers the technology enables. This is a much-needed reform in America. Recall that much of the controversy surrounding NSA surveillance in the US relates to the Department of Justice developing ingenious, strained, and often secret interpretations of laws that greenlighted bulk collection of personal data in ways that surprised even skeptical civil libertarians. But under an ECHR-like framework, experts' surprise itself would be evidence that the law was arbitrary; that the law failed to tell the public what the government can and cannot do with the technology.

The ECHR framework is just one piece of Europe's criminal procedure. Other instruments regulate police investigative practices at the state level, nation states do have oversight mechanisms for intelligence, and in 2016, community law was passed comprehensively regulating how law enforcement agencies, from investigation to prosecution, collect and use personal data. Taken together, this belies the narrative that Europeans "trust the government" and that Eu-

---

[107] *R (Bridges) v. CC South Wales & ors*, Case No: C1/2019/2670.
[108] *Privacy International v. Secretary of State for Foreign and Commonwealth Affairs et al.*, Case No: C-623/17.

ropeans allow police to run roughshod over civil liberties. Americans have few criminal procedure protections as substantively strong as the Europeans, and nothing as comprehensive.

As the military acquires new surveillance techniques that inevitably find their way into the hands of federal and then local law enforcement, the European model would force useful transparency and place limits on power and consequently preserve civil liberties. Short of the European model, the US could create safeguards that require substantive and procedural review before these technologies leave federal government agencies and end up in the local sheriff's office.

The human rights approach has another, more subtle advantage: it can be framed as a positive agenda, as in we are *for human rights*. Technology policy today emphasizes a *negative* approach, one focused on denying China the ability to press its political will on the world through technology. Advancing the cause of human rights, demanding that these rights be respected, gives policymakers a positive frame and a way to reject technologies based on their effects rather than their source.

9.6 Quantum Prediction

Companies developing quantum technologies have identified a number of commercial goals for the technology. Some companies are seeking short-term goals, but Google is aiming for the moonshot of achieving artificial intelligence using quantum computers.[109] Quantum computing is thought to both speed existing machine learning processes but also create the infrastructure for entirely new techniques.[110]

Machine learning may receive a significant advance with quantum computing because if current limitations on encoding quantum information can be overcome, a quantum machine learning process could consider more information than classical approaches. In classical approaches, data scientists deal with so much data that in order to make problems tractable, they either simplify or discard data. Simply put, high-dimensional datasets include too many independent variables to consider. Collapsing datasets makes computing faster or, in some cases, simply possible. For instance, in natural language

---

[109]Google, "Quantum – Google AI" (n.d.).

[110]Sandia National Laboratories and National Nuclear Security Administration, *ASCR Workshop on Quantum Computing for Science* (2015).

processing, in order to make computation of a corpus possible, a data scientist may systematically eliminate all words considered to be "low value" in meaning ("stop words").[111] Similarly, to reduce the problem space, data scientists use stemming and lemmas to collapse words with similar roots into a single concept. Presumably a quantum machine learning approach would have no need for throwing out so much data.

It is not clear if cognition of human experts operates in the same manner as modern machine learning systems, largely because we still have very little understanding of how human cognition works – especially among human experts. It's clear that expert-level human performance requires a combination of innate skill, learning, and thousands of hours of practice. What's not clear is how much of that expert-level performance is based on some kind of memorization and knowledge integration, and how much is based on establishing new neural pathways that can rapidly analyze new patterns.

9.6.1 Product development

Among the most intriguing proposals is the possibility of combining machine learning with quantum simulation of physical objects. The implications of these proposals are profound for product development in materials sciences, pharmaceuticals, and chemicals. Given any goal, such as for a drug that is more targeted and thus has fewer side effects, the combination of quantum sensing and computing could identify treatments that fit the bill. With the quantum sensing approach, scientists will see deeper into molecules. The understanding gained could create a revolution in using structure to target and to choose attributes of a chemical or material that are desired. Once structures are understood, quantum computing, using Grover's algorithm, presumably could search for the optimal candidate structures.[112]

---

[111] Berry et al., *Survey of Text Mining II: Clustering, Classification, and Retrieval* (2008).

[112] Aspuru-Guzik et al. put it nicely: "Imagine that you want to find a potential candidate for a cancer therapy. The user would begin by compiling a list of known compounds that are effective or ineffective for fighting a particular form of cancer. The user then decides a class of molecular features that they believe will be useful for deciding the effectiveness of a drug. Quantum simulation algorithms could then be used to calculate these features for use in a supervised data for a quantum machine learning algorithm. A quantum computer could subsequently use Grover's search to rapidly scan over a database of potential candidate molecules

One can imagine the fantastic outcomes and their knock-on effects from simulation. If one can simulate the chemical basis for energy storage, perhaps a super-efficient battery could be built. Energy then becomes cheaper (because we can store it easily) and the knock-on effects could be that we have more energy capacity and that solar generation and storage become economical for more households. Similar research could be applied to energy transmission efficiency and to countless energy-intensive processes, from creating fertilizer to metals.

What do these capabilities mean for safety regulation? One approach is to trace the requirements for pharmaceutical and chemical safety to current processes, and explore how computer simulation might add, change, or even eliminate requirements. At the highest level, pharmaceuticals go through four levels of review: pre-clinical testing, clinical research, review by the FDA, and finally, surveillance after the drug is in the marketplace. Consider that in the earliest phases of drug development, before humans are involved, developers must answer basic questions surrounding absorption, dosage, and risks surrounding toxicity.

This earliest screening of drugs requires time and labor-intensive explorations, because people are not all alike, and treatments may have different effects on people based on their sex, race, age, body weight, and presence of existing conditions. The Food and Drug Administration specifies procedural rules to ensure good design and to prevent guile.[113] And this is where quantum simulation may offer the best speedups. In addition to basic discovery of promising treatments, the effects of those treatments could be simulated with models of drug absorption and interaction. Once the complex interactions can be modeled and specified on a quantum computer, presumably these models could be run as standalone programs on classical computers. In fact, entire businesses could arise that specialize in creating these models for others to use. A market would exist for creating models based on many different human characteristics going beyond sex and age. One could foresee models for pregnant people, for people with genetic or environmental conditions that may create complications,

---

in search of one that the trained model believes will have therapeutic properties." See Sandia National Laboratories and National Nuclear Security Administration, *ASCR Workshop on Quantum Computing for Science* (2015).

[113]See e.g. FDA, Protocol for and Conduct of a Nonclinical Laboratory Study, 21 CFR 58 (2020).

or even models for single individuals afflicted with cancer or other diseases that have idiosyncratic characteristics.

Later phases of drug development require human experimentation, with all of its complexities and contingencies. The FDA process specifies three rounds of clinical trials with increasing numbers of human subjects. Each phase can take years, and the final phase can involve thousands or even tens of thousands of subjects. It is unclear how quantum simulation might affect these requirements. Perhaps developers could more precisely identify how many people must be tested and whether over-, or under-sampling is called for based on genetic or environmental factors.

Clinical trials elicit side effects from participants, and any patient is now familiar with the lengthy, sometimes conflicting lists of complications that any drug might create. A straightforward counting of adverse event disclosure found that the most popularly prescribed 200 drugs on average have 106 such warnings. One popular drug had 459.[114] How much of this disclosure is noise or risk management instead of useful knowledge about risk? One could imagine quantum machine learning being used to tease out all the conflicting and confusing signals surrounding side effects of medicines. Perhaps there are indeed hundreds of risks from any given drug; finding ways to prioritize these risks could contribute to physicians' risk/reward considerations.

Finally, in the post-market phase, FDA monitors the marketplace for bad outcomes, lack of advertising compliance, and enduring safety and quality risks from manufacturing. Here too one could see quantum simulation providing more efficient oversight. For instance, in the post-market phase, companies making generics may copycat existing treatments, under a different regulatory standard that seeks to ensure that the generic treatment is equivalent in mechanism and effect. One could imagine simulation finding or verifying equivalent treatments. Whether these applications emerge, and whether they could possibly relieve regulatory burden on pharmaceutical makers is a question for another day.

9.6.2 Fairness

Artificial intelligence and machine learning (AI/ML) have raised deep concerns about how data inputs, algorithms, and commercial

---

[114]Duke, Friedlin, and Ryan, "A Quantitative Analysis of Adverse Events and 'Overwarning' in Drug Labeling" (2011).

practices might result in machines that engage in unlawful discrimination or other kinds of unfairness.[115] Because the answers produced by AI/ML systems will be thought to be "smart," users might inadvertently engage in invidious discrimination while laying the moral responsibility with the computer. A rich field known as FAT* (fairness, accountability, and transparency in machine learning, artificial intelligence, and other systems) seeks to create procedural and substantive standards to detect discrimination and other forms of perverse outcomes.[116] A key problem in this space is that there appears to be an inverse relationship between learning power and explainability in modern ML approaches. That is, the most powerful learning systems, because of their complexity, find subtle and unpredictable relationships.[117] Yet this power comes with a price – users may not be able to explain why these relationships occur, these relationships may be specious, and they may correlate with race or other factors that could be perverse.

Of course such transparency does not guarantee fairness, but policymakers will see transparency as an important factor in evaluating machine decision making.

Turning to substantive aspects of fairness, we might see quantum-enhanced learning as inherently disproportionate and powerful when applied to people in many domains. We would not consider it fair for a person to play chess or Go against a supercomputer. But what if we are called to play consumer or investor against adversaries using quantum computing-powered optimization?

In the consumer context, the immense volume of internet traffic and tracking that is collected simply cannot be computed on classical machines. The disconnect between data volume and the ability to process it causes marketers to use abstractions to make sense of consumers, such as profiles that bin consumers into general categories like age, sex, presence of children, and so on. These abstractions are coarse representations of reality, but good enough to target ads. Turning to a quantum computing marketing machine,

---

[115]Calo, "Artificial Intelligence Policy: A Primer and Roadmap" (2018).

[116]See ACM Conference on Fairness, Accountability, and Transparency (ACM FAccT), a computer science conference with a cross-disciplinary focus that brings together researchers and practitioners interested in fairness, accountability, and transparency in socio-technical systems.

[117]Gunning and Aha, "DARPA's Explainable Artificial Intelligence Program" (2019).

individual consumers could come into fuller focus. The fine-grained, second-by-second ways in which we pay attention might be sensed and understood.

We should anticipate such systems to know about our history but also our personality. Lawyers see advertising as a rational information exchange but marketers understand it as a tool that communicates on several levels, including on raw emotion. In a marketplace optimized by quantum computers, sellers might understand our willingness to pay, our strongest preferences, our subjective emotional valences, and the kinds of evidence that cause us to change our minds. Imagine the face-recognizing camera system described above optimized to understand how desperate the consumer is for a product, how the consumer has responded to other offers, how emotion can be invoked to appeal to a certain individual, and whether the consumer is innumerate or otherwise unable to understand common strategic selling techniques such as bundling. Might we see such a marketing machine in the same light as advanced selling techniques targeted at children? Would the standard regulatory approach of labeling (perhaps "quantum ad") be enough to prepare consumers for the kinds of persuasion we may face?[118]

Recall that quantum computing is most likely to be achieved by nation states or dominant technology companies, such as Google. Google reportedly refrained from using user search terms to predict stock movements,[119] apparently because it realized that searches may include material non-public information (which is illegal to use under US law). Google may similarly conclude that quantum trading approaches using search data implicate insider-trading laws. But nation states will not concern themselves with such limitations. In fact, quantum ML might be a seductive tool for the destabilization of other economies. Imagine using quantum optimization in order to identify subtle, inscrutable market effects disadvantageous for Vladimir Putin's oligarchs. Or imagine identifying the kinds of conditions that could poison the chances of a Chinese marketplace competitor, Huawei, from gaining a foothold in telecommunications

---

[118]A core function of advertising law is to help consumers recognize strategic communication so that they can use their own self defenses against deception or other manipulation. Self defense is necessary because there is so much false advertising that regulators could never police it. See Hoofnagle, *Federal Trade Commission Privacy Law and Policy* (2016).

[119]Fortt, "Top 5 Moments From Eric Schmidt's Talk in Abu Dhabi" (2010).

markets. The intelligence community has already found offensive cyber to be a useful, asymmetric, secret tool to undermine adversaries. Won't quantum technology be just as tempting a tool?

The law already remedies many situations where automation or information asymmetry creates imbalances of power. Quantum ML might be a field where such imbalances need transparency forcing, or other remedies, including bans on certain applications.

9.7 Measuring Quantum's Research Output

We conclude the chapter with an attempt to evaluate the impact of policy efforts to date: where is the quantum action?

9.7.1 Academic Publications

To better understand state sponsorship of quantum technologies, this section presents data from the Web of Science to elucidate high-level trends in quantum technology research outputs. The data source is the Web of Science Core Collection, "a curated collection of over 21 000 peer-reviewed, high-quality scholarly journals published worldwide (including Open Access journals) in over 250 science, social sciences, and humanities disciplines."[120]

Quantum Technology's Research Output

We examined statistical data about scientific literature and patents to identify funding and other trends regarding quantum information science.[121] In examining funding sources for 15 130 papers we identified as relevant, Web of Science reports that the National Natural Science Foundation of China (2692) is the dominant funding organization for published research in quantum technologies, followed far behind by the US National Science Foundation (1275). But such a categorization ignores how nations have multifarious routes to funding research. For instance, in addition to the NSF, other major US government supporters of quantum technology research include the

---

[120]Clarivate, "What Is Web of Science Core Collection?" (2021).

[121]A simple text search for "quantum" in titles, abstracts, and keywords returns over 400 000 papers published since 2009. We used two approaches to narrow these results. First, we used a search for publications mentioning the three categories of quantum technologies focused on this book; that returned 15 696 papers ("quantum sen*", $n = 629$; "quantum commun*", $n = 3852$; "quantum compu*", $n = 11 215$). There were 566 duplicate publications appearing in two or more of these searches, resulting in $n = 15 130$ unique publications. Almost all of the literature appears in English.

Department of Defense,[122] the Department of Energy, the National Aeronautics and Space Administration, the National Institute of Standards and Technology, the National Institutes of Health, and the Office of the Director of National Intelligence. In fact, one key observation from this analysis is that the US intelligence community and the US military both have embraced a rich quantum information science research agenda. Furthermore, the Department of Energy is funding quantum technology research in an attempt to promote US superiority in high-performance computing.

In China, many individual provinces have research portfolios in quantum research, supplementing the country's national scientific research organizations. In Europe, individual nations, most prominently Germany, Spain, the Netherlands, and the United Kingdom, have supplemental funding to community-wide efforts. Brazil, Singapore, and Japan also appear prominently. Finally, many private foundations, such as the Alfred P. Sloan Foundation and the Simons Foundation, are active in quantum technology, and their funding contributions may be as consequential as some nations'.

Table 9.6 gives a lower-bound estimate of the number of published papers in quantum technology funded by different nation states. The table is styled as an estimate because funding support data in Web of Science required significant cleaning and some supporters could not be resolved to a country (for instance, some papers are supported by the "Ministry of Education," but many nations have such a body). Also, a single paper can be sponsored by more than one research organization. This table presents two rows for the European Union. The first is EU-community-wide-supported publications plus all the papers funded by individual EU member states (for instance, to recognize the independently funded nation state programs in Germany, the UK, and elsewhere).

Just as patent counting is not an evaluation of patent quality, paper counting is not an evaluation of research quality. Indeed, turning away from the absolute number of papers published to citation met-

---

[122]Funding agencies within the Department of Defense include the Air Force Office of Scientific Research (AFOSR), the Army Research Office (ARO), the Defense Advanced Research Projects Agency (DARPA), the National Security Agency (NSA), and the Office of Naval Research (ONR). The Intelligence Advanced Research Projects Agency (IARPA) was modeled on DARPA, but is organizationally underneath the Office of the Director for National Intelligence, and not part of the US Department of Defense.

Table 9.6. Support for publications on quantum technologies

Nation	Estimated Number of Papers
China	8006
US	6071
European Union (including national support)	5819
EU alone	2520
Japan	1491
Canada	1425
UK	894
Germany	785
Nongovernmental Organizations (Foundations)	618
Australia	598
Brazil	518
Spain	455
Russia	383
France	280
Austria	253
Korea	249
Papers with no data	4641
Total	35 006

rics, among the most-cited research publications, US and European-funded works dominate.

Web of Science tracks the institutions of authors publishing papers. Institution tracking looks for any matching address, so a single paper can have many institutional affiliations. This is especially true in quantum information science, which is inherently interdisciplinary, and frequently research involves collaboration across institutions. Table 9.7 presents the most frequently appearing institutions in quantum technology papers.

Turning to author national affiliations, Web of Science tracks the addresses that appear in published papers, and categorizes them by nation. Since multiple addresses can appear in papers, a single paper can be affiliated with more than one nation. In quantum technologies, the US has the most authors (Table 9.8).

Web of Science also provides high-level categorization of quantum technology publications, revealing the wide variety of disciplines

Table 9.7. Affiliations listed by authors on quantum technology publications

Institution	Number Published
Chinese Academy of Sciences	836
Centre National de la Recherche Scientifique (CNRS)	440
University of Science Technology of China	432
University of California System	411
University of Waterloo	346
US Department of Energy	324
Max Planck Society	307
National University of Singapore	305
Massachusetts Institute of Technology	292
University of Oxford	285
University System of Maryland	281
Tsinghua University	276
National Institute of Standards Technology	243
University of Maryland College Park	238
Russian Academy of Sciences	223
Consiglio Nazionale delle Ricerche (CNR)	218
Harvard University	196
University of Tokyo	195
University of London	180
Beijing University of Posts Telecommunications	177
California Institute of Technology	166
United States Department of Defense	165
Delft University of Technology	157
ETH Zurich	156
University College London	154
(affiliation data missing)	247
Total	7250

Table 9.8. National affiliation of QIS authors

National Affiliation	Authors
US	3973
China	3680
Germany	1451
England	1200
Japan	1114
Canada	1026
Australia	767
India	654
France	630
Italy	618
Russia	453
Spain	448
Switzerland	419
Singapore	383
Austria	370
No regional data	235
Total	17 421

that contribute to the expertise of QIS. This table highlights that many science disciplines, including chemistry, physics, electrical engineering, computer science, and nanoscience, are relevant to the conception and design of quantum technologies. We present this information in Table 9.5

It is important to recognize the limitations of these data. First, the analysis obviously only focuses on published research: research that is classified or simply unpublished is not included. Such omissions are not fatal to our analysis, because the players in quantum technologies today have incentives to publish. Indeed, authors affiliated with or funded by D-Wave, Google, IBM, Microsoft, Lockheed Martin, Rigetti, and Volkswagen, along with scientists at military-affiliated research laboratories, appear in the results. A second, more significant limitation is that paper counting overlooks publication quality. While China appears to be pulling ahead in research output, there are systemic incentive problems documented in some countries' publication practices. China has dramatically increased its scientific scholarly output in the past three decades, in part by giving generous cash awards to authors. A 2017 article found that payments for publication in *Science* or *Nature* came with an average bonus to the first author of $43 783.[123] Lower-tier institutions were willing to pay authors more than higher-tier ones. Publication in the *Journal of the Association for Information Science and Technology* (JASIST) netted the first author on average almost $2500. Given that the average faculty salary for a university professor in China is about $8600, these sums are significant. Payments for publications as a policy were reportedly ended in 2020, but these statistics are clearly influenced by China's former policy.[124]

A third limitation is that some attributes are missing significant data. For instance, in funding organization, about 30 percent of the papers lack any information about the research sponsor: this could be an oversight, or an attempt to hide a significant sponsorship.

Finally, sources of funding are multifarious and referred to in inconsistent ways by authors. As a result, producing these tables required significant data cleaning to address inconsistencies and errors, so unmeasured errors resulting from reporting bias or manipulation may be present.

9.7.2 Quantum Technology's Patent Output

Issued patents are another way to measure the success of research expenditures. A 2017 survey of quantum technologies by *The Economist* reflected a national competition in the patent landscape of quantum technologies.[125] Using data current through 2015, the publication found that the US had by far the most patent applications for quantum computing. However, there was a surge of Chinese applications focusing on communications and cryptography in recent years, with China exceeding the US 367 to 233. Investment in sensing was on par between these superpowers. Other countries with fulsome quantum portfolios included Canada (quantum computing), Germany (sensing), and Japan (quantum computing and cryptography).[126]

---

[123]Quan, B. Chen, and Shu, "Publish or Impoverish: an Investigation of The Monetary Reward System of Science in China (1999–2016)" (2017).

[124]Mallapaty, "China Bans Cash Rewards for Publishing Papers" (2020).

[125]Palmer, "Technology Quarterly: Here, There and Everywhere" (2017).

[126]See also Patinformatics, "Quantum Information Technology Patent Landscape Reports" (2017).

Patents Concerning Qubits and Quantum Entanglement Since 2000

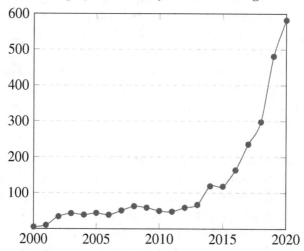

Figure 9.3. Over time, there has been a steady increase in patents published that concern qubits or entanglement. In 2019, 481 such patents were published world-wide. Source: analysis based on Derwent World Patents Index database. For more information, see the sidebar "Who Has Quantum Technology Patents?" on page 453.

Who Has Quantum Technology Patents?

Richard P. Feynman gave a seminal talk, "Simulating Physics with Computers," at the first Conference on Physics and Computation in 1981. Charles Bennett and Gilles Brassard proposed the first quantum cryptography protocol in 1984. David Deutsch formulated a model for a universal quantum computer in a 1985 paper. Peter Shor's RSA-busting algorithm was published in 1994, and Grover's search algorithm in 1996. None of these events resulted in a patent being granted. Starting in the mid-1990s, several large companies were awarded patents in quantum technologies.[a] These entities have more than 30 patents published as of December 2020:[b]

Patent Owner	# Patents
International Business Machines Corp. (IBM)	236
D-Wave Systems, Inc.	157
Intel Corp.	80
Microsoft	74
US Military Branches	68
Northrop Grumman	64
Google	59
Zhejiang Gongshang University	55
Toshiba	55
Lockheed Martin	45
NTT	43
MIT	41
Hewlett-Packard	36
Rigetti	34
South China Normal University	32
Total	1079

[a]In quantum cryptography, British Telecom led the field with 9 patents, while IBM, the University of California, General Electric, NTT, NEC, and the UK and US Secretaries of Defense were also in the mix. In quantum computing, leaders included IBM, Mitsubishi, Silicon Graphics, Hitachi, Lucent, MIT, and the US Air Force.

[b]Based on a search in Derwent World Patents Index for patents published that include the terms "quantum entangle!" or "qubit" since 2000. (The "!" is the Westlaw "root expander" search metacharacter.) The search produces 2650 responses. Responses were cleaned using OpenRefine.

9.8 Conclusion

Our focus in this chapter is at the national level, with primary emphasis on policy options available to the US and Western governments.

While we believe that there is clearly a role for international agreements, and while we are strong advocates of bringing concepts from modern physics into the pre-college curricula, it is at the national policy level – the level of national sovereignty – where policy goals are most likely to be translated into meaningful dollars spent and policies enacted as laws or regulations.

Will Quantum Computers Make Humans Obsolete?

What if humanity realizes the project to build quantum computers and these machines can solve hard problems in the blink of an eye? What then?

As computers can solve hard problems, work that only humans can perform might become trivial. Natural language processing might develop *Star Trek*'s universal translator. As machine text and image generation become more sophisticated, many would be satisfied with less expensive, ubiquitous, quickly generated works of original art and literature that do not require the training and patronage of flesh-based artists.

"Don't worry about such creative destruction," say techno-utopians. "For each job technology obsoletes, another opportunity arises." The flaw with this narrative is that the innovations discussed here are aimed at problems solved using human intelligence and creativity.

Computers need not equal human performance: even if the machines are merely middling, employers interested in saving money, and consumers who value convenience over quality, will satisfice with the mediocre.

Computers will also systematically improve, thanks to the progress of technology. A recent article on the translation industry notes that while the market for human translators continues to expand, much of the work is now "post-editing machine translation."[a] What happens when post-editing is no longer necessary?

In *Homo Deus*, Yuval Noah Harari explains how many who feel secure in their jobs today could be replaced by algorithms.[b] Job security is imperiled by what Harari calls the *great decoupling* of intelligence from consciousness. "For AI to squeeze humans out of the job market it needs only outperform us in the specific abilities a particular profession demands."

Harari observes that our liberal notions of human worth are tied to and justified by the uses of the human body for warfare and for work, an idea echoed by Sun Microsystems founder Bill Joy.[c] When both functions can be performed well enough by computers, what need will the future have for humans?

---

[a]Tirosh, "Top Translation Industry Trends for 2020" (2020).
[b]Harari, *Homo Deus: A Brief History of Tomorrow* (2017).
[c]Joy, "Why The Future Doesn't Need Us" (2000).

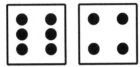

The Quantum Age: Conclusions

W^E are at the cusp of a quantum technological revolution. Quantum sensing, computing, and communication offer some significant improvements on classical technologies, in some cases they create fundamentally new capabilities.

This book begins a conversation on these consequential quantum technologies, as they reshape how companies and government measure and observe, communicate, and make sense of the world through simulations and problem solving.

Many technologies are deployed by companies and governments on society with weak or underconceptualized plans to deal with the technologies' implications. With Quantum Information Science, attempts to cast implementations of quantum technology as entirely novel are inapt. Novelty narratives may be a product of hype and confusion, but regardless of the purpose of their use, arguments of novelty may flummox policy and planning processes.

This book has explored the invention of many technologies, some novel, some not. We have argued that historical analogy is a good guide for analysis of quantum technologies. When it comes to quantum technologies, some of its most promising applications are (dramatic) improvements on classical methods, such as simulation and code-breaking (see Chapter 5). But we experienced similar breakthroughs 80 years ago, when the first digital electronic computers were deployed for the very same purposes: physics simulations and code-breaking. While quantum does offer entirely novel capabilities, such as quantum cryptography and quantum networking (see Chapter 7), we believe that these will not be worth the extra expense and complexity for broad deployment for the foreseeable future.

Quantum technologies are quickly arriving. Even if the most hyped promises in quantum computing are not realized in the next decade, in the near term quantum sensing could shift relationships irrevocably. This book has painted the landscape of quantum's implications – from nation-state concerns of strategic conflict, intelligence gathering, and law enforcement activities; to the concerns of companies that may be subject to industrial policy priorities and restrictions; to the level of the individual who may face institutions with great asymmetries in sensing and sensemaking power. We should start deciding now how these technologies will be used, before others make the choice for us.

We are both optimistic and excited about the potential for quantum technologies to improve our lives. A careful overview of the field suggests the contours of those improvements.

10.1 Quantum Computing Winter Is a Probable Scenario for 2030

Chapter 8 modeled possible scenarios for quantum technologies, in order to motivate a policy discussion. We think it important to seriously consider the likelihood of the quantum winter scenario in the near term. Recall that in our quantum winter scenario, large-scale quantum computers simply cannot be realized in the next decade or two. Nor do applications emerge in quantum simulators or smaller-scale devices that are compelling enough to trigger virtuous cycles. In this scenario, quantum sensing advances because of its maturity and sound economic drivers, mostly from medical, law enforcement, defense and intelligence application. But quantum communications loses steam as cryptanalysis threats fade.

We believe that there are two factors that make quantum winter probable. First, no consensus has emerged on a substrate that will enable large-scale quantum computing. In simple terms, whereas computing vendors rushed to adopt the transistor in the 1950s, there is no similar technology that presents itself for quantum computing. Second, no technologist, no company, no actor in the quantum computing space has implemented an application that is truly game-changing – a reason to use a quantum computer rather than a conventional one. To create a virtuous cycle, quantum computing needs an application that ordinary businesses find worthwhile to invest in.

The most pressing risk of a quantum winter scenario is an unwillingness to recognize the possibility and plan for it.

Specifically, we are not concerned about the private companies pouring investor dollars into quantum computing. These companies will be able to shift more quickly than other institutions if a quantum winter comes. We are concerned that a hard freeze may damage our capacity to evaluate when the thaw is upon us – and that nations that fail to pivot quickly will be significantly disadvantaged.

One signpost of a thaw could be the widespread agreement on a substrate for stable, scalable quantum computing.

10.1.1 Public/Private Scenario

We are hopeful that the public/private research and development scenario is the most likely future for quantum technologies. This scenario is most likely because state-of-the-science developments are being achieved in several nations, sometimes in government/private partnerships, but also by private companies acting alone. Today's private sector does not have the commercial landscape of the 1940s. Large, sophisticated technology companies such as Google and Microsoft have more cash on hand than some nation states, and these companies see billions more in profit from materials science, chemistry, and drug development applications of quantum simulation.

In the public/private scenario, significant breakthroughs and applied development continue to occur in both the public and private sectors – not just in the US, but also in quantum technology powerhouses like Canada, China, Germany, the Netherlands, and the United Kingdom. Unlike strategic technology developments of the past such as the atom bomb and global positioning systems that were only in the reach of governments, today the private sector has both the financial resources and scientific capability to make nation-state level investments and realize accomplishments – as evidenced by the recent achievements of the private outer space industry. Governments might try to limit this innovation with export controls. But, again unlike the development of the atomic bomb, no single country is dominant in quantum technologies, meaning that there are likely to be many sellers of controlled technologies.

Innovators will have high-powered incentives to evangelize quantum technologies and find many uses for their inventions outside defense and intelligence. For all these reasons, we think the quantum technology future is bright, and will be open relative to previous technology revolutions. The public/private quantum scenario is the technology's brightest because of incentive alignment. Quantum technology's greatest contributions to people – and to companies' profit statements – will come not from cryptanalysis but from advances in material science, chemistry, medicine, and every field that could benefit from precision engineering, from consumer durables to manufacture of gadgets.

10.2 Assessing the Next Decade of Quantum Technologies

Whether or not the year 2030 sees us in a quantum winter, we believe that the 2020s will be good times for those involved in the research and business of quantum information technologies.

10.2.1 Prospects for Quantum Sensing

Quantum sensing (see Chapter 2) is already a mature, successful technology. Currently in its first-generation, just one form of quantum sensing – Magnetic Resonance Imaging – has contributed to the treatment of countless people. Other first generation technologies like the atomic clock made it possible to have reliable, worldwide position, navigation and timing devices thanks to GPS.

For the coming decade and perhaps beyond, second generation quantum sensing will be the most exciting class of quantum technology, providing not just improvements on existing methods but new capabilities as well. More exquisite sensing of magnetic and gravitational fields has obvious implications for military, intelligence, and law enforcement, but uses in the private sector will abound: medical imaging technologies that are both more precise and non-invasive; sensing underground deposits of minerals and valuable materials will benefit mining interests; high-precision manufacturing, possibly including futuristic engineering production runs that yield *identical* artifacts because they are assembled at the atomic level.

Contrary to many media and policy narratives, the next novel and troubling threats to privacy will likely come from quantum sensing rather than encryption-cracking quantum computing. Already clever technologists are deploying ever-smaller sensors on satellites and on unmanned aerial vehicles. These technologies will be used to peer into private spaces and the kinds of countermeasures ordinary people possess – window blinds and doors – simply will not provide protection.

Quantum sensing is a precursor technology for both computing and communication. As such, quantum sensing will directly or indirectly benefit from investment in other quantum technologies. Mastery of quantum sensing is necessary for quantum computing, and as that mastery develops, entrepreneurs will likely find many non-computing uses of quantum sensors to benefit society.

10.2.2 Prospects for Quantum Computing

Quantum computing will be the most exciting form of quantum technology, if large-scale devices can be developed. Quantum computing's biggest potential contributions might change life as we know it. The spotlight on cryptanalysis (see Chapter 5) has left these other uses of quantum computers in the shadows, and as these lesser discussed applications are realized, cryptanalysis will be left in the shadowy recesses of government agencies. It will be similar to what happened with electronic computers: yes, there is cool stuff going on behind the curtain, but there will be so much going on in front of the curtain that most of us won't notice.

Richard Feynman's vision of quantum computers – as simulators for physical systems – is not only more likely, but more beneficial for humankind than code-breaking. We can imagine advances in materials science letting firms create products with new properties; advances in solar cells making energy capture more efficient; simulations in chemistry leading to new classes of drugs and improvements on existing ones; and unraveling some of nature's mysteries, like photosynthesis and nitrogen fixation, enabling humans to feed more people. And that's just the beginning! Just like the personal computer revolution, the quantum computing revolution will produce unimagined uses and benefits. Perhaps cryptanalysis will be remembered faintly, like the old artillery tables that drove computing in the 1940s (see Chapter 4). Cryptanalysis' role will be secondary because the process is harder than popularly understood, because countermeasures are already available, and because companies will generate more profit pursuing other uses of quantum computers.

The fundamental technological challenges in realizing quantum computing (see Chapter 6) are more difficult than those faced by classical computing. Classical computing's breakthrough came with the transistor and then the integrated circuit, together a massive improvement on vacuum-tube approaches. Semiconductors enabled decades of scaling in power, miniaturization in size, and reduction in cost. Quantum computing has yet to experience its own transistor revolution because of the fundamental challenge of managing quantum states. Scaling a quantum computer becomes more difficult with each additional qubit; the same constraint has not limited classical computing until recently where quantum effects have complicated the development of 7 nanometer chips.

Quantum computing requires a basic science breakthrough similar to the invention of the transistor. That breakthrough must enable the management of an enormous number of quantum states, coherence over long periods, and the ability to measure the managed states. The basic science breakthrough may lie in photonic approaches, or in the topological qubit, or ion traps, but we believe that it is unlikely to occur in superconducting media currently used to make the largest quantum computers. Until scaling is possible, many of the most discussed applications of universal quantum computing simply cannot be realized. Instead, scientists will build special purpose devices that benefit from fantastic computational power, but only perform limited experiments, like the analog devices of early classical computing.

10.2.3 Prospects for Quantum Communications

Europe and China have embraced a focus on quantum communications in both of its forms, quantum key distribution (QKD) and in quantum networking/internet (see Chapter 7). Because these nations have substituted for the market, quantum communications will receive a boost that normal business drivers would not produce. In effect, nations will subsidize the development and marketization of quantum communications, at least in the form of QKD.

Defense against the future is the driving rationale for QKD adoption. If one's secrets must remain hidden for 10, 25, or 50 years, one must have a strategy to address growing computational power from adversaries. QKD, because it is information-theoretically secure rather than relying upon number theory for security, should provide protection against future attackers with large quantum computers. Today many working systems use QKD for distributing keys but AES-256 for actually encrypting data. Although this is likely to be safe, AES-256 *could* be cracked at some point in the future, even using classical or quantum approaches. As the speed of QKD improves, the time that each AES-256 key is used will decrease. At some point there may be no need for AES-256 at all.

Post-quantum cryptography is an alternative to QKD that uses computationally-secure algorithms that are believed to be resilient against quantum computers. But reliance on post-quantum cryptography may be misplaced; clever scientists could discover a new algorithm that unscrambles ciphertext quickly, or perhaps quantum computers scale massively, so much so that brute force can undo the cryptography. The switch to post-quantum cryptography is essential, but conversion to QKD requires an analysis of institutions' risk appetite and the time value of their secrets. For many companies, operations plans may need only be secret for a business cycle, but for governments, decades-long secrecy requirements may justify extra precaution.

The prospects for quantum internet are weaker than for QKD. It is not clear to us why institutions would adopt quantum internet given implementation complexities. One answer lies in network reliance, or rather the lack of it. The classical Internet is akin to the shared, "party lines" of the early telephone network. Many strangers can listen in. Interception and copying is easy. We use encryption to shield our content, yet encryption cannot prevent revealing forms of investigation based on network metadata – who is talking to whom, how often, and when. Many people use the word *trust* to describe what really is *reliance* on networks, with their unknowable operators, paths, and vulnerabilities. That is, they *trust* the network not to violate their security policy, because they have no mechanism for *assuring* that the network does not. The network is *trusted*, even though it may not be *trustworthy*.

Quantum internet likely takes the majority of SIGINT opportunities out of the equation, making communications end-to-end secure. Operators of a quantum internet need still worry about side channel attacks on endpoint devices and against the people who use them. Availability can be compromised by attacks on the fiber itself, although free-space systems have no such problem. Operators will have to discover countermeasures against tampering and use physical isolation for quantum repeaters. But if the quantum internet is developed, users can deny adversaries the ability to capture their communications *and* deny adversaries access to metadata analysis on communications. Adversaries will not know when or with whom communication is taking place. These metadata-denying advantages may be the driving rationale behind investment in China and the European Union, in a kind of technological revanche against the "golden age" of SIGINT. Quantum internet would actually bring about intelligence agencies' greatest fear, the notion that communications could "go dark" and not be available for analysis.

10.3 Law and Policy Priorities for the Quantum Age

Chapter 9 presents a full list of policy issues raised by quantum technologies. Our approach recognizes that innovators sometimes present technologies as entirely novel, flummoxing the public and policymakers about potential regulatory implications. Recognizing that quantum technologies are mostly improvements on classical methods, and that many others have implications that are predictable, we draw upon lessons from the history of technology to elucidate likely development cycles and challenges to governance.

If limited to just five challenges and approaches, we think the following are the most significant:

Innovation policy   Quantum computing is still in a pre-transistor-revolution phase in its development. To realize scalable, fault-tolerant quantum computing will require an enormous and decades-long commitment of investment in basic research. The US, after a period where policymakers looked to private technology giants to assume more of the responsibility for basic research, now invests billions in QIS research. From the Apollo Space Program to the GPS constellation to the Internet itself, the US government has been a humble driver of innovations that devolve to the general public, accruing to the benefit of all, and in the process, educating and training legions of people. The government stands as a counterexample to the over-hyped, popular narrative of the lone inventor who saves the day. The lone inventor narrative is particularly unlikely in quantum technologies, because of the need for multidisciplinary expertise. We are more likely to realize scalable quantum computing with healthy government patronage, more likely to avoid private-company winner-take-all stratagems, and once quantum computing arrives, government programs are more likely to incubate the people necessary to lead a quantum computing revolution.

Immigration   To build the expertise and multidisciplinary talent, among the quickest solutions is a liberal immigration policy. Approaches that ease the burdens with visiting, studying in, and staying in quantum technology hubs will create advantages. We recount how most PhDs in computer science and engineering are "non-resident aliens" in the US, and suggest that liberal immigration policy could let us keep more of those highly trained people in America. The anti-immigration, even xenophobic emanations from the US government during the Trump administration pushed scientists to Canada, Germany, and the Netherlands, countries with high standards of living and major quantum technology centers. We risk a brain drain unless we create a more welcoming environment and ease the burdens to permanent residence in the US.

Strategic competition Similarly, to realize the quantum age, nations should invest in parallel, enabling technologies. Outer space programs are especially critical in this regard. Nations that have space programs will be able to enjoy quantum sensing and communications capabilities in ways that nations limited to terrestrial deployment cannot. Also, we will realize more quantum technology innovation if inventors can rely on and integrate existing components in their products. A visible example comes from Jian-Wei Pan and Chao-Yang Lu's optical Jiuzhang quantum computer (see Chapter 6, p. 250), a close inspection of which reveals it to be constructed of many components from American optics maker ThorLabs. The US needs to carefully weigh the benefits from levying export controls on more quantum technology precursors against the risks that such innovation will occur anyway, but with components manufactured by foreign, state-supported competitors.

Human futures Through no fault of their own, people are inheriting a world where the traditional sources of human value, as worker, thinker, and fighter, will narrow thanks to automation. Even those on the top of the pile, like the computer programmer, are the focus of intense automation efforts. With our American conception of human value so tied to our economic outputs, the fuse on our incentive and reward system shortens with every step technologists make in automation. No one is safe from automation.

The European campaign to enshrine and expand basic human rights could be an effective hedge strategy for technological futures. Embracing a positive rights system (a right *to* some good, such as education or a basic income, rather than a negative system that is concerned with freedoms *from* government) might help us transition to a world where technology itself has narrowed the workplace.

We ought to be having conversations now about our technical-economic trajectories. Ideas that might seem esoteric now, such as universal basic income, might be the only economic future for most people.

The social benefit scenarios from quantum technologies will be life-changing. But in a highly stratified economy such as ours, those benefits could both be realized and still leave people in a system more feudal than free.

Civil liberties   We assess that the greatest threats to civil liberties in the near term will come from quantum sensing rather than quantum computers. As sensing devices are miniaturized and mounted on aerial and satellite platforms, quantum-equipped actors will see more than others, and in some cases, into private spaces.

Nation states should adopt technology-neutral legal frameworks[1] to address advances in quantum sensing that will create new capabilities to peer into private spaces and technological protections.

Chapter 9 discusses one legal approach, the European human-rights-based framework for addressing technological invasions of privacy by law enforcement. Applied with care, the European model is flexible enough to both anticipate new practices and subject them to substantive limits. Under the European model, governments must seek legal authorization to use investigative methods, those methods must be necessary for a specific law enforcement purpose, and the methods must be proportionate. The effect of these high-level principles is to require governments to disclose their surveillance methods, and to limit the creep of powerful technologies into general criminal deterrence efforts, while allowing aggressive techniques when a credible and specific threat arises. There are now case-law examples of European courts limiting new technologies, such as face recognition, and preventing new technologies from being used for general criminal deterrence, and even for general terrorism deterrence.

---

[1]Not because technology is neutral, but rather because so many US limits on surveillance are keyed to specific technologies or to interference associated with physical touching. A technology-neutral approach would abstract away from the specific technology used and provide legal certainty about acceptable conduct (Koops et al., "Should ICT Regulation Be Technology-Neutral?" (2006)).

Turning to technological countermeasures, it is prudent for institutions to switch now to post-quantum encryption algorithms. Privacy law also suggests several interventions that make sense now, such as limiting data hoarding so that these are not captured decades from now and decrypted.

⌣

We are at the cusp of a quantum technology revolution. We hope this book anticipates the social challenges presented by quantum sensing, computing, and communications technologies. It is now up to policymakers and innovators to pursue normative goals for how the quantum age will be realized.

Appendices

Quantum Science

Here in the back of the book we provide more information about the quantum realm and the weirdness of quantum effects. These two appendices introduce the building blocks of quantum technologies: the atom, quantum sizes, light, and quantum speeds.

A

Introduction to the Quantum Realm

Q
UANTUM MECHANICS describes nature at very small scales – at
the atomic and subatomic levels, but quantum effects have been
observed in large molecules as well. The idea that our everyday
world is made from small particles, which we call *atoms*, dates
to the ancient Greeks. Today we think of atoms as small spherical
objects that are ten-billionths of a meter in diameter. This measurement $(10 \times 10^{-10}\,\text{m})$ is so common that it has a special name – the
Angstrom – and a special symbol, Å. Objects that are angstrom-sized
behave very differently than objects the size of everyday objects like
tennis balls and automobiles. Because humans grow up looking at
and manipulating everyday objects, most people do not have any
intuition about how angstrom-sized objects behave until they have
been educated in modern physics. While Democritus of Abdera came
up with the idea of atoms more than 2500 years ago and our understanding of chemistry has evolved over many centuries, our understanding of quantum mechanics was developed mostly over the past
125 years.

Information theory concerns how information is stored, communicated and quantified. Although humans have been storing and communicating information for thousands of years, our mathematical
understanding of what information actually is dates to a paper by
Claude Shannon from 1948, "A Mathematical Theory of Communication."[1] Among other things, the paper introduced the term "bit,"

---

[1]Shannon, "A Mathematical Theory of Communication" (1948).

short for *binary digit*, as the fundamental unit of information. Much of what is known about the nature of information – including codes, compression and encryption – was first worked out in the 1940s.

As its name implies, QIS combines these two disciplines. Broadly, QIS is the study of approaches that combine knowledge of how quantum effects can be used to measure, sense, communicate, and compute.

This appendix is the first of two that are intended to provide an introduction to quantum mechanics for policymakers who may be generally knowledgeable about our technological world, but who (realistically) did not progress beyond algebra and introductory physics in high school or college. This chapter explains quantum scale and starts an exploration as to why effects at the quantum scale are so radically different from humans' day-to-day experience. Appendix B explores more of quantum mechanics and shows how that theory applies to information science. Readers who want to jump directly to a functional understanding of what quantum technologies enable should review Part 1, "Quantum Technologies."

A.1 The Quantum World: A Brief Introduction

Albert Einstein, Niels Bohr, Max Planck, Werner Heisenberg, and others led what is known as the "first quantum revolution" when they created quantum mechanical theory in the early twentieth-century. Their work was sparked not by a desire to understand things that were very small, but to explain phenomena that could be measured in the world of the 1890s. In short order, they realized that explaining these phenomena required rethinking their understanding of matter, energy, and even time. To do so, they used a combination of physical experiments that were carefully constructed so that their results depended upon the interaction of quantum forces.

The experiments and their interpretations made by these physicists had profound consequences. Fission and fusion bombs are quantum weapons. Other quantum devices powered by the first quantum revolution include the atomic clock, lasers, the transistor, and medical imaging technology, such as magnetic resonance imaging (MRI).

The pace of QIS innovation is increasing, so much so that Jonathan Dowling and Gerald Milburn have labeled the current age the "second quantum revolution." In this second revolution, technologies leverage the special physics of the very small to measure physical phenomena and time more precisely (quantum metrology), to create imagery or otherwise sense phenomena invisible to ordinary sight (quantum sensing), to communicate information, including more secure encryption keys (quantum communications), and to engage in computing (quantum computing).[2]

Writing in 2003, Dowling and Milburn attributed the second revolution to the need for miniaturization and to the potential performance enhancements that QIS provided over technologies governed by classical physics. Today, miniaturization and performance continue to be important driving factors, but other political imperatives and technology developments have emerged to contribute to the second revolution.

Quantum theory seems perplexing because humans have no experience of the subatomic world in daily life. Quantum physics is counterintuitive and difficult to grasp; unfortunately, this means that the label *quantum* frequently becomes a smokescreen for claptrap. When learning about QIS, it is important to distinguish reasoned discussion of quantum effects from quantum fiction designed to entertain, confuse, or distract.

Quantum fiction is readily seen in Hollywood movies where a superhero might pass her hand through a wall, explaining "well, we are mostly made up of empty space" and then perhaps adding a throwaway explanation that her hand is making use of "quantum tunneling." This seems reasonable, because the atomic nucleus is in fact tiny compared to the size of atom and quantum tunneling is a real phenomenon. Quantum tunneling appears to allow particles to skip through barriers, it is the basis of scanning tunneling microscopy, and it presents a fundamental limit for how small the features of integrated circuit transistors might actually get. But quantum tunneling only happens at the subatomic scale. In real life, a superhero cannot phase through a wall because the electrons in the hero's hand repel the electrons in the wall. That is why the quantum fiction in superhero movies relies on computer graphics to add the visual effects, rather than relying on quantum physics.

A.2   Terminology, Size, and Frequency

This section introduces the terminology of modern physics and conveys a sense of the sizes involved.

---

[2]Dowling and Milburn, "Quantum Technology: The Second Quantum Revolution" (2003).

A.2.1 The Atom

Hydrogen is the simplest atom, with a single negatively charged electron orbiting around a positively charged nucleus that contains a single proton. Because it has a single proton, the hydrogen atom is said to have an atomic weight of 1. Hydrogen gas is a molecule that consists of two hydrogen atoms: the two nuclei each consists of a single proton, and the two electrons are shared between them, forming what chemists call a *covalent bond*. (Although the chemical formula for elemental hydrogen is H_2, it is sometimes written H:H to emphasize that the two electrons are shared.)

A small fraction of the hydrogen on the planet has both a positively charged proton and a neutrally charged *neutron* in its nucleus: this kind of hydrogen is called deuterium and it has an atomic weight of 2. Water made from deuterium is called heavy water and played a role in the German atomic bomb program in World War II because it can be used as a moderator to sustain a nuclear chain reaction, a critical step in producing plutonium. It is also used in medical research, to measure food intake and energy balance.

A third kind of hydrogen, called tritium, has two neutrons; tritium is highly radioactive because the atom's nucleus has twice as many neutrons as protons. (Nuclei become unstable if the neutron/proton ratio is more than 1.5:1.) Tritium has been used in self-illuminating mechanical watch dials and as a tracer in medical diagnosis. It is also an ingredient in certain kinds of nuclear weapons.

Quantum mechanics describes electrons, protons and neutrons with mathematical equations that define a probability distribution. Instead of thinking of electrons whizzing around the nucleus like planets around the Sun, think instead of an electron cloud surrounding the nucleus, like a swarm of bees buzzing around a hive. But even that analogy is flawed. Mathematically, the hydrogen's electron is better described with equations that describe a wave centered on the atomic nucleus, like the vibrating surface of a bell that has been struck. The equation describes how the electron's energy changes when it absorbs light.

Quantum mechanics also describes how hydrogen atoms resemble a spherical shell with a diameter of approximately 1.1 Å centered on the hydrogen nucleus. The nucleus is described with a similar equation, except that it has a diameter of 0.000 017 Å (1.7×10^{-5} Å, or 1.7×10^{-15} m).

Hydrogen's electrons can be anywhere, no space within the atom is strictly empty. Look at an even smaller scale, and even the "empty space" within the atom – as well as the empty space between atoms – may be filled with observable space–time fluctuations – quantum foam – in which mass and energy is created and destroyed in a manner that is consistent with the Heisenberg Uncertainty Principle.

A.2.2 Quantum Sizes

The previous section uses measurements like 1.1 Å and 10×10^{-15} m without much reflection; this section attempts to provide a better understanding of quantum sizes.

In the classic 1977 short movie *Powers of Ten and the Relative Size of Things in the Universe and the Effect of Adding Another Zero*, the noted twentieth-century designers Charles and Ray Eames take the viewer on a voyage through 46 orders of magnitude. [3]

When the movie starts, the field of view is 1 square meter (1.09 yards) and shows a man and woman at rest on a blanket on the western shore of Lake Michigan. The field of view then zooms out, a factor of 10 every 10 seconds. At 20 seconds, the field of view is 100 meters, showing the entire field, at 30 seconds the 1000 m field of view shows several blocks, and so on.

Scientists and engineers commonly use exponents to describe large numbers. The measurement 1000 m can be written as 10^3 m or as 1 km (1 kilometer). The notation 10^3 literally means "the number ten multiplied by itself three times," or $10 \times 10 \times 10$. Scientific notation is useful for measurements like 10^{10} m (the distance the Earth travels through space in about four days), 10^{20} m (the scale of the structure of the Milky Way Galaxy and its rich brotherhood of stars) and 10^{24} m, the maximum scale shown in the Eames movie. Today we believe that the size of the observable universe is 93 billion light years, or 3.6×10^{25} meters. Many computer programs use the letter "E" to represent scientific notation, so the reader may encounter the measurement written as "3.6E25m."

The second half of the Eames film returns to the couple in Burnham Park and then zooms off in the other direction, everything in the frame getting smaller by a factor of 10 every 10 seconds. Twenty seconds into the second half, the field of view is 10^{-2} m across (also called 1 centimeter, or one hundredth of a meter). The frame shows

---

[3] The nine-minute Eames film is online www.eamesoffice.com/the-work/powers-of-ten/.

Understanding Negative Exponents

Mathematicians define the number 10^0 as 1, which might seem strange. *How can you multiply anything by itself zero times? Should not the answer be zero, not one?* This is an example of a model that works well in one domain, but fails when applied to another.

Because the addition of exponents can be defined in terms of multiplication, subtracting exponents is defined as division. Just as 1000 divided by 100 is 10, 10^3 divided by 10^2 is 10^1. That is, $10^3 \div 10^1 = 10^{3-2} = 10^1 = 10$. It then follows that 10^0 is 1. (This also works if you think of the exponent x in the equation 10^x as the number of 1 followed by x zeros, which shows the advantage of having a superior mental model.)

Negative exponents extend this idea in the other direction. The number 10^{-1} is the same as the number 0.1 or $\frac{1}{10}$. More generally, $10^x = \frac{1}{10^x}$.

---

a patch of skin on the man's hand. At 50 seconds the scale is 10^{-5} m, or 10 micrometers (also called *microns*), which is the size of a white blood cell. At 10^{-10} m (1 Å) the screen fills with the electron shell of a hydrogen atom. At 10^{-15} m, which the film calls a "fermi" (a unit of measure named for Enrico Fermi but not widely used), the screen shows a proton and a neutron, two of the building blocks of matter. The film stops at 10^{-16} m, which the narrator explains is the scale of quarks, electrons, and positrons.

As the preceding paragraphs demonstrate, one of the reasons policy specialists find it difficult to digest quantum literature is that the same measurements can be described many different ways. For example, the nitrogen atoms used for quantum sensing discussed in this book have a diameter of 1.12 Å, but that measurement might alternatively be reported as a radius of 56 picometers (pm), 0.056 nanometers (nm) or 5.6×10^{-11} m.

In addition, scientists typically describe the measurement of a sphere with its radius (the distance from the center to the surface), rather than the diameter, because the equations that describe the properties of circles and spheres are simpler when based on r (the radius) rather than d (the diameter). But for people who think of atoms as tiny tennis balls, the concept of radius can be confusing, because we tend to think of tennis balls as spheres which have an official *diameter* of 6.54–6.86 cm (as defined by the International Tennis Federation) – and not as spheres that have an official radius of 3.27×10^{-2} m to 3.43×10^{-2} m. (Of course, both measurements are exactly the same.)

At quantum scales, nature is probabilistic and objects have behaviors reminiscent of both waves and particles. This differs from how objects behave at the scale of real tennis balls, rackets and courts. The way these objects behave guides our intuition and, as a result, shaped the development of what is called classical physics. In our ordinary lives, one can determine how objects will act by knowing their mass, inertia, and so on. At a quantum scale, reality is governed by probability. That is, one can make predictions about the location of subatomic particles but these predictions are probabilities. As such, quantum science is as unsettling as is it profound.

A.2.3 Light

The fundamentals of light are a focus of early education. For example, many students in high school will learn that sunlight is actually a mixture of all the colors in the rainbow, and what we call a "rainbow" is actually drops of rain acting like a prism, splitting sunlight into its component colors. On the other hand, students who take theater class will learn that what looks like white light can be produced by mixing red, green, and blue light together. (That works because most people's eyes have three kinds of color-sensing cells, sensitive to red, green, and blue light respectively.) Light can be filtered by color: shine white light into a red filter, and red light comes out the other side. But light of one color cannot be changed into another color: pass red light into a blue filter, and nothing comes out.

For many years scientists were confused about the fundamental nature of light: some scientists, like Isaac Newton, thought light was made out of tiny objects he called *corpuscles*. Other scientists like Thomas Young thought that light was actually some kind of wave traveling through some kind of medium variously called the *ether*, also written *aether*, *æther* (or even αιθηρ if you happen to write in ancient Greek).

If you use a prism to produce a rainbow, you will discover that some of the Sun's energy extends on both ends beyond the familiar red-orange-yellow-green-blue-indigo-violet (a.k.a. ROYGBIV) color chart. Place your hand to the left of the red and your hand will grow

Through a Glass Darkly...

QIS forces us to understand just how limited human perception is. As humans, we indeed see the world through a glass darkly. Consider what we see of the world – the visible light spectrum (see below). The quantum realm exists mostly outside the world of human experience. QIS and resulting quantum technologies are counterintuitive because there are few, fleeting moments when humans see quantum effects. Our entire experience is based on the physics of relatively large objects.

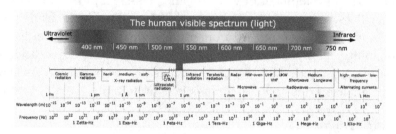

Will better understanding the physics of the small change how we perceive our own greatness, and even relevance? The same passage of the Bible that gives us the lovely metaphor of *seeing through a glass darkly*, a reference to the imperfect mirrors of antiquity, explains the concept that our perception of god is incomplete: *For we know in part, and we prophesy in part. But when that which is perfect is come, then that which is in part shall be done away.*

Quantum physics is replacing the imperfect mirror of classical mechanics. That imperfect mirror allowed us to leave many questions to prophesy. As the quantum physics mirror is perfected, what will be left to prophesy?

Sidebar 1. Visible light is a tiny part of the electromagnetic spectrum. Image CC BY-SA 4.0 by Wikimedia user Horst Frank.

warm from infrared light, which is too far red for the eye to see. Sir William Herschel discovered this effect in 1800, although he used a thermometer and not his hand.

Place your hand to the right of the violet and you will get a suntan, and then a burn, from the ultraviolet light. Johann Wilhelm Ritter discovered ultraviolet this way in 1801, although Ritter used silver chloride, a chemical that turns black when it is exposed to sunlight (and was the basis of the wet chemistry used in photography for nearly two centuries). Ritter discovered that silver chloride turned black faster with blue light than with red light – and faster still when placed past the violet edge of a spectrum created with sunlight and a glass prism.

What is called visible light is actually just a tiny part of the *electromagnetic spectrum*, which includes radio waves, microwaves, infrared rays, visible light, ultraviolet light, X-rays, and gamma rays (see Table A.2). As humans, we perceive the reality of the physical world through a glass darkly.

A.2.4 Quantum Speeds

Light is one of the best understood quantum objects, in part because of its simplicity, in part because it is the easiest to study, and in part because humans can observe it. In ancient times people thought that light moved instantly, but in the 1670s the Danish astronomer Ole Rømer determined light must have a finite speed based on his observations of eclipses of Jupiter's moon Io. Rømer's estimation of the speed of light was about 220 000 kilometers per second, roughly three-quarters the actual value of 299 792 km/s in vacuum.

All colors of light travel at the same speed, as do radio waves, microwaves, and gamma rays. It turns out that a beam of monochromatic light also has a wavelength and a frequency. They are related by this equation:

$$c = \lambda f \tag{1}$$

where:

c = the speed of light (roughly 300 000 km/s)
λ = the wavelength of light
f = the frequency of the light, measured in cycles per second (Hz).

That is, the wavelength times the frequency is equal to the speed of light. Since the speed of light is constant (it is a sort of universal speed limit), light with small wavelength has a high frequency, and light with large wavelength has a small frequency.

Table A.1. The electromagnetic spectrum.

Kind of light	Wavelength	Frequency
Longwave radio	3×10^8m	10^0Hz = 1Hz
AM radio waves	3×10^2m	10^6Hz = 100Kilohertz
FM radio waves	3×10^0m = 3m	10^8Hz = 100Megahertz
Microwaves	3×10^{-2}m = 3cm	10^{10}Hz = 10Gigahertz
Near Infrared	3×10^{-6}m = 3μm	10^{14}Hz = 10Terahertz
Visible	380–740nm	405–790 Terahertz
Ultraviolet	3×10^{-8}m = 30nm	10^{16}Hz = 1Petahertz
X-rays	3×10^{-10}m = 3Å	10^{18}Hz = 100Petahertz
Gamma rays	3×10^{-14}m	10^{22}Hz

Table A.2. The visible electromagnetic spectrum.

Color	Wavelength	Frequency
Violet	380–450 nm	680–790 THz
Blue	450–485 nm	620–680 THz
Cyan	485–500 nm	600–620 THz
Green	500–565 nm	530–600 THz
Yellow	565–590 nm	510–530 THz
Orange	590–625 nm	480–510 THz
Red	625–740 nm	405–480 THz

Table A.1 shows the wavelengths and frequencies for various kinds of light. Notice that invisible light does not follow the convention of falling within even powers of 10. Unfortunately visible light does not fit neatly into this table; our eyes evolved to perceive light in the relatively tiny range of light that has wavelengths of 380nm to 740nm, as shown in Table A.1.[4]

This exposition reveals an important point: the quantum realm and almost all of its effects happen outside human perception. We perceive the world through a glass darkly – a glass that only reveals

---

[4]Not every human's eyes work the same way. There is a version of so-called "color blindness" in which the blue cones are sensitive in the ultraviolet, and there are some humans who have four color receptors, a condition called tetrachromacy. Such people, who are exceedingly rare, see a richer pallet of colors.

visible light. The quantum world and quantum effects typically require special equipment to perceive and to manipulate. As a result, we have little day-to-day experience with the quantum world, and its attributes are thus counterintuitive and take work and study to learn.

This appendix introduced the basics of the quantum realm. The appendix covered the reasons why quantum information science is exciting, the relative sizes of quantum phenomena, fundamental properties of light, and the idea that everything has wave- and particle-like properties. This foundation is necessary for the next appendix, which turns to quantum effects: wave mechanics, the uncertainty principle, polarization, entanglement, superposition, and the "cat state."

B

Introduction to Quantum Effects

W HAT are quantum effects, how does one build an intuitive sense of them, and what do quantum effects mean? The roots of these important questions are found in wave mechanics. The previous appendix began the exploration of the quantum world with a review of quantum sizes, measurement, and the properties of light. This appendix builds on that knowledge by summarizing the history and debates of wave mechanics, which was developed at the start of the twentieth century. The appendix then introduces three quantum effects that flow from wave mechanics: uncertainty, entanglement, and superposition. These three quantum effects form the basis of the quantum computing, communication, and sensing technologies discussed later in this work.

B.1 Wave Mechanics

What are quantum effects and what do they mean? Consider Richard Feynman[1] (pronounced Fine-man), the American physicist who was also a great popularizer of science. Feynman was critical of attempts to understand the meaning of quantum mechanics. As he made clear in numerous public speeches and lectures, quantum mechanics is a set of mathematical equations that explain experiments and observed

---

[1] Feynman shared the 1965 Nobel Prize in physics with Sin-Itiro Tomonaga and Julian Schwinger "for their fundamental work in quantum electrodynamics with deep-ploughing consequences for the physics of elementary particles." Nobel-Prize.org, "The Nobel Prize in Physics 1965" (2019).

phenomena. "I think I can safely say that nobody understands quantum mechanics," was one of his more memorable quotations.

Yet, despite these recommendations, physicists and the public alike are thirsty for some kind of intuitive understanding of what these quantum equations *mean*. Such an understanding is especially important for this book, since our goal is to provide insight into quantum information science and its implications without delving into the underlying physics and math. The remainder of this section, describes four critical *observations* that are the basis of quantum physics and are critical for grasping what is special and different about quantum technologies.

B.1.1 Quantum Swirls

What happens at the quantum domain doesn't stay in the quantum domain: quantum effects are visible all around us if one knows where and how to look. Perhaps the most obvious evidence is what physicists in the early twentieth century called the *wave–particle duality*. This duality indicates that the physical building blocks of reality – mass and energy – result in effects at the macro-scale that are reminiscent of both waves and particles. This confounded physicists for a time, as they assumed things like light and matter had to be *either* discrete particles *or* waves oscillating in some kind of medium. The birth of quantum physics resulted from the realization (and the corresponding mathematics) that light and matter are neither waves nor particles, that there is no medium, and that tiny microscopic objects don't behave like tennis balls.

The swirl of colors in a soap bubble (Figure B.1) illustrates a quantum process at work. The colors are created by interference between two wave fronts: the light reflecting off the front side and the back side of the soap film. This demonstrates the wave-like properties of light. Different colors are caused by light with different wavelengths, unquestionably demonstrating that light is a wave. Such wave-like behavior is not limited to light: similar effects can be observed in tiny "particles" of matter (such as electrons), and even in large organic molecules.[2]

On the other hand, if you take light from the Sun and shine it on a piece of metal, you'll discover that the Sun's ultraviolet light – the same kind of invisible light responsible for sunburn – can dislodge

---

[2]Gerlich et al., "Quantum Interference of Large Organic Molecules" (2011).

Figure B.1. The colorful swirls in a soap bubble are the result of constructive and destructive interference of light reflecting against the inside and outside soap film walls. The changing distance between the two walls at different points in the bubble simultaneously results in constructive interference of some colors and destructive interference of others. As a result, the soap film seems to possess different colors at different points. Image CC-BY-SA Wikimedia user Werner100359.

Figure B.2. In this illustration of a young child jumping rope, the movement of the rope describes a circular wave. The rope is the wave's medium, the rope's wavelength twice the child's arm-span, and the frequency is the number of times per second that the rope passes under the child's legs. The wave's amplitude is distance from the line between the child's hands and the rope's midpoint at the child's ankle.

electrons from the surface of the metal, producing a slight voltage, while light from the red end of the spectrum can't. This is called the photoelectric effect.

What is odd about the photoelectric effect, though, is that whether or not light produces electricity when it hits the metal depends entirely on the light's color – its wavelength or frequency – and not the light's brightness or intensity.

There are two numbers that describe a wave propagating through a medium: the wave's amplitude and its frequency. The amplitude is how much the wave displaces the medium from its resting state, also called its ground state. The frequency is how many times per second the wave causes the medium to oscillate. (See Figure B.2.)

Classical physics says that the energy transferred by a wave is proportional to its amplitude. If light were a wave, its brightness

On Quantum and Elementary Particles

In this section we've used the imprecise phrase *quantum particles* to describe very small particles, whereas most texts would probably have used the term *elementary particles*, meaning the smallest particles that are the building blocks of matter.

Electrons, protons, and neutrons were once called *elementary particles* because they were thought to be the fundamental building blocks of matter. Today, most physicists subscribe to the *Standard Model* which describes the hundred-or-so subatomic particles out of which the universe is thought to be made. Under the Standard Model, the term *elementary particle* is reserved for *leptons* and *quarks*. Electrons are leptons, whereas protons and neutrons are made up of quarks. Protons in particular are made up of two Up and one Down quarks, while the neutron is made up of two Down and one Up quarks. Quarks and leptons are both called *fermions*. There are 24 kinds of fermions: the six quarks (named up, down, strange, charm, bottom, and top), six leptons (the electron, electron neutrino, muon, muon neutrino, tau particle, and tau neutrino), and, for each lepton, its antiparticle.

The photon is neither lepton nor quark: it is a boson, which is the name used for particles that follow Bose–Einstein statistics. The key difference between fermions and bosons is that fermions obey the Pauli exclusion principle, which means that two fermions cannot be in the same place and in the same state at the same time, while any number of bosons can be packed together. Light is a boson (and in particular, a *gauge boson*), which is why many photons can be packed together in a laser. Likewise helium is a boson (it's actually called a *composite boson*), which allows it to form a superfluid when it is cooled close to absolute zero.

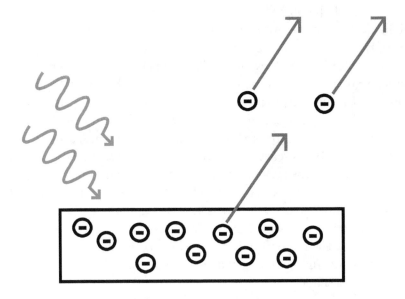

Figure B.3. The photoelectric effect results when light striking metal causes electrons to eject. Einstein explained the effect by saying that the energy of light was carried in individual particles, which are now called *photons*, and that the energy of those particles is proportional to the frequency of the light, with photons from higher-frequency light having more energy. Image CC-BY-SA Wikimedia user Wolfmankurd.

would be the wave's amplitude, and its color would be its frequency. Einstein's explanation required that light be viewed as a stream of particles, not waves, with the energy of each particle depending, perplexingly, on that particle's "wavelength."

B.1.2 Light: Newton Thought It Was a Particle

The nature of light was a centuries-long riddle for scientists. Just what is light, and how does it work? Why are some things different colors? Why is there color during the day but not at night? Teasing out which of the observed effects were due to the underlying nature of light, and which were due to the nature of the observer, took literally thousands of years of scientific work.

In 1704, Sir Isaac Newton published his treatise *Opticks*, in which he showed how the fundamental nature of light could be revealed through the use of prisms and mirrors. In that book Newton also promoted what was called the *corpuscular theory of light*, the idea that light was made up of tiny particles called *corpuscles*, a concept first proposed by Descartes in 1637. Newton's work on light bolstered the corpuscular theory, since light travels in straight lines and reflects from mirrors at right angles, like billiards bouncing off a pool table's bumpers. Waves traveling across the surface of a bath tub or lake just don't act that way. Furthermore, Newton argued that if light were a wave, then one would see interference fringes in the boundary between light and darkness that arise when an object with a sharp edge casts a shadow.

For all of Newton's prestige, the corpuscle theory really didn't do a good job explaining why light has color. But the real nail in the theory's coffin was the discovery that light in fact did produce interference patterns.

B.1.3 Light: It Acts Like a Wave

By the end of the eighteenth century, physicists had a basic understanding of waves from observing their behavior in water. For example, physicists understood that waves traveled through some kind of medium, causing it to cycle up and down.

Recall from the illustration of the child jumping rope, the height of a water wave is its amplitude, while the distance between the peaks is the wavelength. The frequency is the number of times per second that the rope passes over the child's head. The frequency and the wavelength of a wave are inversely related.

Interference happens between waves when two (or more) waves meet and pass through each other: where the wave peaks align, the interference is *constructive:* the peaks add together, increasing their intensity. Where a peak aligns with a trough, the interference is *destructive*, and the waves cancel each other out. You can readily perceive this effect with sound by having a colleague stand a few feet away from you with two tuning forks. If your colleague strikes both forks and holds them a foot apart, you will perceive the sound to be louder and quieter as you approach or retreat from your friend's position. The change in volume is caused alternately by constructive and destructive inteference of the sound waves, which are now known as compression waves in the medium of air.[3]

In 1801, the British scientist Thomas Young devised an experiment that established beyond a doubt that light has wave-like properties. In the experiment (see Figures B.6 and B.7), a stream of light travels through two slits in a black plate. Young reasoned that if light were made out of tiny ball-like particles (Newton's "corpuscles"), the particles passing through each slit would produce a slightly larger rectangular line on the screen. And indeed, that's what happens if the slits are large. But when the slits are small, an interference pattern emerges, showing that light has wave-like properties.

At the time, Young and others assumed this meant that light was actually a kind of wave, like sound, and not a kind of particle as Newton had hypothesized. (Full-length books have been devoted to the two-split experiment[4] the complexity of which will not be fully conveyed here.)

Of course, once you know what to look for, interference shows up in all kinds of places: put a lightly curved watch glass on a piece of white paper and illuminate it from above, and you will observe a bull's-eye pattern of rainbows (if illuminated with white light), or light and dark circles (if illuminated with monochromatic light). These circles are called Newton's rings (Figure B.4) and they are an interference fringe; they allow you to make precise measurements

---

[3]The invention of the vacuum pump in 1650 by Otto von Guericke and the discovery of air pressure was a major driver of the scientific and engineering revolutions that were to follow. For an excellent history of vacuum science, see Grant, *Much Ado About Nothing: Theories of Space and Vacuum From The Middle Ages to The Scientific Revolution* (2008).

[4]Ananthaswamy, *Through Two Doors at Once: The Elegant Experiment That Captures The Enigma of Our Quantum Reality* (2018).

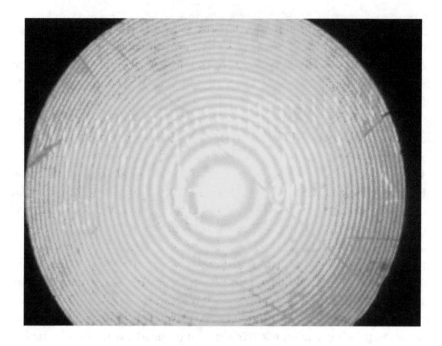

Figure B.4. Newton's rings observed through a microscope. The pattern is created with a 20cm convex lens illuminated from below by a monochromatic sodium lamp. The scale has 100μm increments. Image CC-0 by Wikicommons user Warrencarpani.

regarding changes in distance or pressure between the glass and the paper.

Physicists have repeatedly made good use of light's wave-like properties since 1801 – and they continue to do so to this day.

Consider the use of the Doppler Effect, which is the term that physicists use to describe the apparent upwards shift in frequency when the distance between a wave emitter and an observer is decreasing, and the corresponding apparent decrease in frequency when that distance is increasing. If an emergency vehicle with a blaring siren approaches and then speeds past you on a street, the siren's wail will be heard at a higher pitch as the vehicle approaches and passes a listener, and then at a lower pitch as the vehicle recedes. This change in pitch was first characterized by the Austrian physicist Christian Doppler in 1842. The shift is caused because the decreasing distance between the vehicle and the listener effectively results in the peaks of each sound wave hitting the listener's eardrum faster than they would if there was no relative motion between the two. Likewise, when the vehicle is receding, the sound waves are effectively stretched out.

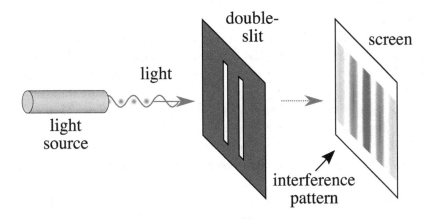

Figure B.5. Thomas Young's dual-slit experiment demonstrated that light has wave-like properties. In the double-slit experiment, light from an emitter travels through two slits and forms an interference pattern on the screen, just as waves passing through two holes in a water break cause interference on a lake. Image CC-BY-SA by NekoJaNekoJa with author edits.

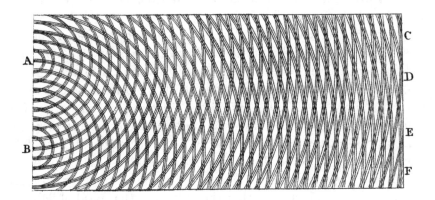

Figure B.6. This is a drawing from Thomas Young's notebook showing how light rays traveling from two point sources A and B result in constructive and destructive interference. In the dual-slit experiment, each slit can be thought of as a stack of point sources that emit light solely in the horizontal plane. Image public domain.

In 1929, US astronomer Edwin Powell Hubble showed that the light from stars and distant galaxies is also shifted red, implying that every galaxy in the night sky is moving away from us. This was the first evidence that the universe is expanding, which is indirect evidence for the Big Bang theory. Red shift measurement depends on the fact that light has wave-like properties, and the fact that light's behavior at macroscopic scales almost perfectly approximates the behavior of a wave moving through some kind of medium. (Indeed, in 1873, James Clerk Maxwell published his theory of electromagnetism with his now-famous set of equations that beautifully described the behavior of light, basing that description on the core idea that light was a wave.)

Physicists can also use the wave-like way that light casts interference patterns to make precise measurements of distance using a technique known as *interferometry*. The technique works by splitting coherent, monochromatic light from a single source into two beams which reflect off two different mirrors and are then recombined: if the distances are exactly the same, then the peaks from one path precisely match the peaks from the other, the interference is positive, and the resulting light is the same brightness as the original source before splitting. But if one path extends precisely one-half of a wavelength longer, then the peaks from one path line up with the troughs from the other, and the interference is destructive: the two beams cancel each other out.

In 2015, a pair of 2000 m-long L-shaped interferometers at the Laser Interferometer Gravitational-Wave Observatory (LIGO) were able to detect minute distortions in the curvature of space–time resulting from gravity waves generated by the collision of two neutron stars in galaxy NGC 4993, 144 million light-years from Earth. The collision was also detected by Virgo interferometer operated by the European Gravitational Observatory, with scientists at both labs winning the 2017 Nobel Prize in physics as a result. As an aside, not only did the experiment demonstrate the existence of gravitational waves, it also showed that they travel at the speed of light, as predicted by Einstein's general theory of relativity.

B.1.4 Light: How Can It Possibly Be a Wave?

It turns out that light can't be a wave for two very basic reasons: there's no medium to vibrate, and light comes in discrete, countable units – something that waves just don't do.

Waves on the surface of a lake result from bulges and troughs in water, while sound waves result from compression of air. Water and air are both a *medium* that transmits wave energy. Given that light was obviously a wave phenomenon, scientists of the nineteenth century[5] wanted to better understand the medium that light was moving through. They called that medium the *Luminiferous aether*, or simply *aether* (or even *ether*).

Physicists knew that light could travel through air, water, glass and even a vacuum, so aether had to be everywhere, penetrating everything. And since the Earth was rotating, light traveling in the direction of the Earth's motion should be impacted by the aether differently than light traveling at right angles. This created an opportunity for measurement. In 1887, Albert A. Michelson of the Case School of Applied Science and Edward W. Morley of Western Reserve University, both in Columbus, Ohio, built a massive interferometer in the basement of a university dormitory to measure this effect of the earth's movement through the aether.[6] They failed to find any effect, demonstrating that there was no aether for light to be traveling through.

Another problem with the wave theory of light is that it can't mathematically explain the amount of light emitted by objects when they heat up. If you have ever worked with a furnace, or even an electric stove, you know that when things like metal rods get hot, they tend to glow – first red, then orange, and eventually bright white. This is called *black-body radiation*, because the color of the light is independent of the color of the object being heated – it even comes off objects that are pure black.

In the late-nineteenth century, physicists started measuring the light coming off of hot objects and then trying to develop mathematical models to explain their measurements. Based on the wave theory of light, the amount of ultraviolet radiation coming off a hot piece of metal should have been significantly higher than the amount of blue or green light – but it was significantly less. In fact, predictions based on Maxwell's equations indicated that as the frequency of light steadily increased, the amount of light coming off should steadily in-

---

[5] Although much of the work to detect the aether took place in the nineteenth century, theories regarding the aether date back to Robert Boyle and Christiaan Hyugen Huygen, *Traité de la Lumière* (1690) in the seventeenth century.

[6] The Case School of Applied Science and Western Reserve University merged in 1967 to form Case Western Reserve.

crease as well, such that an infinite amount of light was coming off with light that had an infinite frequency. Clearly that wasn't happening. This mismatch between theory and observation was called *the ultraviolet catastrophe.*

In 1900, the German physicist Max Planck published a mathematical theory that properly predicted radiation emitted by black bodies. The theory assumes that the light emitted by black-body radiation is *quantized* at specific levels. Planck didn't go so far as to say that quantization was inherent in all kinds of light. Still, this work earned Planck the 1918 Nobel Prize in Physics, "in recognition of the services he rendered to the advancement of Physics by his discovery of energy quanta."

Five years later, Einstein built upon Planck's work and suggested that light itself was quantized, and not merely the energy levels at which light is radiated from black bodies. With this leap of intuition, Einstein was able to explain the aforementioned photoelectric effect. Einstein's 1905 explanation of how it works[7] was experimentally confirmed by Robert Millikan in 1915 at the University of Chicago.[8] It was for this work that Einstein was awarded the 1921 Nobel Prize in Physics "for his services to Theoretical Physics, and especially for his discovery of the law of the photoelectric effect."

Thus, the inescapable conclusion of more than a century's worth of physics research is that light is both a wave and a particle – or, more accurately, that physicists can construct experiments in which light has observable effects that appear similar to the wave-like effects that physicists can observe in sound waves, and the particle-like properties that physicists can observe in objects like tennis balls.

Before the invention of quantum mechanics, some physicists called this the "wave–particle duality," a name that has unfortunately persisted to this day (although the authors will try not to use that phrase elsewhere in this book). Einstein explained it this way in 1938:

> It seems as though we must use sometimes the one theory and sometimes the other, while at times we may use either. We are faced with a new kind of difficulty. We have two contradictory pictures of reality; separately nei-

---

[7]Einstein, "Über Einen Die Erzeugung Und Verwandlung Des Lichtes Betreffenden Heuristischen Gesichtspunkt (On The Production and Transformation of Light From a Heuristic Viewpoint)" (1905).

[8]American Physical Society, "Robert A. Millikan" (n.d.).

ther of them fully explains the phenomena of light, but together they do.[9]

In fact, as far as light goes, quantum theory explains virtually *all* observations that humans have ever made. The one exception is that quantum theory does not explain the curvature of space–time, which clearly affects the way that light bends around massive gravitational objects like stars and black holes. But with the exception of gravity, the quantum theory of light appears to be complete.

The word *photon* itself was coined by Gilbert N. Lewis in a 1926 letter to *Nature*.[10]

It's Not Just Light: Everything Has Both Wave-like and Particle-like Properties

This apparent combination of both wave-like and particle-like effects is not confined to light: all matter has wave-like properties, from tiny particles of matter like electrons, to much larger molecules, to planets and stars. More to the point, these waves can even be measured – at least in the case of electrons and molecules.

In 1924, Louis-Victor de Broglie derived an equation that relates the wavelength of any object (λ) to momentum[11] (p) and Planck's constant (h). That equation is:

$$\lambda = \frac{h}{p} \tag{1}$$

de Broglie's equation implied that *everything* has a measurable wavelength (or, if you prefer, a measurable frequency). When scientists went out to measure these waves, they found them ... with precisely the wavelength that de Broglie's equation predicts. The first confirmation came from Bell Labs in 1927,[12] when slow-moving electrons hitting crystalline nickel were shown to refract (at the quantum level, the arrangement of atoms in crystalline nickel looks like a lot of ridges or slits). The idea that matter has wave-like properties was so radical, and the confirmation was precise, that the Nobel committee awarded de Broglie the 1929 Prize in Physics "for his discovery of the wave nature of electrons."

---

[9]Einstein and Infeld, *The Evolution of Physics: The Growth of Ideas From Early Concepts to Relativity and Quanta* (1938).

[10]Lewis, "The Conservation of Photons" (1926).

[11]Recall that the momentum of an object is its mass times its speed.

[12]Davisson and Germer, "Reflection of Electrons by a Crystal of Nickel" (1928).

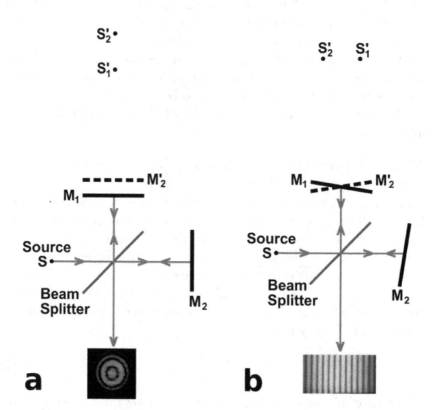

Figure B.7. A Michelson interferometer uses a source of light and a beam splitter to make precise measurements of the relative angles and distances of the two mirrors M_1 and M_2. The bull's-eye pattern results from rings of constructive and destructive interference between the convex lens and a plane of glass, with each band representing an increase in separation between the planes of glass equal to an additional wavelength of light. If light were actually a wave traveling through the aether, then the interference pattern would be smeared when the direction of the earth's movement when the movement was aligned with an axis of the lens; the resulting pattern would be a series of bars, rather than a bull's-eye. Michelson and Morley never observed such a pattern: this was taken as proof that the Earth is not moving through an aether medium. Image CC-BY Wikimedia user Stigmatella_aurantiaca.

In 1999, researchers at the University of Vienna demonstrated diffraction patterns from C_{60} "buckyballs" (fullerenes), which have a diameter of roughly 0.7 nm, meaning that even large molecules have observable wave-like properties. Larger objects, like books, cars, and people, have wavelengths, although they are tiny, even by quantum standards – that's because an object's wavelength is inversely proportional to its mass. In the case of a 58 g regulation tennis ball being served at the 263 km/h,[13] the fastest serve on record, p is 0.058 kg times 4383 m/s, giving a wavelength of 2.6×10^{-37} m, which is 22 orders of magnitude smaller than the diameter of a proton. That's the reason why the wave-like nature of particles typically isn't visible in our day-to-day classical world: the wavelengths are so small that they can be easily ignored.

In 2018, scientists at Hitachi demonstrated a version of the dual-slit experiment using an advanced device that can detect individual electrons and show them on a screen.[14] In the experiment, individual electrons are accelerated in a vacuum to 50 000 volts, which is 40% the speed of light. The electrons then pass on either side of an electron biprism (a very thin wire with a negative charge) and then smash into a detector. The team posted a video on YouTube showing the screen as each additional electron arrives. Since just 10 electrons travel through the device each second, there is no way for the electrons to interfere with one another – most of the time, there is no electron in the device. And indeed, as the first few electrons appear on the screen, they appear to be randomly placed. But after the experiment runs for 20 minutes, a clear pattern of bright and dark bars appears on the screen. This is the characteristic interference pattern of the dual-slit experiment.

So what's going on?

The Hitachi experiment shows that the electrons are arriving at the detector in accordance with a probability distribution. The bright bars are where electrons have a high probability of appearing; the dark bars are where the probability is low. By measuring the distances, it's possible to calculate the wavelength that would cause such a pattern to appear: it is the same wavelength that is revealed by the de Broglie equation.

---

[13] en.wikipedia.org/wiki/Fastest_recorded_tennis_serves
[14] www.hitachi.com/rd/portal/highlight/quantum/doubleslit/index.html

Paradoxically, the only way to make sense of this experiment is to let go of the classical notions that electrons are little particles that travel along paths predetermined by the mass, charge and momentum. Instead, think of the electron gun, the two slits, and the detector as a single system. The release of each electron, its acceleration to 50 000 volts, its travel through the slits, and its detection are not four distinct events, but a single action that takes place in space and time, transferring a tiny bit of mass from the electron emitter and a tiny bit of energy from the acceleration plates to the detector. This transfer of mass and energy can be described by a single equation that, when solved, provides the probability that is different for different points of the single electron detector.

If you crave a classical explanation for what is happening, consider a gambler who is rolling a pair of dice. With each role, there is a chance that the sum of the two dice will be 2, or 12, or any number in between. There's no way for the gambler to predict the next roll of the dice, but the gambler knows that, over time, a roll of 7 is the most likely. Likewise, in the Hitachi experiment, there's no way to predict the location of each electron, but over time the pattern of light and dark bars will clearly emerge.

Light and Matter: whatever it is, it's described by Schrödinger's wave equation and Heisenberg's matrices.

In 1925, the 24-year-old Werner Heisenberg was working as an assistant to Max Born at the Institute of Theoretical Physics at the University of Göttingen. There Heisenberg developed a mathematical formulation based on matrix math that accurately described the interactions between light and matter that scientists had been able to precisely measure up to that point.[15] The following year, Erwin Schrödinger developed what is now called the Schrödinger Wave Equation which does the same thing, but which is based on partial differential equations. The two formulations are in fact mathematically equivalent, although it is sometimes easier to use one formulation, and sometimes easier to work with the other. It is these systems of equations that are called *quantum mechanics.*

---

[15]While the phrase "the interaction of light and matter" may sound quite grandiose, most of these interactions are simply what happens when electrons in atoms absorb a photon and jump to a higher energy level, and when electrons drop back down to a lower energy level, emitting a photon.

For many people, philosophers and physicists alike, the challenge of quantum mechanics comes when trying to ascribe meaning or an "interpretation" to these equations. Our recommendation here is the same as Richard Feynman's: don't try to ascribe sense or meaning to the equations, just accept that they accurately predict experimental observations. Leave it at that.

For example, if you set up the equation to describe the position of an electron around a hydrogen atom, you can then take the value of function at any point in 3-dimensional space, square the value of the function and then take the absolute value (that is, if the number is negative, make it positive), and the result predicts the density of the electron cloud at that point over the course of many observations of many different atoms. This is called the Born rule, named after Max Born, who suggested the relationship in 1926. Viewing $|\Psi^2|$ as a probability is clean mathematically, but it raises many problems philosophically.

The first problem has to do with the formulation of squaring the number and then taking the absolute value. One has to do this because the function itself is a complex, vector function. That is, at any point (x,y,z) the function evaluates to a number with two components, one that is a real number (such as 0.5 or -0.2) and one that is a complex number (such as $0.25i$). Recall that i is the number that, when multiplied by itself, produces -1. That is, $i^2 = -1$ or $i = \sqrt{-1}$. This is why it is necessary to both square the wave equation and to take its absolute value: because probabilities have to be positive. (More exactly, the value of the function is actually multiplied by its complex conjugate.)

So what does the wave equation actually *mean*? It turns out that we do not really know. The Born rule produces the right answers, but we do not know why. Specifically, we do not know why the rule works, and we do not know what it means philosophically about the nature of reality. This is what Feynman meant when he said "I think I can safely say that nobody understands quantum mechanics." Feynman was making a point about epistemology.

Put another way, the wave equation accurately describes quantum phenomena observed in experiments. But from an epistemological viewpoint, no one has any first-hand *knowledge* what these equations *actually mean*. Only one of our senses can perceive quantum events directly – specifically, the dark-adapted human eye can perceive individual photons. But that's about it. When it comes to electrons, protons, atoms, or even molecules, our senses are limited to indirect measurements. When it comes to air pressure, we do not perceive air molecules pounding against our skin as the result of Brownian Motion. In fact, we do not perceive air pressure at all, which is why its existence was unknown for most of human history.

This kind of empirical relativism is fundamentally unsatisfying to many, and as a result there have been many efforts to interpret the meaning of the wave equation into words that make sense to humans. There is also an ongoing effort in theoretical physics called *Quantum Reconstruction* that seeks to derive the Born rule, as well as other seemingly arbitrary aspects of quantum mechanics, from a significantly smaller set of fundamental postulates. Physicist John Wheeler advocated this approach in 1983, arguing that there should be laws of physics that emerge from mathematics, what he called "law without law":

> [A]ll of Physics in my view, will be seen someday to follow the pattern of thermodynamics and statistical mechanics, of regularity based on chaos, of "law without law." Specially, I believe that everything is built higgledy-piggledy on the unpredictable outcomes of billions upon billions of elementary quantum phenomena, and that the laws and initial conditions of physics arise out of this chaos by the action of a regulating principle, the discovery and proper formulation of which is the number one task.[16]

For many people, this is ultimately what is most unsettling about quantum mechanics: in practically every other field of science and social science, scientists base their theories on clear, consistent mental models. They perform mental experiments to see how those models work. They then put math to the models, and finally, collect data to see if observed phenomena agree with the models. That's the basic process started a thousand years ago in ancient Arabia, when the scientist Ibn al-Haytham conducted experiments in optics and used the results of his experiments to prove one theory of vision and disprove another.[17] It's the approach that Newton used to create his laws of motion, it's the basic process of economics.

---

[16]Wheeler, "On Recognizing 'Law Without Law,' Oersted Medal Response at The Joint APS–AAPT Meeting, New York, 25 January 1983" (1983).

[17]al-Haytham, *Book of Optics* (1011).

However, this approach is different from the approach the Pythagoreans used to invent mathematics, that Aristotle used to explain the world, and that Einstein used to create his theory of relativity. In those cases, people sought to create an intellectual framework that was internally consistent. Indeed, when Einstein's assistant Rosenthal-Schneider asked him what he would have done if the 1919 transit of Mercury across the Sun did not confirm the General Theory of Relativity, Einstein replied, "Then I would feel sorry for the good Lord. The theory is correct."[18]

B.2 Quantum Effects 1: Uncertainty

In early 1926, Heisenberg was invited to give a talk on the matrix mechanics in Berlin. In the audience were Max Planck (who won the 1918 Nobel Prize in Physics "in recognition of the services he rendered to the advancement of Physics by his discovery of energy quanta"), Max Theodor Felix von Laue (who won the 1914 Nobel Prize in Physics "for his discovery of the diffraction of X-rays by crystals"), Walther Hermann Nernst (who discovered the third law of thermodynamics and had won the 1920 Nobel Prize in Chemistry "in recognition of his work in thermochemistry"), and Albert Einstein (who as previously noted had won the 1921 Nobel Prize in Physics for the photoelectric effect).

This assemblage of some of the world's foremost physicists must have been quite intimidating to the 25-year-old Heisenberg! He could probably not have imagined at the time, but in just six years he would earn the 1932 Nobel Prize in Physics, "for the creation of quantum mechanics, the application of which has, *inter alia*, led to the discovery of the allotropic forms of hydrogen."

Einstein invited Heisenberg to come back to his house after the lecture, and the two discussed the fundamental relationship between theory and experimental observation.[19] According to Heisenberg,[20] Einstein argued that a physicist must start with a theory, and from that decide what observations are possible (and presumably which experiments to perform). Heisenberg, in contrast, said that one must start with what is observed during the course of an experiment. If

---

[18]Batten, "Subtle Are Einstein's Thoughts" (2005).

[19]This was the first time that Heisenberg was to meet Einstein, but not the last: the two had a lifelong relationship which Heisenberg wrote about in his posthumously published book, *Encounters with Einstein*.

[20]Heisenberg, *Encounters with Einstein* (1983).

nothing can be observed, then, from the point of view of physics, there is nothing to explain.

This difference in opinion between Einstein and Heisenberg proved to be foundational, influencing how the two would view physics for decades to come.

It's important to realize that the word *observation* here has two meanings, one very specific, the other quite general. The specific meaning is quite literally something that a person (presumably a physicist) can observe, or more accurately, perceive. An observation might be a flash of light, the sound of an explosion, or even the movement of a dial. The second meaning of observation is more general: since scientific instruments have lights and dials, the word *observation* really means anything that can be measured scientifically. And since sensitive scientific instruments can detect a single electron or photon, this really means anything that can interact with an atom or an atomic particle in some detectable manner. If something cannot be detected, then there is no reason to explain it with a theory – indeed, it is not possible to explain with a theory, because there is (by definition) no way to prove if the theory is right or wrong.

Heisenberg returned to Copenhagen and continued to develop quantum mechanics, where he discovered another curious aspect of the theory: according to his math, it should not be possible to precisely determine the position and the speed of an object simultaneously. This was not a consequence of poor instrumentation, it was a result of the underlying physics. This is because the act of measuring something requires interacting with that thing. For example, if you wish to measure the size of a coin, you can put the coin against a ruler, but then you need to bounce light off the coin and into your eye so that you can observe the coin's dimensions. And each time a photon bounces off the coin, there is a physical consequence. Heisenberg called this the *indeterminacy principle*; today it is commonly called the Heisenberg Uncertainty Principle.

To understand the uncertainty principle, let's follow Heisenberg's thought processes. Let's say that one wants to describe the quantum state of a silver coin. To start, one would need to note the precise position of every silver atom that the coin contains. To do this, one could use a microscope that bounced light off each atom on the atom's surface to carefully establish each atom's position. One could capture this bounced light and slowly measure the *state* of the entire object.

rear:

A. Piccard	E. Henriot	P. Ehrenfest	E. Herzen
Th. de Donder	E. Schrödinger	J. E. Verschaffelt	W. Pauli
W. Heisenberg	R. H. Fowler	L. Brillouin	

middle:

P. Debye	M. Knudsen	W.L. Bragg	H. A. Kramers
P. A. M. Dirac	A. H. Compton	L. de Broglie	M. Born
N. Bohr			

Front:

I. Langmuir	M. Planck	M. Curie	H.A . Lorentz
A. Einstein	P. Langevin	Ch.-E. Guye	C. T. R. Wilson
O. W. Richardson			

Figure B.8. Fifth Solvay Conference, Brussels, October 24–29, 1927. Photograph by Benjamin Couprie, Institut International de Physique Solvay.

This is the thought experiment that Heisenberg devised in 1927, although to be accurate, Heisenberg's thought experiment involved finding the location of a single electron using an optical microscope, rather than identifying all of the atoms in a coin. But the basic idea is the same.

By 1927, it was well established that light is quantized – it was six years after Einstein received his Nobel Prize, after all. So Heisenberg's microscope has to be using photons of some sort. What kind of photons should the microscope use to measure an electron?

The year 1927 also marked the fifth invitation-only conference of the International Solvay Institute for Physics and Chemistry, which is noted for its groundbreaking discussions of quantum theory. Of the conference's 29 invited attendees, 17 were or became winners of the Nobel Prize, including Niels Bohr, Albert Einstein, Marie Curie, Paul Dirac, Werner Heisenberg, and Erwin Schrodinger. The conference photo (Figure B.8) has been compared with the Bennett photo from the 1981 Physics of Computation Conference (Figure 4.9).

Traditional light microscopes use visible light. Referring back to Table A.2, visible light has wavelengths between 380 nm and 740 nm. Photons of those sizes are great for looking at things like red blood cells, which have a diameter of roughly $7\mu m$ (7000nm) – i.e. roughly 10 times the size of the wavelength of red photons. But those photons are way too big for looking at individual atoms, let alone an individual electron. Recall that nitrogen atoms so important for quantum sensing have a radius of 0.056nm.

A microscope works by using lenses to focus the light passing through different parts of the object to different parts of the resulting image: this is only possible because the wavelength of the light is much smaller than the size of the object under study. If you want to measure the position of individual atoms, you need to use photons with wavelengths that are roughly the same size as an atom. Looking again to Table A.1, one can see that taking pictures of atoms requires using X-rays – and that's a problem.

Since the energy of a photon is proportional to its frequency ($E = hf$), the energy is inversely proportional to its wavelength ($f = c/\lambda$, so $E = hc/\lambda$). Those photons with the atom-sized wavelength are called *X-rays*, and each one packs so much energy that it can whack an atom far, far away from the point of impact.[21]

Now the stage is set for Heisenberg's discovery of the uncertainty principle. It turns out that there was no way to precisely and simultaneously measure an object's position and its momentum at the atomic level: light that could precisely determine the position of an atom would result in significant energy transfer to the atom causing it to move. Light that was weak enough so that there would be no significant transfer of energy has too large a wavelength to make precise measurements. That is, as position uncertainty decreased, momentum uncertainty had to increase, and *vice versa*. Heisenberg crunched through the math, and arrived at his famous equation, which can be written as:

$$\Delta x \Delta p_x \geq \hbar \tag{2}$$

Where Δx is the uncertainty in position in the x dimension, Δp_x is uncertainty in momentum in the x dimension, and $\hbar$ is the value

---

[21]Such impacts and energy transfer are the reason that X-rays cause cancer. Of course, even ultraviolet light, with a wavelength of just 30nm, is still powerful enough to damage genes within cells and cause cancer: it just takes a longer cumulative exposure.

of Planck's constant divided by 2π, known as the reduced Planck constant. It has a value of 1.05×10^{-34} joule seconds. The joule is a measure of energy; a food calorie has roughly 4200 joules, so $\hbar$ is truly a tiny quantity by our day-to-day standards, which is the reason why we tend not to notice the inherent measurement uncertainty in the world around us.

Heisenberg's key insight – and his fundamental point of disagreement with Einstein – is that it doesn't make sense to theorize aspects of the electron, such as its position and its momentum, unless there is an actual way to measure these aspects. So it is meaningless to say that the electron *has* a precise position and momentum. As Heisenberg wrote:

> If one wants to be clear about what is meant by "position of an object," for example of an electron...then one has to specify definite experiments by which the "position of an electron" can be measured; otherwise this term has no meaning at all.[22]

This kind of relationship between position and momentum is called *complementarity*, and there are many other instances of it in quantum physics. Perhaps the most relevant for quantum information science is the polarization of light, which turns out to be critical for quantum cryptography.

B.3 Quantum Effects 2: Polarization

Polarization is a fundamental property of light that many people are familiar with in their day-to-day experience, thanks to the widespread availability of sunglasses made from polarized filters. Polarization is also the basis of the liquid crystal displays on many computer screens and watches, which is why such displays sometimes turn black if you look at them through a pair of polarizing sunglasses.

Light polarization was discovered in 1669 by Erasmus Bartholinus (1625–1698), a Danish physicist, physician, and mathematician.[23] Bartholinus noticed that when light bounces off a crystal of calcite (also known as calc-spar or Icelandic Spar), there are two reflections, as if there are two kinds of light. In fact, there are.

---

[22]Heisenberg, "Über Den Anschaulichen Inhalt Der Quantentheoretischen Kinematik Und Mechanik" (1927).

[23]Horváth, *Polarization Patterns in Nature: Imaging Polarimetry with Atmospheric Optical and Biological Applications* (2003).

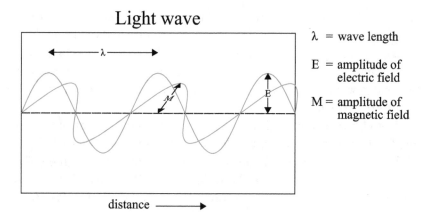

Figure B.9. Here, linearly polarized light is illustrated as a transverse wave with electric and magnetic fields oscillating at right angles to the direction of propagation. Image CC BY-SA by Wikimedia user Gpvos.

As discussed above, light can be described as a transverse wave, as shown in Figure B.9. If you look at the figure, you'll see that there are actually two light waves moving in the direction of the arrow: one wave that ripples up and down, and a second wave that ripples left and right. This diagram is more realistic than one might imagine: at the most fundamental quantum level, light from the Sun, a fire, or a hot stove is actually a mixture of two kinds of light: that is, light that is polarized vertically, and light that is polarized horizontally. We can say that this light is *disorganized*, but it's more common to say that it is not polarized. (This is similar to saying that white light is not colored light, when in fact, white light is actually made up of light of many colors.)

The blue light from the sky on a cloudless day is strongly polarized if you measure it in some directions but not others; it's likely that some birds that fly long distances use this fact to navigate.[24] Light that reflects off flat water tends to be horizontally polarized, and dragonflies make use of this because their eyes can detect the imbalance. Polarized sunglasses do the same: whereas traditional sunglasses absorb both kinds of polarized light, polarized sunglasses are positioned so that vertically polarized light can pass through while horizontally polarized light is blocked. Such sunglasses do a great

---

[24]Horváth, *Polarization Patterns in Nature: Imaging Polarimetry with Atmospheric Optical and Biological Applications* (2003).

Polaroid's First Product Wasn't a Camera

Large polarizing films and filters became cheaply available after Edwin Land (1909–1991) invented a way to attach crystals that polarized to film such that all of the crystals lined up. Crystals that polarized light were well known when Land became interested in the topic as an undergraduate at Harvard University: such crystals had been widely used since the 1850s in polarizing kaleidoscopes, entertaining toys which were commonly found in middle-class houses. Scientists wanted to produce large polarizing crystals to make it easier to use polarized light in microscopy and for experiments, but the crystals were fragile and resisted attempts to grow them large. Land's breakthrough discovery in 1928 was that he could grow many tiny crystals and then force them to line up by squeezing a colloidal suspension of the crystals through long narrow slits.[a] Land left Harvard, perfected the technique, returned to Harvard, then established the Land–Wheelwright Laboratories in 1932 with his Harvard physics instructor, George Wheelwright, and quit Harvard again. (Land never graduated from Harvard, a fate that would befall other notable entrepreneurs who enrolled as undergraduates but never managed to pull their diploma over the finish line.) The company was renamed the Polaroid Corporation after its primary product in 1937, although it would eventually become better known for its developments in instant photography, electronics, optics, and mechanical engineering.

---

[a]Robson, "Profile Edwin H. Land" (1984).

job cutting glare from water, roads, and even other cars: they also let people on boats to see better beneath the surface of the water, which is great for fishing.

For outdoor photography, a polarizing filter attached to the front of a camera will preferentially dim the polarized light from the blue sky compared to the clouds, which has the result of intensifying the clouds and producing spectacular photos (see Figure B.10). Years ago these filters were commonly mounted on a rotatable annulus, so that the photographer could turn the filter as appropriate to maximize the intensity of the clouds while turning the sky to a deep blue. These days, it's more common to purchase filters that can create

Circular Polarizers and 3D Glasses

Circular polarizers are also responsible for the revolution in 3D movies. The key behind the illusion of depth in these movies is that each eye is presented with a slightly different view, something called a *stereoscopic image*. The brain is sensitive to the slight differences between the two images, which creates the illusion of depth. The first 3D movies were black-and-white affairs, with one image projected using red light, the other blue. Viewers wore cardboard glasses with red and blue filters, such that each eye only saw one image. (Red and blue were chosen because they are at opposite ends of the visible light spectrum, which makes it easier to create highly efficient filters that pass one kind of light while blocking the other.) This technique was invented in 1915 and used in comic books and in movies from the 1950s through the 1980s.

Polarized light makes it possible to project 3D movies in color. The early systems used two linear polarizers, typically placed at 45° and 135°. The problem with these systems was that moviegoers had to sit up straight: any tilt of the head would ruin the 3D effect. That's why modern 3D systems use circular polarization: one eye receives light that's polarized in a clockwise direction, the other in a counterclockwise direction. Rather than use two projectors that need to be precisely aligned, it's common to use a single projector with an electrically controlled liquid crystal filter that can rapidly switch polarizations, so that alternating frames go to the left and right eyes.

light that is circularly polarized: it gives the photographer a little less control, but it's easier to use because the photographer doesn't need to worry about orientation. (See the sidebar "Circular Polarizers and 3D Glasses" on page 508 for more information.)

The polarization of light holds an important place in quantum information science because it is the underlying phenomenon on which quantum key distribution, also known as quantum cryptography, is based. It is also one of those quantum effects that are visible at the macroscopic scale and with our human senses.

Here is a simplification of the mathematics of polarization: every photon is polarized in one of two directions, and those directions are determined by how the polarization is *measured*. So if we are mea-

Figure B.10. The effects of a polarizing filter on the sky in a photograph. The picture on the right uses the filter. Image CC BY-SA by Wikimedia user PiccoloNamek.

suring the polarization of light with a linear polarizing filter that is horizontally aligned, the photons that pass through the filter are said to be horizontally polarized while those that do not are vertically polarized. If we are sitting in a 1980s 3D movie, the photons that enter our right eye might be polarized at 45°, while those that go in our left would be polarized at −45° or 135°. And if we are in a modern 3D movie, then the photons that go into the right eye may be circularly polarized in the clockwise direction, while those that go in the left eye may be polarized in the counterclockwise direction. No matter how you measure it, countless scientific experiments have confirmed that there is apparently just one *bit*[25] of polarization state within the photon: the photon can either be aligned with your polarization measurement, or it can be opposed to it. That's because, at the quantum level, polarization is simply the manifestation of something called angular momentum. You can think of circularly polarized photons as tiny spinning corkscrews zipping off at the speed of light in some particular direction.

B.3.1 Six Experiments with Quantum Polarization

With these concepts of polarization, the next section introduces six experiments that you can do yourself. You will need three linear

---

[25] A bit is a binary digit, colloquially thought of as a 0 or a 1, or as the values "false" and "true." Claude E. Shannon (1916–2001), the "father" of information theory, attributes the word to the American mathematician John W. Tukey (1915–2000), although the word was in usage before Claude gave it a precise mathematical definition in 1948. See Garfinkel and Grunspan, *The Computer Book* (2018). Bits are discussed on p. 86.

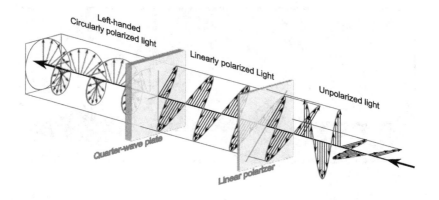

Figure B.11. This illustration shows differences in the waves among disorganized, linearly polarized, and circularly polarized light moving from right to left. Disorganized, or unpolarized, light, characterized by a mixture of polarizations, appears on the right side. The first file manipulated the waves into linearly polarized light (center); the second to circularly polarized light (left). Image public domain by Wikimedia user Dave3457; see Wikipedia for a more detailed explanation.

polarizing filters. (Don't use a circular polarizing filter: you won't get the same results.)

Experiment 1: A Single Linear Polarizing Filter Take a single sheet of a linear polarizing filter and look through it at an incandescent light bulb (if you can find one), a burning candle (be careful not to catch the filter on fire!), or a red-hot stove. All of these objects emit black-body radiation with roughly equal amounts of photons polarized in each direction. (If you don't have any of those, just use a white wall.) You'll see that the filter decreases the intensity roughly by half, but you shouldn't see anything special (Figure B.12, left pane). We will call this the ↔ direction, or a 0° rotation.

What's happening here is that light that has linear polarization that's aligned with the filter passes, while light that is not aligned with the polarizer does not pass. If you use a light meter, you'll see that roughly half of the light is blocked.

Experiment 2: Two Linear Polarizing Filters at (0°, 0°) Now take two linear polarizing filters (Figure B.12, right pane), hold them at the same angle, and look through both of them together. You'll see that the light passes through, and it's about the same strength as when passing through a single filter. Schematically, this is ↔ ↔, or

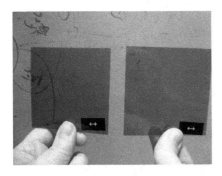

Figure B.12. Two linear polarizing filters with their polarization angles aligned (experiment 1, left), and overlapping (experiment 2, right) at ↔.

two filters at 0° rotation each. Using the logic introduced in Experiment 1, the light that makes it through the first filter is polarized in the ↔ direction, so it can pass through the second filter as well.

Experiment 3: Two Linear Polarizing Filters at (0°, 90°)
Rotate one of the filters 90°, so that one filter is ↔ and the other is ↕ (that is, 0° and 90°, as in Figure B.13, right pane). Position the filters so that you can look through either ↔ filter or both the ↔ and the ↕ filter at the same time. You'll see that the combination of the two filters blocks nearly all the light. Using the logic introduced in Experiment 1, the light that makes it through the first filter is polarized in the ↔ direction, and when it arrives at the second filter it can't pass.

Experiment 4: Two Linear Polarizing Filters at (0°, 45°)
Experiment 3 observed the interaction of light and two filters, one at 0°, one at 90°. If instead the filters are oriented at 0° and 45°, that is, at ↕ and ↖, there is still no surprise (Figure B.13, left pane). Roughly half of the light (50%, or 0.5) passes through the first filter, and roughly half of that light passes through the second. So the combination of the 0° and the 45° filter lowers the light to 25% or .25 of its original intensity.

Experiment 5: Three Linear Polarizing Filters at (0°, 90°, 45°) Now take three filters and arrange them as ↕ ↔ ↖. You will see the same lack of light passing through the three filters as you saw with the two ↕ ↔ filters. There are no surprises here. Only light

Figure B.13. Two overlapping linear polarizing filters with a 90° angle between their polarization angles (experiment 3, left) and a 45° angle (experiment 4, right). Notice that at 90°, no light comes through, whereas at 45° roughly half of the light comes through.

Figure B.14. Three overlapping linear polarizing filters in two different orientations. In both cases the rearmost filter is at 0°. On the left (experiment 5) the middle filter is at 90°, and the one closest to the camera is at 45°. Notice that the filter at 0° combined with the filter at 90° blocks all of the light; the filter at 45° has no effect. On the right (experiment 6), the middle filter is at 45°, and the one closest to the camera is at 90°. Notice that the triangle showing where the 0° and 90° filter overlap is actually darker than the four-sided shape in the middle where the filters are stacked at 0°, then 45°, then 90°.

that is polarized in the up-down direction passes through the first filter. That light can't pass through the second filter. The third filter is present, but it doesn't do anything. See Figure B.14.

Experiment 6: Three Linear Polarizing Filters at (0°, 45°, 90°) Given the results of Experiment 5, what happens if one reverses the order in which light passes through the 90° and the 45°

filter? That is, what happens if the light encounters the filters as a stack of ↕ ↘ ↔ (0°, 45°, 90°)? See Figure B.14 right pane.

Before answering the question about experiment 6, you'll note that what was happening in Experiment 4, when the light that passed through the 0° filter suddenly encountered the 45° filter, went unexplained. Why would roughly half of the light make it through, and is it roughly half, or is it exactly half?

Polarization can be thought of as the direction of oscillation of the transverse wave, or as the angular momentum (or spin) of each photon. So the light that passes through the first filter is oriented at 0° (↕). When this light hits the filter oriented at 45°, it has a 50% chance of passing through and a 50% chance of being absorbed.[26] But now the light passing through the second filter has a polarization of 45°, so when this light hits the third filter, there is once again a 50% chance that the light will pass through and a 50% chance of it being absorbed. As a result, when the filters are at 0°, 45°, and 90°, the amount of light passing through the first filter is 50%, the amount of light passing through the second is 25%, and the amount of light passing through the third is 12.5% of the original.

Once the photon passes through the first 0° filter, it is absolutely certain that it will pass through a second 0° filter and be blocked by a 90° filter. But if the photon encounters a 45° filter before the 90° filter, then all bets are off: the photon might pass, or it might be blocked by the 45°, and if it passes through, then it might be blocked by the 90°, or it might pass through. This is a direct result of the photon only having a single bit of internal state to represent the direction of its angular momentum: it's either polarized horizontally or vertically, it's polarized at 45° or −45°, or it's spinning clockwise or counterclockwise. One set of measurements gives no information about the other set of measurements.

B.4 Quantum Effects 3: Entanglement

This section turns to the phenomenon known as quantum entanglement.

Entangled particles are particles that are somehow linked on the quantum level, even though they are physically separated with no

---

[26]The amount of light passing through can actually be calculated using the Born Rule as $\cos(\theta)^2$ where θ is the angle between the polarization of the first filter and the second filter. Note that $\cos(45 \deg)^2 = .5$.

way to communicate. Entanglement has no direct analog in the classical world, and it is so strange that Einstein labeled it "spooky actions at a distance."[27] One way to think of it is that entangled particles are part of a system, where measuring any part of the system reveals information about other parts.

When particles are entangled, measurement of one causes the other to act in a predictable fashion. Entanglement appears to violate relativity, because measurement appears to cause the other particle to react instantly, superluminally, even when the particles are separated by great distances. Spooky action occurs without sending information through physical space. Sometimes it is said that entanglement enables communication at faster-than-light speeds, but this is impossible, as discussed in the sidebar "Alas, Faster-than-light Communication Is Not Possible" on page 301.

One of the simplest systems of entangled particles is a pair of photons released when a high-speed laser pulse strikes a special kind of crystal that has been *pumped* into a high-energy state. Spontaneous parametric down conversion (SPDC) is a common method to create entangled photons. In SPDC, the laser pulse, striking a beta barium borate crystal, causes the crystal to release two complementary photons traveling in opposite directions. Because they are created together, angular momentum is conserved, which means that if one photon is spinning clockwise as it zips through space, then the other must be spinning counterclockwise. So far, so good.

Recall that each photon's angular momentum is related to how it will interact with a linear polarizing filter. If one photon will pass through a polarizing filter that's oriented at 0°, then the other one will pass through a filter oriented at 90°. If one photon passes through a filter oriented at 45°, the other one will pass through a filter oriented at −45°. So if one sets up two filters, one for each photon, and the filters are oriented at 0° and 90°, then the entangled photons will either pass through *both* of the filters, or they will pass through *neither* of the filters. On the other hand, if both of the filters are oriented 0°, then one of the entangled photons will pass through one of the filters and the other entangled photon will be absorbed.

What happens if instead of placing the filters at 0° and 90°, the two filters are placed at 0° and 45°? Unlike polarization at 0° and

---

[27]Einstein and Born, *The Born–Einstein Letters: Correspondence between Albert Einstein and Max and Hedwig Born From 1916–1955, with Commentaries by Max Born* (1971).

90°, or clockwise and counterclockwise, polarization at 0° and 45° are unrelated at the quantum level. Just as half of the light that passed through the 0° filter could pass through a 45° filter, if the first photon encounters a 0° filter and the second encounters a 45° filter, then each photon will have a 50% chance of passing through to the other side: there will be no correlation between the two measurements.

This is the essence of entanglement: it is also the essence of quantum key distribution. And it was profoundly disturbing to Albert Einstein, Boris Podolsky, and Nathan Rosen, who identified the problem when they were working together at the Institute for Advanced Study in Princeton, NJ in 1934, and published their classic paper on the topic in 1935: "Can Quantum-Mechanical Description of Physical Reality Be Considered Complete?" – known as the EPR paper.

Simply put, here is the paradox that the EPR paper identifies: because they are moving in opposite directions at the speed of light, there is no way for the two photons to communicate with each other. Nothing, after all, can move faster than the speed of light. If the first photon hits a polarizing filter at 0° and the second hits a filter at 90°, only one of them will pass through. But if they hit filters at 0° and 45°, then each photon has a 50% chance of passing through. Run a lot of experiments in which two entangled photons hit a pair of filters at 0° and 45°, and roughly a quarter of the time neither photon will pass, a quarter of the time the A photon will pass, a quarter of the time the B photon will pass, and a quarter of the time both photons will pass. This happens even if the orientation of the polarizing filters is set after the entangled photons are created.

How does each photon know the orientation of both filters at the time of impact?

The challenge here is that the two entangled photons are described by a single wave function. This made no sense to the scientists. What holds the photons together? If the photons were in some kind of communication, it would need to be faster than the speed of light, and that would violate Relativity. On the other hand, by 1934 wave mechanics was well enough developed that it had accurately predicted the outcome of every experiment designed to test it: wave mechanics was clearly correct. Therefore, the EPR paper argued, the description of reality provided by quantum mechanics must not be complete – there must be more to the description of each entangled photon than its wave function. Or, as the paper states it:

From this follows that either (1) *the quantum-mechanical description of reality given by the wave function is not complete* or (2) *when the operators corresponding to two physical quantities do not commute the two quantities cannot have simultaneous reality.* For if both of them had simultaneous reality – and thus definite values – these values would enter into the complete description, according to the condition of completeness. If then the wave function provided such a complete description of reality, it would contain these values; these would then be predictable. This not being the case, we are left with the alternatives stated.[28] (emphasis in original)

In his March 1947 letter to Born, Einstein put his objection into more colorful language:

I cannot make a case for my attitude in physics which you would consider at all reasonable. I admit, of course, that there is a considerable amount of validity in the statistical approach which you were the first to recognise clearly as necessary given the framework of the existing formalism. I cannot seriously believe in it because the theory cannot be reconciled with the idea that physics should represent a reality in time and space, free from spooky actions at a distance.[29]

Entanglement is a powerful technique that is central to quantum computing, metrology (the study of measurement), sensing, and communication. In quantum computing, entanglement is used to create coordinated ensembles of particles. Operating together, these ensembles may provide faster computing in a quantum computer. In metrology and sensing, an entangled photon can illuminate an object while the linked particle can be measured to learn about the target. In communication, entanglement can be used to create random sequences of bits that can be used as encryption keys for securely exchanging information even in the presence of surveillance. As will be seen in

---

[28]Einstein, Podolsky, and Rosen, "Can Quantum-Mechanical Description of Physical Reality Be Considered Complete?" (1935).

[29]Einstein and Born, *The Born–Einstein Letters: Correspondence between Albert Einstein and Max and Hedwig Born From 1916–1955, with Commentaries by Max Born* (1971), p. 158.

Chapter 7, in 2017, Chinese researchers maintained entangled photons at 1200 km using a satellite that communicated with two base stations. As *Science* explained it, "Spooky action achieved at record distance."[30]

The EPR paper argues that there must be some deeper theory from which the probabilistic quantum theory could be derived. That theory would presumably assign to particles like photons and electrons additional state that would be described by new variables, and from those variables the observed probabilities could be derived. From the point of view of this underlying, more complete, and utterly hypothetical theory, there would be no randomness. Today this is called the "hidden variable theory."

In 1964 physicist John Stewart Bell developed a hypothesis that would need to be true for any explanation of quantum mechanical results based on hidden variables.[31] In the years that followed, experiments were designed that could prove or disprove the hypothesis: these were called *Bell tests*. In the intervening years, these experiments have been carried out with ever-increasing precision and levels of exactness. The conclusion of this line of work is now clear: entanglement exists. Entangled particles are somehow linked. There are no hidden variables.

B.5 Quantum Effects 4: Superposition

Let us go back to our experiments with light and linear polarizing filters. Recall that if a photon passes through the first filter at 0°, it will pass through a second filter at 0°, but it only has a 50% chance of passing through a filter at 45°, and it has a 0% chance of passing through a filter at 90°.

One of the reasons that Schrödinger's wave equation (described more fully below) was such a breakthrough is that it gave physicists a mathematical approach for describing this situation. Once the equation is written down it's then possible to solve for the amount of light that passes through the second filter. If p is the fraction of light that passes through the second filter after passing through the first, and θ is the angle between the two filters, then the equation is $p = \cos^2(\theta)$, where cos is the trigonometric cosine function that evaluates to 0 at 0°, 1 at 90°, and $\sqrt{0.5}$ at 45°. The function is squared in line with Born's rule.

---

[30]Popkin, "Spooky Action Achieved at Record Distance" (2017).
[31]Bell, "On The Einstein Podolsky Rosen Paradox" (1964).

Man Plays Dice with Einstein's Words

Einstein never said, or wrote, one of the most famous quotations attributed to him – that God does not play dice with the Universe. In his December 4, 1926 letter to Max Born, Einstein actually wrote:

"Die Quantenmechanik ist sehr achtunggebietend. Aber eine innere Stimme sagt mir, daß das noch nicht der wahre Jakob ist. Die Theorie liefert viel, aber dem Geheimnis des Alten bringt sie uns kaum näher. Jedenfalls bin ich überzeugt, daß der nicht würfelt."[a]

The Hebrew University of Jerusalem, Israel owns the copyright on the letter. In the 2005 publication of Einstein and Born's letters, the German was thus translated:

"Quantum mechanics is certainly imposing. But an inner voice tells me that it is not yet the real thing. The theory says a lot, but does not really bring us any closer to the secret of the "Old One." I, at any rate, am convinced that He is not playing at dice."[b]

Here's another translation, with commentary:

"Even 'God does not play dice,' arguably Einstein's most famous quote, isn't quite his words. It derives from a letter written in German in December 1926 to his friend and sparring partner, theoretical physicist Max Born. It is published in the new volume of Einstein's papers, in which the editors comment on its 'varying translations' since the 1920s. Theirs is: 'Quantum mechanics ... delivers much, but does not really bring us any closer to the secret of the Old One. I, at any rate, am convinced that He does not play dice.' Einstein does not use the word 'God' (Gott) here, but 'the Old One' (Der Alte). This signifies a "personification of nature," notes physicist and Nobel laureate Leon Lederman (author of The God Particle, *1993).[c]*

---

[a]Einstein, Born, and Heisenberg, *Albert Einstein Max Born, Briefwechsel 1916–1955: Mit Einem Geleitwort von Bertrand Russell (Deutsch)* (2005).
[b]Einstein and Born, *The Born–Einstein Letters 1916–1955: Friendship, Politics and Physics in Uncertain Times* (2005).
[c]Robinson, "Did Einstein Really Say That?" (2018).

If you are wondering why p was used in the above paragraph for the fraction of light passing through, rather than f, that's because the equation really isn't about the *fraction* of light passing through: it provides the *probability* that any particular photon that passes through the first filter will pass through the second. This probability (which ignores the probability that the photon will be absorbed by the substrate on which the polarizing material rests) holds true in general for any pair of polarizing filters.

- This is why 12.5% of the light that enters a sandwich of three polarizing filters at 0°, 45°, and 90° will pass through: 50% will pass through the first filter at 0°, 50% of that light will pass through the filter at 45°, and then 50% of that light will pass through the filter at 90°.

- This is also why 0% of light will pass through a sandwich of filters at 0°, 90°, and 45°: 50% of the light will pass through the first filter at 0°, then 0% will pass through the filter at 90°. And that's that. There's no more light. If there was light leaving the filter at 90°, 50% of it would pass through the filter at 45°. But there isn't any light, so nothing passes through.

The word *superposition* can be used to describe what's happening here at the quantum level. In quantum mechanics, the Schrödinger wave equation allows any wave to be described as a combination[32] of any other waves. Physicists and engineers can use this property to describe physical systems with simplified wave equations that focus on the particular quantum phenomena on which they are focusing, or they can write exceedingly complex wave equations with many terms to consider more possibilities (or simply to impress their friends and intimidate their rivals).

To get a better understanding of what might be happening in the case of the three polarized filters, each photon approaching a polarizing filter can be described as a superposition of two photon possibilities: the possibility that the photon will travel through the filter, and the possibility that the photon will be absorbed. If these are the only two possible outcomes – that is, if one ignores the possi-

---

[32]In quantum mechanics, the waves are actually represented as *linear functions* of other waves, which means that waves can be added or subtracted in any proportion, but cannot be multiplied or divided.

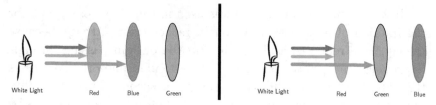

Figure B.15. Photons in the white light from the candle are represented here as colored lines containing photons from the red, green, and blue regions of the visible spectrum. These photons are all at different energy levels, with the blue photons having roughly twice as much energy per-photon as the red photons. When the white light encounters the red filter, only the red photons can pass through: the rest are absorbed, which is why the world looks red when you look through the filter. The red photons, in turn, are blocked by the blue filter (left) or the green filter (right). Thus, no light can pass through any combination of red, green, and blue filters, no matter which order the light encounters the filters. (Illustration credit: Simson Garfinkel)

bility that the photon might be reflected off the surface of the filter – then these two possibilities must sum to 1:

$$1 = p_{\text{pass}} + p_{\text{absorb}} \tag{3}$$

Recall that the probability of a the photon passing through was $\cos^2(\theta)$. So another way of writing this equation is:

$$1 = \cos^2(\theta) + p_{\text{absorb}} \tag{4}$$

which is equal to:

$$1 = \cos^2(\theta) + \sin^2(\theta) \tag{5}$$

From a wave mechanics point of view, this is actually a summation of two wave equations: one that represents the probability that the photon will pass through the filter, and the other representing the probability that the photon will be absorbed.

The remainder of this section will discuss why behavior of the three polarizing filters at 0°, 45°, and 90° seems so strange, by explaining what's happening at the quantum level if colored filters were used instead of polarizing ones.

Most of us have a clear understanding of how light passes through colored glass as a result of our day-to-day experiences and from color theory. White is made up of all the colors of the rainbow. Red light passes through red glass and blue light passes through blue glass (Figure B.15). This is why blue things look black through a red filter, and red things look black through a blue filter. Old-style 3D movies and comic books were based on this basic optics.

Polarized light doesn't work this way. Although it's tempting to think that polarizing filters act like colored gels, except that they let through light that is aligned as little arrows ($\leftrightarrow$ for a filter at 0°, $\updownarrow$ for a filter at 90°), that's not what is happening. If it was, then only a tiny bit of light could possibly make it through a polarizing filter set at 0° – not only would the light at 90° be absorbed, but so too would the light at 45° be blocked.

Color and polarization are different, because individual photons really do have individual color – a photon's color is directly related to its wavelength, which is a real thing that you can measure in many different ways. A photon's polarization, in contrast, is a superposition of wave functions. Those wave functions are determined by the photon's angular momentum, or spin.

Candles emit a stream of photons in every direction. Any individual photon's spin is going to be in one direction or the other, but overall the numbers will be equal because angular momentum is conserved. So when one of these photons hits that first polarizing filter, it has a 50% chance of traveling through, and a 50% chance of being absorbed.

If that photon travels through the filter, its polarization is now aligned with the crystals out of which the filter was built. When that photon comes to a filter that's 45° out of alignment, there is only a 50% chance that the photon will properly interact with the crystals in the second filter and pass through. But if it does, its polarization is now aligned with the second set of crystals. If you want a classical, non-wave-equation way of thinking about this, you can pretend that the second filter *turned* the photons that successfully passed through (Figure B.16). If you want a quantum mechanical explanation, you could say that the wave function describing photons on the left side of the 45° filter describes a superposition of photons that can pass through the filter and those that cannot; likewise the wave function that describes photons on the right side of the filter is a superposition of those that *did* pass through the filter and ... well, and nothing. But *that* wave function can itself be described as a superposition of photons that can pass through a filter at 0° and those that can pass through a filter at 90°.

Figure B.16. Photons in the white light from the candle are represented here with black lines that represent the stream of photons leaving the candle. Each photon contains two possible polarizations (or two possible angular momenta). When this stream of photons hits a linear polarizing filter at 0°, only 50% of the photons can pass through. These photons have now been measured to have a linear polarization of 0°. If these photons interact with a linear polarizing filter that has a 45° offset (left), 50% of the photons can pass, because $\cos^2(45 \deg) = 0.5$. Alternatively, if these photons interact with a linear polarizing filter that has a 90° offset (right), none of the photons can pass, because $\cos^2(90 \deg) = 0$.

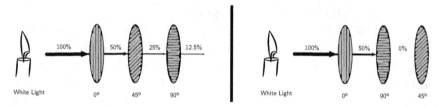

Figure B.17. In this example, light is directed to travel through three polarizing filters. On the left, the light passes through the filters that are set at the angles 0°, 45°, 90°, which means that the photons encounter two transitions of 45°, one after the other. Each transition reduces the amount of light that passes by 50%. On the right the light is set to pass through the filters that are set at the angles 0°, 90°, 45°, which means that the photons should first encounter a transition of 90° and then one of −45°. However the first transition blocks all of the light.

B.6 The Cat State

The experiments that we've presented in this appendix and the explanations for the somewhat paradoxical results are much simpler (and correspondingly less accurate) than you are likely to find in other books on quantum information science, let alone on quantum mechanics. Although many quantum devices are based on polarized light, they don't measure whether photons are transmitted or absorbed: instead, they send the photon into a crystal that either reflects or transmits the light depending on its phase, and then they use two sensors, each capable of detecting a single photon.

Complex two and four-beam systems are not discussed in this appendix because simplifying the presentation enables one to demonstrate with experiments using three low-cost and mass-produced linear polarizing filters, while reading this very book. This approach can give one an intuitive feel for the strangeness of quantum mechanics – a strangeness that arises because the behavior of tiny quantum particles is so very different than the behaviors we observe at the macroscopic scale.

This book also intentionally avoids any discussion of what the Schrödinger wave equation actually *means* – whether there is a wave function that collapses when it is measured or observed, as the so-called Copenhagen Interpretation of quantum mechanics holds, or whether the universe literally splits in two, as is held by the alternative many-worlds interpretation. And no time is devoted here to the vastly less popular *pilot wave* theory, first proposed by de Broglie, then rediscovered by David Bohm in 1952, which holds that the probabilistic interactions are themselves an illusion, and that the wave function describes a real wave that pushes around real particles. In pilot wave theory, the pilot wave is the wave described by the Schrödinger Wave Equation; a second equation called the Guiding Equation describes how the wave moves. Pilot wave theory does require hidden variables, but they are global: the entire wave function is instantaneously affected by every other particle in the universe.

Schrödinger and Einstein were both fundamentally dissatisfied with a theory of reality that depended so intimately upon the role of the observer. To that end, they created a thought experiment that today is referred to as Schrödinger's cat. The idea was to imprison a cat and a radioactive source in a box that has no contact with the outside world. There is a 50% chance that the radioactive material will decay within an hour and that the decay will be detected by a machine that's also in the box. If the decay is detected, the machine kills the cat – either by releasing poison gas (in Schrödinger's version) or by blowing up some explosive (Einstein). So at the end of the hour, the cat is either alive or it's dead. But since quantum mechanical events aren't actually settled until they are *observed* (at least, that's the story according to the Copenhagen interpretation), then the cat is *both alive and dead* until someone opens the box and checks on it. Unless the cat is also an observer, in which case it either observes that it's alive, or it's dead.

Perhaps the cat is both alive and dead: perhaps the universe has split in two, and there are really two cats. That's the many-worlds interpretation of quantum mechanics. Or perhaps there is a pilot wave, determined by all of the particles in the universe, and the radioactive material either did decay or it didn't, and it was either detected or it wasn't, and the cat is either alive or it's dead, no matter if there is an observer or not. That's the pilot wave theory at work.

In quantum theory, some scientists use the cheeky term "cat state" to describe an object that simultaneously exists in two opposing states. Humanity currently lacks the scientific tools to test these multiple competing interpretations of reality, and because all of them are equally compatible with the quantum devices that are being created in labs today and likely to be created for the foreseeable future, you, dear reader, can choose the ultimate nature of your own physical reality.

Bibliography

23andMe (n.d.). "Choosing Which Reports to View" (). Last accessed March 26, 2021. customercare.23andme.com/hc/en-us/articles/212195308-Choosing-Which-Reports-to-View.

4iQ (May 2020). "2020 4iQ Identity Breach Report."

Aaronson, Scott (March 2005). "Guest Column: NP-Complete Problems and Physical Reality." *SIGACT News* 36.1, pp. 30–52. doi.org/10.1145/1052796.1052804.

— (March 2008). "The Limits of Quantum." *Scientific American* 298.3, pp. 62–69.

— (2013). *Quantum Computing since Democritus.* Cambridge: Cambridge University Press.

Abbott, B. P. et al. (February 2016). "Observation of Gravitational Waves From a Binary Black Hole Merger." *Physical Review Letters* 116.061102. link.aps.org/doi/10.1103/PhysRevLett.116.061102.

Acin, A. and L. Masanes (2016). "Certified Randomness in Quantum Physics." *Nature* 540.7632, pp. 213–219.

Adleman, Leonard (1994). "Molecular Computation of Solutions to Combinatorial Problems." *Science* 266.5187, pp. 1021–1024. science.sciencemag.org/content/266/5187/1021.

— (2011). "Pre-RSA Days: History and Lessons." In: *ACM Turing Award Lectures.* New York: Association for Computing Machinery, pp. 2002. doi.org/10.1145/1283920.1961904.

Adleman, L. M., R. L. Rivest, and A. Shamir (September 1983). *Cryptographic Communications System and Method.* US Patent No. 4,405,829. Patent filed September 14, 1977. www.google.com/patents/US4405829.

Aggarwal, Vinod K. and Andrew W. Reddie (2018). "Comparative Industrial Policy and Cybersecurity: a Framework for Analysis." *Journal of Cyber Policy* 3.3, pp. 291–305.

Agrawal, Manindra, Neeraj Kayal, and Nitin Saxena (June 2004). "Primes Is in P." *Annals of Mathematics* 160. Godel Prize, Fulkerson Prize, pp. 781–793. www.microsoft.com/en-us/research/publication/primes-is-in-p/.

Aliberti, Keith and Thomas Bruen (2006). "Quantum Computation and Communication." *Army Logistician* 38.5, pp. 42–48.

Alper, Alexandra (December 2019). "US Finalizing Rules to Limit Sensitive Tech Exports to China, Others." *Reuters*. www.reuters.com/article/us-usa-tech-china-exclusive-idUSKBN1YL1B8.

Alpert, Carol Lynn, Emily Edwards, and James Freericks (May 13, 2020). *Key Concepts for Future QIS Learners*. Workshop Output. qis-learners.research.illinois.edu/.

Altman, Ehud et al. (December 14, 2019). "Quantum Simulators: Architectures and Opportunities." arxiv.org/abs/1912.06938.

Alzar, Carlos L. Garrido (December 13, 2019). "Compact Chip-Scale Guided Cold Atom Gyrometers for Inertial Navigation: Enabling Technologies and Design Study." *AVS Quantum Science*. avs.scitation.org/doi/10.1116/1.5120348.

American Physical Society (n.d.). "Robert A. Millikan" (). Last accessed December 25, 2019. www.aps.org/programs/outreach/history/historicsites/millikan.cfm.

Ananthaswamy, A. (2018). *Through Two Doors at Once: The Elegant Experiment That Captures The Enigma of Our Quantum Reality*. New York: Penguin Publishing Group. books.google.com/books?id=bGNKDwAAQBAJ.

Anqi, Huang et al. (2018). "Implementation Vulnerabilities in General Quantum Cryptography." *New Journal of Physics* 20.10.

Apple Computer (November 10, 2020). "Apple Unleashes M1." www.apple.com/newsroom/2020/11/apple-unleashes-m1/.

Applegate, David L., Robert E. Bixby, Vašek Chvátal, William Cook, et al. (2009). "Certification of an Optimal TSP Tour through 85,900 Cities." *Operations Research Letters* 37.1, pp. 11–15. www.sciencedirect.com/science/article/pii/S0167637708001132.

Applegate, David L., Robert E. Bixby, Vašek Chvátal, and William J. Cook (2006). *The Traveling Salesman Problem*. Princeton, NJ: Princeton University Press.

Aramon, Maliheh et al. (April 2019). "Physics-Inspired Optimization for Quadratic Unconstrained Problems Using a Digital Annealer." *Frontiers in Physics* 7, pp. 48.

Arias, Elisa Felicitas and Gérard Petit (2019). "The Hyperfine Transition for The Definition of The Second." *Annalen Der Physik* 531.5, pp. 1900068. onlinelibrary.wiley.com/doi/abs/10.1002/andp.201900068.

Arkell, Harriet (September 10, 2014). "Death From above without Warning: 70 Years after The First One Fell, Interactive Map Reveals Just Where Hitler's V2 Rockets Killed Thousands of British Civilians in Final Months of WW2." *Daily Mail.* www.dailymail.co.uk/news/article-2750353/Interactive-map-reveals-hundreds-sites-Hitler-s-V2-rockets-killed-thousands-British-civilians-final-months-WW2.html.

Arnbak, A. M. and N. A. N. M. van Eijk (2012). *Certificate Authority Collapse: Regulating Systemic Vulnerabilities in The HTTPS Value Chain.* Arlington, VA: Telecommunications Policy Research Conference.

Arute, Frank et al. (2019). "Quantum Supremacy Using a Programmable Superconducting Processor." *Nature* 574.7779, pp. 505–510. doi.org/10.1038/s41586-019-1666-5.

Ashida, Yuya et al. (2019). "Molybdenum-Catalysed Ammonia Production with Samarium Diiodide and Alcohols or Water." *Nature* 568.7753, pp. 536–540. doi.org/10.1038/s41586-019-1134-2.

Bailey, Dennis (2004). *The Open Society Paradox: Why The 21st Century Calls for More Openness – Not Less.* Washington, D.C.: Brassey's.

Baloo, Jaya (June 30, 2019). "KPN's Quantum Journey, Cyberweek 2019, Tel Aviv, Israel." YouTube video. www.youtube.com/watch?v=ePZFBxX3DnY.

Barker, Elaine and John Kelsey (March 2007). *Recommendation for Random Number Generation Using Deterministic Random Bit Generators (Revised).* Last accessed May 30, 2020. National Institute of Standards and Technology. csrc.nist.gov/publications/detail/sp/800-90/revised/archive/2007-03-14.

Barlow, John Perry (1996). "A Declaration of The Independence of Cyberspace." *Journal.* homes.eff.org/~barlow/Declaration-Final.html.

Barzanjeh, Shabir et al. (February 2015). "Microwave Quantum Illumination." *Physical Review Letters* 114 (8). pp. 080503. link.aps.org/doi/10.1103/PhysRevLett.114.080503.

Batten, Alan H. (September 26, 2005). "Subtle Are Einstein's Thoughts." *Physics World* 18.9, pp. 16. physicsworld.com/a/s ubtle-are-einsteins-thoughts/.

Bayer, Ronald and Sandro Galea (2015). "Public Health in The Precision-Medicine Era." *New England Journal of Medicine* 373.6, pp. 499–501.

Bell, John Stewart (1964). "On The Einstein Podolsky Rosen Paradox." *Physics* 1.3, pp. 195–290. cds.cern.ch/record/111654/files /vol1p195-200_001.pdf.

Bellovin, Steven M. (June 1, 2000). "Wiretapping The Net." www.n ae.edu/7430/WiretappingtheNet.

ben-Aaron, Diana (April 1985). "Weizenbaum Examines Computers and Society." *The Tech* 105 (16).

Ben-Atar, Doron S. (2004). *Trade Secrets: Intellectual Piracy and The Origins of American Industrial Power.* New Haven, CT: Yale University Press. books.google.com/books?id=dHD1m AEACAAJ.

Benioff, Paul (1980). "The Computer As a Physical System: A Microscopic Quantum Mechanical Hamiltonian Model of Computers As Represented by Turing Machines." *Journal of Statistical Physics* 22.5, pp. 563–591.

— (1982a). "Quantum Mechanical Hamiltonian Models of Discrete Processes That Erase Their Own Histories: Application to Turing Machines." *International Journal of Theoretical Physics* 21.3, pp. 177–201.

— (1982b). "Quantum Mechanical Models of Turing Machines That Dissipate No Energy." *Physical Review Letters* 48.1581.

Bennett, C. H. (November 1973). "Logical Reversibility of Computation." *IBM Journal of Research and Development* 17.6, pp. 525–532. doi.org/10.1147/rd.176.0525.

Bennett, C. H. and G. Brassard (1984). "Quantum Cryptography: Public Key Distribution and Coin Tossing," pp. 175–179. resear cher.watson.ibm.com/researcher/files/us-bennetc/BB84highest.pdf .

Bennett, Charles H. et al. (1993). "Teleporting an Unknown Quantum State via Dual Classical and Einstein-Podolsky-Rosen Channels." *Physical Review Letters* 70.13, pp. 1895–1899.

Beranek, Leo (2011). "Founding a Culture of Engineering Creativity." In: *A Culture of Innovation: Insider Accounts of Computing and Life at BBN: A Sixty Year Report, 18 October 1948 to*

1 July 2010. East Sandwich, Massachusetts: Waterside Publishing. www.cbi.umn.edu/hostedpublications/pdf/CultureInnovation _bbn.pdf.

Berger, Abi (2002). "Magnetic Resonance Imaging." *BMJ Clinical Research Ed.* 324.7328, pp. 35.

Bernstein, Daniel J. (2009). "Introduction to Post-Quantum Cryptography." In: *Post-Quantum Cryptography*. Berlin, Heidelberg: Springer, pp. 1–14.

Bernstein, Daniel J. and T. Lange (2017). "Post-Quantum Cryptography." *Nature* 549.7671, pp. 188–194.

Berry, Michael W. et al. (2008). *Survey of Text Mining II: Clustering, Classification, and Retrieval*. Springer.

Berthiaume, André and Gilles Brassard (1994). "Oracle Quantum Computing." *Journal of Modern Optics* 41.12, pp. 2521–2535. doi.org/10.1080/09500349414552351.

Bierhorst, Peter et al. (2018). "Experimentally Generated Randomness Certified by The Impossibility of Superluminal Signals." *Nature* 556.7700, pp. 223–226.

Biological Nitrogen Fixation: Research Challenges – A Review of Research Grants Funded by the US Agency for International Development (1994). Washington, DC: The National Academies Press. www.nap.edu/catalog/9288/biological-nitrogen-fixation-res earch-challenges-a-review-of-research-grants.

Black, John, Martin Cochran, and Trevor Highland (2006). "A Study of The MD5 Attacks: Insights and Improvements." *Fast Software Encryption, 13th International Workshop, FSE 2006*. Vol. 4047. Lecture Notes in Computer Science. Berlin, Heidelberg: Springer, pp. 262–277. iacr.org/archive/fse2006/40470265 /40470265.pdf.

Blakeslee, Sandra (June 1, 1984). "Nuclear Spill at Juarez Looms As One of The Worst." *New York Times*. www.nytimes.com/198 4/05/01/science/nuclear-spill-at-juarez-looms-as-one-of-worst.ht ml.

Blakley, George Robert (1979). "Safeguarding Cryptographic Keys." *International Workshop on Managing Requirements Knowledge*. IEEE Computer Society. Los Alamitos, CA: IEEE Computer Society, pp. 313–318.

Blanchette, Jean-François (2012). *Burdens of Proof: Cryptographic Culture and Evidence Law in The Age of Electronic Documents*.

Cambridge, MA: The MIT Press. www.jstor.org/stable/j.ctt5vjp
dh.

Boaron, Alberto et al. (2018). "Secure Quantum Key Distribution
Over 421 Km of Optical Fiber." *Physical Review Letters* 121.19,
pp. 190502.

Boneh, Dan and Matthew K. Franklin (2001). "Identity-Based Encryption From The Weil Pairing." *Advances in Cryptology –
Proceedings of CRYPTO 2001*, pp. 213–229.

Bongs, Kai et al. (2019). "Taking Atom Interferometric Quantum
Sensors From The Laboratory to Real-World Applications." *Nature Reviews Physics* 1.12, pp. 731–739. doi.org/10.1038/s42254-
019-0117-4.

Bonneau, J., C. Herley, et al. (2012). "The Quest to Replace Passwords: A Framework for Comparative Evaluation of Web Authentication Schemes." *2012 IEEE Symposium on Security and
Privacy*. San Francisco, CA, pp. 553–567.

Bonneau, J., Cormac Herley, et al. (June 2015). "Passwords and
The Evolution of Imperfect Authentication." *Communications
of The Association for Computing Machinery* 58.7, pp. 78–87.
doi.org/10.1145/2699390.

Boothby, Bill (2017). "Space Weapons and The Law." *International
Law Studies* 93, pp. 179–214.

Boudot, Fabrice et al. (February 28, 2020). "Factorization of RSA-
250." *Cado-Nfs-Discussion*. lists.gforge.inria.fr/pipermail/cado-nfs
-discuss/2020-February/001166.html.

Braithwaite, Matt (July 7, 2016). "Experimenting with Post-Quantum
Cryptography." *Google Security Blog*. security.googleblog.com/2
016/07/experimenting-with-post-quantum.html.

Brassard, G. (2005). "Brief History of Quantum Cryptography: a
Personal Perspective." *IEEE Information Theory Workshop on
Theory and Practice in Information-Theoretic Security, 2005*.
Updated version located at arxiv.org/abs/quant-ph/0604072v1,
pp. 19–23.

Bratley, Paul and Jean Millo (1972). "Computer Recreations: Self-
Reproducing Programs." *Software – Practice and Experience* 2,
pp. 397–400. onlinelibrary.wiley.com/doi/abs/10.1002/spe.438002
0411.

Briegel, H. J. et al. (1998). "Quantum Repeaters: The Role of Imperfect Local Operations in Quantum Communication." *Physical Review Letters* 81.26, pp. 5932–5935.

Brown, I. David (2016). *The Chemical Bond in Inorganic Chemistry: The Bond Valence Model, 2nd ed.* Oxford University Press.

Brumfiel, Geoff (February 2016). "US Navy Brings Back Navigation by The Stars for Officers." *National Public Radio.*

— (2019). "India Claims Successful Test Of Anti-Satellite Weapon." *National Public Radio: All Things Considered.*

Buchanan, Ben (2020). *The Hacker and The State: Cyber Attacks and The New Normal of Geopolitics.* Cambridge, MA: Harvard University Press.

Buchner, M. et al. (2018). "Tutorial: Basic Principles, Limits of Detection, and Pitfalls of Highly Sensitive SQUID Magnetometry for Nanomagnetism and Spintronics." *Journal of Applied Physics* 124.16.

Budker, Dmitry and Michael Romalis (April 1, 2007). "Optical Magnetometry." *Journal.* escholarship.org/uc/item/1c79s7vb.

Bunch, Bryan H. (2004). *The History of Science and Technology: A Browser's Guide to The Great Discoveries, Inventions, and The People Who Made Them, From The Dawn of Time to Today.* Boston, MA: Houghton Mifflin.

Bundesministerium für Bildung und Forschung (January 2020). "Die Zweite Quantenrevolution Maßgeblich Mitgestalten."

Bureau International des Poids et Mesures (July 2017). "50th Anniversary of The Adoption of The Atomic Definition of The Second." Last accessed October 9, 2020. www.bipm.org/en/news/full-stories/2017-07-definition-second.html.

— (n.d.). *The International System of Units.* www.bipm.org/en/publications/si-brochure/.

Bush, V. (1931). "The Differential Analyzer. A New Machine for Solving Differential Equations." *Journal of The Franklin Institute* 212.4, pp. 447–488. www.sciencedirect.com/science/article/pii/S0016003231906169.

Bush, V., F. D. Gage, and H. R. Stewart (1927). "A Continuous Integraph." *Journal of The Franklin Institute* 203.1, pp. 63–84. www.sciencedirect.com/science/article/pii/S0016003227900970.

C4ADS (April 16, 2019). "Above Us Only Stars: Exposing GPS Spoofing in Russia and Syria." www.c4reports.org/aboveusonlystars.

Calo, Ryan (2018). "Artificial Intelligence Policy: A Primer and Roadmap." *Bologna Law Review*. bolognalawreview.unibo.it/articl e/view/8670.

Capra, Fritjof (1975). *Tao of Physics: an Exploration of The Parallels between Modern Physics and Eastern Mysticism*. 2nd ed. New Science Library. Boston, MA: Shambhala.

Caro, Jose et al. (2011). "GPS Space Segment." *Navipedia*. Last edited June 19, 2018; last accessed October 18, 2020. gssc.esa.in t/navipedia/index.php/GPS_Space_Segment.

Carreyrou, John (2018). *Bad Blood: Secrets and Lies in a Silicon Valley Startup*. New York: Alfred A. Knopf.

Catalog Technologies, Inc. (n.d.). "Catalog" (). Last accessed January 12, 2021. www.catalogdna.com.

Center for Long Term Cybersecurity (2019). "Cybersecurity Scenarios 2025."

Chang, C. W. et al. (2019). "Quantum-Enhanced Noise Radar." *Applied Physics Letters* 114.11, pp. 112601.

Chang, Weng-Long, Minyi Guo, and M. S. Ho (2005). "Fast Parallel Molecular Algorithms for DNA-Based Computation: Factoring Integers." *IEEE Transactions on NanoBioscience* 4.2, pp. 149–163.

Chen, Yu-Ao et al. (2021). "An Integrated Space-To-Ground Quantum Communication Network Over 4,600 Kilometres." *Nature* 589.7841, pp. 214–219.

Cheng, Kai-Wen and Chien-Cheng Tseng (June 5, 2002). "Quantum Plain and Carry Look-Ahead Adders." *ArXiv.org Quantum Physics*. arxiv.org/abs/quant-ph/0206028.

Chiang, Ted (2019). *Exhalation*. New York: Alfred A. Knopf.

Cho, A. (2020). "The Short, Strange Life of Quantum Radar." *Science* 369.6511, pp. 1556–1557.

Chou, C. W. et al. (2010). "Optical Clocks and Relativity." *Science* 329.5999, pp. 1630–1633. science.sciencemag.org/content/329/59 99/1630.

Church, Alonzo (1936). "An Unsolvable Problem of Elementary Number Theory." *American Journal of Mathematics* 58.2, pp. 345–363. www.jstor.org/stable/2371045.

Chwala, A. et al. (2012). "Full Tensor SQUID Gradiometer for Airborne Exploration." *ASEG Extended Abstracts* 2012.1, pp. 1–4.

Clarivate (2021). "What Is Web of Science Core Collection?" clariv
ate.libguides.com/woscc/basics.

Clark, David D. (2019). *Designing an Internet*. Cambridge, MA:
MIT Press.

Clymer, A. Ben (1993). "The Mechanical Analog Computers of
Hannibal Ford and William Newell." *IEEE Annals of The History of Computing* 15.2, pp. 19–34.

Cocks, Clifford (2001). "An Identity Based Encryption Scheme
Based on Quadratic Residues." *Proceedings of The 8th IMA
International Conference on Cryptography and Coding*. Berlin,
Heidelberg: Springer.

Congressional Research Service (2020). "US Research and Development Funding and Performance: Fact Sheet."

Cook, William J. (2012). *In Pursuit of The Traveling Salesman*.
Princeton, NJ: Princeton University Press.

Copeland, B. Jack, ed. (2005). *Alan Turing's Automatic Computing Engine*. Oxford University Press.

Copeland, René (October 17, 2017). "The International Quantum
Race." *The Coming Quantum Revolution: Security and Policy
Implications*, Video Conference proceedings.

Coppersmith, D. (May 1994). "The Data Encryption Standard
(DES) and Its Strength against Attacks." *IBM Journal of Research and Development* 38.3, pp. 243–250.

Cory, David G., Amr F. Fahmy, and Timothy F. Havel (1997).
"Ensemble Quantum Computing by NMR Spectroscopy." *Proceedings of The National Academy of Sciences* 94.5, pp. 1634–
1639. www.pnas.org/content/94/5/1634.

Cox, K. C. et al. (2018). "Quantum-Limited Atomic Receiver in
The Electrically Small Regime." *Physical Review Letters* 121.11,
pp. 110502.

Crane, Leah (June 24, 2020). "Honeywell Claims It Has Built The
Most Powerful Quantum Computer Ever." *New Scientist*. www
.newscientist.com/article/2246940-honeywell-claims-it-has-built-th
e-most-powerful-quantum-computer-ever/.

Creery, Madison (June 26, 2019). "The Russian Edge in Electronic
Warfare." *Georgetown Security Studies Review*. georgetownsecuri
tystudiesreview.org/2019/06/26/the-russian-edge-in-electronic-wa
rfare/.

Cross, A. W. et al. (2019). "Validating Quantum Computers Using
Randomized Model Circuits." *Physical Review A* 100.3.

Curtis E. LeMay Center for Doctrine Development and Education (March 13, 2019). "Introduction to Targeting." www.doctrine.af .mil/Portals/61/documents/Annex_3-60/3-60-D01-Target-Intro.p df.

Cybersecurity and Infrastructure Security Agency (January 5, 2021). "Joint Statement by The Federal Bureau of Investigation (FBI), The Cybersecurity and Infrastructure Security Agency (CISA), The Office of The Director of National Intelligence (ODNI), and The National Security Agency (NSA)." www.cis a.gov/news/2021/01/05/joint-statement-federal-bureau-investigat ion-fbi-cybersecurity-and-infrastructure.

D-Wave Systems Inc. (May 29, 2019). "Quantum Experiences: Applications and User Projects on D-Wave." www.youtube.com/wa tch?v=NTnu1UiFXVo.

— (September 29, 2020). "D-Wave Announces General Availability of First Quantum Computer Built for Business." www.dwavesys .com/press-releases/d-wave-announces-general-availability-first-qu antum-computer-built-business.

Dam, Kleese K. van (February 5, 2020). *From Long-Distance Entanglement to Building a Nationwide Quantum Internet: Report of The DOE Quantum Internet Blueprint Workshop.* Tech. rep. BNL-216179-2020-FORE. United States.

Danzig, Richard (May 2018). "Technology Roulette: Managing Loss of Control As Many Militaries Pursue Technological Superiority." *CNAS.* www.cnas.org/publications/reports/technology-roulette.

Dasgupta, Subrata (2014). *It Began with Babbage: The Genesis of Computer Science.* Oxford University Press.

Dattani, Nikesh S. and Nathaniel Bryans (2014). "Quantum Factorization of 56153 with Only 4 Qubits." arxiv.org/abs/1411.675 8.

Davies, William (2020). *This Is Not Normal: The Collapse of Liberal Britain.* Verso. www.versobooks.com/books/3628-this-is-not-normal.

Davisson, Clinton J. and Lester H. Germer (1928). "Reflection of Electrons by a Crystal of Nickel." *Proceedings of The National Academy of Sciences of The United States of America* 14 (4), pp. 317–322.

Dean, Jeffrey (November 10, 2010). "Building Software Systems At Google and Lessons Learned." Talk at Stanford University. www.youtube.com/watch?v=modXC5IWTJI.

Defense Advanced Research Projects Agency (2020). "Quantum Sensing and Computing." www.darpa.mil/attachments/Quantum SensingLayout2.pdf.

Degen, C. L., F. Reinhard, and P. Cappellaro (July 25, 2017). "Quantum Sensing." *Reviews of Modern Physics* 89.3, pp. 035002.

Department of Commerce, Bureau of Industry and Security (November 19, 2018). "Review of Controls for Certain Emerging Technologies." *Federal Register* 83.

Department of Homeland Security, US Coast Guard (2018). "GPS Problem Reporting." navcen.uscg.gov/?pageName=gpsUserInput.

Deutsch, David (1985). "Quantum Theory, The Church–Turing Principle and The Universal Quantum Computer." *Proceedings of The Royal Society of London. A. Mathematical and Physical Sciences* 400.1818, pp. 97–117.

Didion, Joan (2003). *Where I Was From.* New York: Knopf.

Diffie, Whitfield and Martin E Hellman (1976). "New Directions in Cryptography." *IEEE Transactions on Information Theory* 22.6, pp. 644–654.

— (1977). "Special Feature Exhaustive Cryptanalysis of The NBS Data Encryption Standard." *Computer* 10.6, pp. 74–84.

Dinolt, George et al. (September 2010). *Parallelizing SHA-256, SHA-1 MD5 and AES on The CellBroadbandEngine.* Tech. rep. NPS-CS-10-11. Monterey, CA: Naval Postgraduate School. calh oun.nps.edu/handle/10945/551.

Director of National Intelligence (2019). "What Is Intelligence?" www.dni.gov/index.php/what-we-do/what-is-intelligence.

DiVincenzo, David P. (1997). "Topics in Quantum Computers." In: *Mesoscopic Electron Transport,* ed. Lydia L. Sohn, Leo P. Kouwenhoven, and Gerd Schön. Dordrecht: Springer Netherlands, pp. 657–677. doi.org/10.1007/978-94-015-8839-3_18.

— (September 2000). "The Physical Implementation of Quantum Computation." *Fortschritte Der Physik* 48.9–11, pp. 771–783. dx.doi.org/10.1002/1521-3978(200009)48:9/11%3C771::AID-PROP771%3E3.0.CO;2-E.

Doudna, Jennifer A. and Emmanuelle Charpentier (2014). "The New Frontier of Genome Engineering with CRISPR-Cas9." *Science* 346.6213, pp. 1258096.

Dowling, J. P. and G. J. Milburn (2003). "Quantum Technology: The Second Quantum Revolution." *Philosophical Transactions of The Royal Society of London A: Mathematical, Physical and Engineering Sciences* 361.1809, pp. 1655–1674.

Doyle, Leonard (April 28, 1993). "Business Spy War Erupts between US and France: Paris Forced to Come Clean on Hi-Tech Dirty Tricks." *The Independent.* tinyurl.com/4jxv6fzr.

Duke, J., J. Friedlin, and P. Ryan (2011). "A Quantitative Analysis of Adverse Events and 'Overwarning' in Drug Labeling." *Archives of Internal Medicine* 171.10, pp. 944–946.

Dworkin, Morris J. et al. (November 26, 2001). *Advanced Encryption Standard (AES).* Tech. rep. FIPS-197. Gaithersburg, MD: National Institute of Standards and Technology. www.nist.gov /publications/advanced-encryption-standard-aes.

Dyakonov, Mikhail (2019). "When Will Useful Quantum Computers Be Constructed? Not in The Foreseeable Future, This Physicist Argues. Here's Why: The Case against: Quantum Computing." *IEEE Spectrum* 56.3.

— (2020). *Will We Ever Have a Quantum Computer?* Berlin: Springer.

Einstein, Albert (1905). "Über Einen Die Erzeugung Und Verwandlung Des Lichtes Betreffenden Heuristischen Gesichtspunkt (On The Production and Transformation of Light From a Heuristic Viewpoint)." *Annalen Der Physik (1900) (Series 4)* 322.6, pp. 132–148. www.gsjournal.net/Science-Journals/Essays/View/2 490;%20www.gsjournal.net/Science-Journals/Essays/View/2491 ;%20www.zbp.univie.ac.at/einstein/einstein1.pdf.

Einstein, Albert and Max Born (1971). *The Born–Einstein Letters: Correspondence between Albert Einstein and Max and Hedwig Born From 1916–1955, with Commentaries by Max Born.* London: Macmillan.

— (2005). *The Born–Einstein Letters 1916–1955: Friendship, Politics and Physics in Uncertain Times.* London: Macmillan.

Einstein, Albert, Max Born, and Werner Heisenberg (2005). *Albert Einstein Max Born, Briefwechsel 1916–1955: Mit Einem Geleitwort von Bertrand Russell (Deutsch).* Munich: Langen Müller, pp. 52.

Einstein, Albert and Leopold Infeld (1938). *The Evolution of Physics: The Growth of Ideas From Early Concepts to Relativity and Quanta.* Cambridge: Cambridge University Press.

Einstein, Albert, Boris Podolsky, and Nathan Rosen (May 15, 1935). "Can Quantum-Mechanical Description of Physical Reality Be Considered Complete?" *Physical Review* 47, pp. 777–780.

Ekerå, Martin and Johan Håstad (2017). "Quantum Algorithms for Computing Short Discrete Logarithms and Factoring RSA Integers." *International Workshop on Post-Quantum Cryptography*. Berlin: Springer, pp. 347–363.

Ekert, Artur K. (August 1991). "Quantum Cryptography Based on Bell's Theorem." *Physical Review Letters* 67 (6), pp. 661–663. link.aps.org/doi/10.1103/PhysRevLett.67.661.

Electronic Frontier Foundation, ed. (1998). *Cracking DES: Secrets of Encryption Research, Wiretap Politics and Chip Design*. en. San Francisco, CA: Electronic Frontier Foundation.

Elfving, V. E. et al. (2020). "How Will Quantum Computers Provide an Industrially Relevant Computational Advantage in Quantum Chemistry?" arxiv.org/abs/2009.12472.

Elliott, Chip and Henry Yeh (July 2007). *DARPA Quantum Network Testbed*. AFRL-IF-RS-TR-2007-180. Cambridge, MA: BBN Technologies. www.dtic.mil/docs/citations/ADA471450.

Ellis, James, Clifford Cocks, and Malcolm Williamson (1975). *Public-Key Cryptography*. Classified reports (titles uncertain) at Government Communications Headquarters (GCHQ), Cheltenham, UK. Work declassified in 1997. Awarded the 100th IEEE Milestone Award for the first discovery (albeit long secret) of public-key cryptography. www.gchq.gov.uk/Press/Pages/100th-IEEE-milestone-award.aspx.

European Commission (March 11, 2020). "A New Circular Economy Action Plan." ec.europa.eu/environment/circular-economy/.

European Commission, High Level Steering Committee, DG Connect (June 28, 2017a). "Quantum Technologies Flagship Final Report." digital-strategy.ec.europa.eu/en/library/quantum-flagship-high-level-expert-group-publishes-final-report.

— (2017b). "Quantum Technologies Flagship Intermediate Report."

Fagaly, Robert (2014). "SQUID Magnetometers." In: *Measurement, Instrumentation, and Sensors Handbook, Second Edition: Electromagnetic, Optical, Radiation, Chemical, and Biomedical Measurement*. Boca Raton, FL: CRC Press, pp. 1–14.

Faley, M. I. et al. (December 6, 2017). "Superconducting Quantum Interferometers for Nondestructive Evaluation." *Sensors* 17.12, pp. 2798.

Farrell, Henry and Abraham L Newman (2019). "Weaponized Interdependence: How Global Economic Networks Shape State Coercion." *International Security* 44.1, pp. 42–79.

Feynman, Richard P. (December 29, 1959). "There's Plenty of Room at The Bottom: An Invitation to Enter a New Field of Physics." Lecture given at the American Physical Society. Published in *Engineering and Science*, February 1960, pp. 22–36.

— (1982). "Simulating Physics with Computers." *International Journal of Theoretical Physics* 21.6, pp. 467–488. doi.org/10.1007/BF02650179.

— (February 1985a). "Quantum Mechanical Computers." *Optics News* 11.2, pp. 11–20.

— (1985b). "Tiny Computers Obeying Quantum Mechanical Laws." In: *New Directions in Physics: The Los Alamos 40th Anniversary Volume*. Boston, MA: Academic Press.

— (1986). *Foundations of Physics* 16.6, pp. 507–531.

Finke, Doug (2021). "Qubit Count." *Quantum Computing Report*. quantumcomputingreport.com/scorecards/qubit-count/.

Flamini, Fulvio, Nicolo Spagnolo, and Fabio Sciarrino (2018). "Photonic Quantum Information Processing: a Review." *Reports on Progress in Physics* 82.1, pp. 016001.

Fortt, Jon (March 11, 2010). "Top 5 Moments From Eric Schmidt's Talk in Abu Dhabi." *Fortune*.

Fowler, Austin G. et al. (September 2012). "Surface Codes: Towards Practical Large-Scale Quantum Computation." *Physical Review A* 86.3. dx.doi.org/10.1103/PhysRevA.86.032324.

Frank, M. P. (2002). "The Physical Limits of Computing." *Computing in Science Engineering* 4.3, pp. 16–26.

Fredkin, Ed, Rolf Landauer, and Tom Toffoli (1982). "Physics of Computation." *International Journal of Theoretical Physics* 21 (12), pp. 903–903. doi.org/10.1007/BF02084157.

Fredkin, Edward F. (July 5, 2006). "Oral History of Ed Fredkin." Interviewed by Gardner Hendrie. archive.computerhistory.org/resources/access/text/2013/05/102630504-05-01-acc.pdf.

Fredkin, Edward F. and Tommaso Toffoli (April 1982). "Conservative Logic." *International Journal of Theoretical Physics* 21.3–4, pp. 219–253.

— (2001). "Design Principles for Achieving High-Performance Sub-micron Digital Technologies." In: *Collision-Based Computing*. Berlin, Heidelberg: Springer-Verlag, pp. 27–46.

Fujiwara, Masazumi et al. (2020). "Real-Time Nanodiamond Thermometry Probing in Vivo Thermogenic Responses." *Science Advances* 6.37.

Gaithersburg, MD: National Institute of Standards and Technology (2018). "NIST Jump-Starts Quantum Information." www.nist.gov/topics/physics/introduction-new-quantum-revolution/nist-jump-starts-quantum-information.

— (June 5, 2019). "Second: The Future." www.nist.gov/si-redefiniti on/second/second-future.

Gamberini, Sarah Jacobs and Lawrence Rubin (2021). "Quantum Sensing's Potential Impacts on Strategic Deterrence and Modern Warfare." *Orbis*. www.sciencedirect.com/science/article/pii/S0030438721000120.

Gardner, Martin (October 1970). "The Fantastic Combinations of John Conway's New Solitaire Game 'Life'." *Scientific American* 223, pp. 120–123.

— (August 1977). "Mathematical Games: A New Kind of Cipher That Would Take Millions of Years to Break." *Scientific American* 237.2, pp. 120–124. www.nature.com/scientificamerican/jour nal/v237/n2/pdf/scientificamerican0877-120.pdf.

Garfinkel, Simson L. (1994). *PGP: Pretty Good Privacy*. Sebastapol, CA: O'Reilly & Associates.

— (October 11, 1995). "1985–1995: Digital Decade. MIT's Computing Think Tank Chronicles The Electronic Age." *San Jose Mercury News*. simson.net/clips/1995/95.SJMN.MediaLab.pdf.

— (2000). *Database Nation: The Death of Privacy in The 21st Century*. Seabastopol, CA: O'Reilly.

— (May 2005). "Quantum Physics to The Rescue: Cryptographic Systems Can Be Cracked. And People Make Mistakes. Take Those Two Factors out of The Equation, and You Have Quantum Cryptography and a New Way to Protect Your Data." *CSO Magazine*.

Garfinkel (aut.), Simson L. and Hal Abelson (ed.) (1999). *Architects of The Information Society*. Cambridge, MA: MIT Press.

Garfinkel, Simson L. and Rachel H. Grunspan (2018). *The Computer Book*. New York, NY: Sterling Milestones.

Garfinkel, Simson L. and Philip Leclerc (2020). "Randomness Concerns When Deploying Differential Privacy." *Proceedings of The 19th Workshop on Privacy in The Electronic Society.* WPES'20. New York: Association for Computing Machinery, pp. 73–86. doi.org/10.1145/3411497.3420211.

Garroway, A. N. et al. (2001). "Remote Sensing by Nuclear Quadrupole Resonance." *IEEE Transactions on Geoscience and Remote Sensing* 39.6, pp. 1108–1118.

Gellman, Barton (2020). *Dark Mirror: Edward Snowden and The American Surveillance State.* London: The Bodley Head.

Gely, Mario F. et al. (2019). "Observation and Stabilization of Photonic Fock States in a Hot Radio-Frequency Resonator." *Science* 363.6431, pp. 1072–1075. science.sciencemag.org/content/3 63/6431/1072.

Gerlich, Stefan et al. (2011). "Quantum Interference of Large Organic Molecules." *Nature Communications* 2.1, pp. 1–5.

Gershenfeld, Neil A. and Isaac L. Chuang (1997). "Bulk Spin-Resonance Quantum Computation." *Science* 275.5298, pp. 350–356. science .sciencemag.org/content/275/5298/350.

Gibney, Elizabeth (2017). "New Definitions of Scientific Units Are on The Horizon." *Nature* 550.7676, pp. 312–313.

— (2019). "The Quantum Gold Rush." *Nature* 574.7776, pp. 22–24.

Gidney, Craig and Martin Ekerå (2019). "How to Factor 2048 Bit RSA Integers in 8 Hours Using 20 Million Noisy Qubits." arxiv .org/abs/1905.09749.

Giurgica-Tiron, Tudor et al. (2020). "Low Depth Algorithms for Quantum Amplitude Estimation." arxiv.org/abs/2012.03348.

Goldberg, Ian, David Wagner, and Eric Brewer (1997). "Privacy-Enhancing Technologies for The Internet." *Proceedings IEEE COMPCON 97. Digest of Papers.* IEEE. San Jose, CA, pp. 103–109.

Goodin, Dan (December 3, 2019). "New Crypto-Cracking Record Reached, with Less Help Than Usual From Moore's Law." *Ars Technica.* arstechnica.com/information-technology/2019/12/new-crypto-cracking-record-reached-with-less-help-than-usual-from-mo ores-law/.

Google (n.d.). "Quantum – Google AI" (). ai.google/research/teams /applied-science/quantum-ai/.

Google LLC (2021). "Encryption at Rest." Last accessed January 1, 2021. cloud.google.com/security/encryption-at-rest.

Goucher, Adam P. (2012). "Antikythera Mechanism." demonstratio ns.wolfram.com/AntikytheraMechanism/.

"GPS Navigation: From the Gulf War to Civvy Street" (November 2, 2018). www.sciencemuseum.org.uk/objects-and-stories/gps-navigation-gulf-war-civvy-street.

Grant, Edward (2008). *Much Ado About Nothing: Theories of Space and Vacuum From The Middle Ages to The Scientific Revolution*. Cambridge: Cambridge University Press.

Greenspan, Donald (1982). "Deterministic Computer Physics." *International Journal of Theoretical Physics* 21.6, pp. 505–523.

Greve, Frank (April 18, 1993). "Boeing Called A Target Of French Spy Effort." *The Seattle Times*. archive.seattletimes.com/archive /?date=19930418&slug=1696416.

Grier, David Alan (2007). *When Computers Were Human*. Princeton, NJ: Princeton University Press.

Grumbling, Emily and Mark Horowitz (2019). *Quantum Computing: Progress and Prospects*. Washington, DC: National Academies Press. doi.org/10.17226/25196.

Guha, S. and B. Erkmen (2009). "Gaussian-State Quantum-Illumination Receivers for Target Detection." *Physical Review A* 80, pp. 052310.

Gunning, D. and D. W. Aha (2019). "DARPA's Explainable Artificial Intelligence Program." *AI Magazine* 40.2, pp. 44–58.

Gurobi Optimization, LLC (2019). "Air France Tail Assignment Optimization." www.gurobi.com/wp-content/uploads/2019/09/Ai r-France-Case-Study.pdf.

Gwinner, Jan et al. (2020). "Benchmarking 16-Element Quantum Search Algorithms on IBM Quantum Processors." arxiv.org/abs /2007.06539.

Hagestad, William T. (2012). "Chinese IW Capabilities." In: *21st Century Chinese Cyberwarfare*. Cambridgeshire, UK: IT Governance Publishing, pp. 137–146. www.jstor.org/stable/j.ctt5hh5nz .16.

Halder, Matthäus et al. (2007). "Entangling Independent Photons by Time Measurement." *Nature Physics* 3.10, pp. 692–695. doi.o rg/10.1038/nphys700.

Hambling, David (2017). "China's Quantum Submarine Detector Could Seal South China Sea." *New Scientist* 22.

Harari, Y. N. (2017). *Homo Deus: A Brief History of Tomorrow.* New York: Harper Collins.

Hardesty, Jasper (2014). "Safety, Security and Dual-Use Chemicals." *Journal of Chemical Health and Safety* 22.5, pp. 3–16.

Harris, Mark (October 4, 2018). "D-Wave Launches Free Quantum Cloud Service." *IEEE Spectrum.* spectrum.ieee.org/dwave-launch es-free-quantum-cloud-service.

Harris, Robert G. and James M. Carman (1984). "Public Regulation of Marketing Activity: Part II: Regulatory Responses to Market Failures." *Journal of Macromarketing* 4.1, pp. 41–52. doi.org/10.1177/027614678400400105.

al-Haytham, Ibn (1011). *Book of Optics.*

Heidari, Hadi and Vahid Nabaei (2019). "SQUID Sensors." In: *Magnetic Sensors for Biomedical Applications.* New York: John Wiley and Sons, pp. 163–212.

Heisenberg, Werner (March 1927). "Über Den Anschaulichen Inhalt Der Quantentheoretischen Kinematik Und Mechanik." *Zeitschrift Für Physik* 43.3, pp. 172–198. doi.org/10.1007/BF01 397280.

— (1983). *Encounters with Einstein.* Princeton, NJ: Princeton University Press.

Heller, Nathan (December 18, 2017). "The Digital Republic: Is Estonia The Answer to The Crisis of Nation-States?" *New Yorker*, pp. 84–93.

Heuer Jr., Richards J. and Randolph H. Pherson (2015). *Structured Analytic Techniques for Intelligence Analysis.* 2nd ed. Thousand Oaks, California: CQ Press.

Hill, Kashmir (July 24, 2013). "Blueprints of NSA's Ridiculously Expensive Data Center in Utah Suggest It Holds Less Info Than Thought." *Forbes.* www.forbes.com/sites/kashmirhill/2 013/07/24/blueprints-of-nsa-data-center-in-utah-suggest-its-stora ge-capacity-is-less-impressive-than-thought.

— (2021). "Your Face Is Not Your Own." *New York Times Magazine.* www.nytimes.com/interactive/2021/03/18/magazine/facial-r ecognition-clearview-ai.html.

Hillis, W. Daniel (1982). "New Computer Architectures and Their Relationship to Physics or Why Computer Science Is No Good." *International Journal of Theoretical Physics* 21.3, pp. 255–262.

— (1989). "Richard Feynman and The Connection Machine." *Physics Today* 42.2. physicstoday.scitation.org/doi/10.1063/1.881196.

Hillis, Daniel and Brian Silverman (1978). "Original Tinkertoy Computer." *Computer History Museum.* www.computerhistor y.org/collections/catalog/X39.81.

Hlembotskyi, Vladyslav et al. (2020). "Efficient Unstructured Search Implementation on Current Ion-Trap Quantum Processors." arx iv.org/abs/2010.03841.

Ho, Chi-Tang, Xin Zheng, and Shiming Lib (March 2015). "Tea Aroma Formation." *Food Science and Human Wellness*, pp. 9–27. www.sciencedirect.com/science/article/pii/S221345301500018 X.

Hoffman, David (2009). *The Dead Hand: The Untold Story of The Cold War Arms Race and Its Dangerous Legacy.* New York: Anchor.

Holland, J., J. M. Smith, and M. Schuchard (2019). "Measuring Irregular Geographic Exposure on The Internet." *ArXiv.*

Hollingham, Richard (September 7, 2014). "V2: The Nazi Rocket That Launched The Space Age." *BBC.* www.bbc.com/future/arti cle/20140905-the-nazis-space-age-rocket.

Holtmaat, Anthony et al. (2009). "Long-Term, High-Resolution Imaging in The Mouse Neocortex through a Chronic Cranial Window." *Nature Protocols* 4.8, pp. 1128–1144.

Homer (2018). *The Odyssey,* ed. Emily R. Wilson. New York: W.W. Norton and Company.

Honeywell (June 2020). "The World's Highest Performing Quantum Computer Is Here." www.honeywell.com/us/en/news/2020 /06/the-worlds-highest-performing-quantum-computer-is-here.

Hoofnagle, Chris Jay (2016). *Federal Trade Commission Privacy Law and Policy.* New York: Cambridge University Press. assets .cambridge.org/97811071/26787/cover/9781107126787.jpg.

Hoogstraaten, Hans et al. (August 2012). *Black Tulip Report of The Investigation into The DigiNotar Certificate Authority Breach.* Tech. rep. PR-110202, pp. 101. www.researchgate.net /publication/269333601_Black_Tulip_Report_of_the_investigati on_into_the_DigiNotar_Certificate_Authority_breach.

Horváth, Gábor (September 7, 2003). "Polarization Patterns in Nature: Imaging Polarimetry with Atmospheric Optical and Biological Applications." DSc. Thesis. Budapest: Loránd Eötvös University.

House, Don Robert (2001). "A Synopsis of Teletype Corporation History." simson.net/ref/2001/house-teletype-corp-synopsis.pdf.

House, Tamzy J. et al. (1996). *Weather As a Force Multiplier: Owning The Weather in 2025*. Tech. rep. Montgomery, AL: Air War College, Maxwell Air Force Base.

Housley, R. (July 2003). *Use of The RSAES-OAEP Key Transport Algorithm in Cryptographic Message Syntax (CMS)*. RFC 3560 (Proposed Standard). Internet Engineering Task Force. www.ietf.org/rfc/rfc3560.txt.

Hughes, Richard J. et al. (2013). "Network-Centric Quantum Communications with Application to Critical Infrastructure Protection." arxiv.org/abs/1305.0305.

Hull, Isaiah et al. (December 2020). "Quantum Technology for Economists." *Sveriges Riksbank Working Paper Series*.

Huygen, Christiaan (1690). *Traité de la Lumière*. (Treatise on Light). The Hague, Netherlands: chez Pierre Vander Aa marchand libraire. archive.org/details/bub_gb_kVxsaYdZaaoC/page/n4.

Hwang, J. Y., J. B. Chang, and W. P. Chang (2001). "Spread of 60Co Contaminated Steel and Its Legal Consequences in Taiwan." *Health Phys* 81.6, pp. 655–60.

Independent Working Group on Missile Defense (2009). *Missile Defense, The Space Relationship, and The Twenty-First Century: 2009 Report*. Published for the Independent Working Group by the Institute for Foreign Policy Analysis. www.ifpa.org/pdf/IWG2009.pdf.

Information Technology Laboratory (September 2013). *Supplemental ITL Bulletin for September 2013*. csrc.nist.gov/csrc/media/publications/shared/documents/itl-bulletin/itlbul2013-09-supplemental.pdf.

International Trade Administration (2021). "US Export Controls." Last accessed March 6, 2021. www.trade.gov/us-export-controls.

Jaques, Samuel et al. (2019). "Implementing Grover Oracles for Quantum Key Search on AES and LowMC." Report 2019/1146. eprint.iacr.org/2019/1146.

Jenks, W. G., S. S. H. Sadeghi, and J. P. Wilkswo Jr. (1997). "Review Article: SQUIDs for Nondestructive Evaluation." *Journal of Physics D: Applied Physics* 30, pp. 293–323. citeseerx.ist.psu.edu/viewdoc/summary?doi=10.1.1.145.5200.

Jernigan, Carter and Behram F. T. Mistree (September 22, 2009). "Gaydar: Facebook Friendships Expose Sexual Orientation." *First Monday*. firstmonday.org/article/view/2611/2302.

Joint Chiefs of Staff (2020). *DOD Dictionary of Military and Associated Terms.* fas.org/irp/doddir/dod/jp1_02.pdf.

Jones, Sam (May 2014). "MoD's 'quantum Compass' Offers Potential to Replace GPS." *Financial Times.*

Jordan, Stephen P. (May 2008). "Quantum Computation beyond The Circuit Model." PhD thesis. Massachusetts Institute of Technology, Cambridge, MA. arxiv.org/pdf/0809.2307.pdf.

— (February 1, 2021). "Quantum Algorithm Zoo." Last accessed February 15, 2021. quantumalgorithmzoo.org.

Joy, Bill (April 2000). "Why The Future Doesn't Need Us." *Wired.* www.wired.com/2000/04/joy-2/.

Juma, Calestous (2016). *Innovation and Its Enemies: Why People Resist New Technologies.* New York: Oxford University Press.

Jun Han et al. (2012). "ACComplice: Location Inference Using Accelerometers on Smartphones." *2012 Fourth International Conference on Communication Systems and Networks (COMSNETS 2012)*, pp. 1–9.

Jurcevic, Petar et al. (2020). "Demonstration of Quantum Volume 64 on a Superconducting Quantum Computing System." arxiv.org/abs/2008.08571.

Juzeliunas, E., Y. P. Ma, and J. P. Wikswo (2004). "Remote Sensing of Aluminum Alloy Corrosion by SQUID Magnetometry." *Journal of Solid State Electrochemistry* 8, pp. 435–441.

Kadrich, Mark (2007). *Endpoint Security.* Boston, MA: Addison-Wesley Professional.

Kahn, David (1996). *The Codebreakers: The Comprehensive History of Secret Communication From Ancient Times to The Internet.* New York: Scribner.

Kan, Shirley (April 2007). *China's Anti-Satellite Weapon Test.* New York: Congressional Research Service. fas.org/sgp/crs/row/RS22652.pdf.

Kania, Elsa B. and John Costello (2018). *Quantum Hegemony? China's Ambitions and The Challenge to US Innovation Leadership.* Washington, DC: Center for a New American Security.

Katyal, Sonia (September 14, 2017). "Why You Should Be Suspicious of That Study Claiming A.I. Can Detect a Person's Sexual Orientation." *Slate.*

Keats, Jonathon (September 27, 2011). "The Search for a More Perfect Kilogram." *Wired.* www.wired.com/2011/09/ff-kilogram/.

Kerr, Orin S. (2001). "The Fourth Amendment in Cyberspace: Can Encryption Create a Reasonable Expectation of Privacy?" *Connecticut Law Review* 33.2, pp. 503–534.

— (2015). "The Fourth Amendment and The Global Internet." *Stan. L. Rev.* 67, p. 285.

Khan, I. et al. (2018). "Satellite-Based QKD." *Optics and Photonics News* 29.2, pp. 26–33.

Kim, Donggyu et al. (2019). "A CMOS-Integrated Quantum Sensor Based on Nitrogen–vacancy Centres." *Nature Electronics* 2.7, pp. 284–289. doi.org/10.1038/s41928-019-0275-5.

King, Gilbert (June 6, 2012). "Fritz Haber's Experiments in Life and Death." *Smithsonian Magazine.* www.smithsonianmag.com /history/fritz-habers-experiments-in-life-and-death-114161301/.

Knuth, Donald E. (December 1970). "Von Neumann's First Computer Program." *ACM Computing Survey* 2.4, pp. 247–260. dl.a cm.org/doi/10.1145/356580.356581.

Koblitz, Neal (January 1987). "Elliptic Curve Cryptosystems." *Mathematics of Computation* 48.177, pp. 203–209.

Koblitz, N. and A. Menezes (2016). "A Riddle Wrapped in an Enigma." *IEEE Security Privacy* 14.6, pp. 34–42.

Koh, John S., Steven M. Bellovin, and Jason Nieh (2019). "Why Joanie Can Encrypt: Easy Email Encryption with Easy Key Management." *Proceedings of The Fourteenth EuroSys Conference 2019*. Dresden: Association for Computing Machinery. doi .org/10.1145/3302424.3303980.

Kohnfelder, Loren M. (1978). "Towards a Practical Public-Key Cryptosystem." Undergraduate thesis, Massachusetts Institute of Technology, Cambridge, MA. dspace.mit.edu/handle/1721.1 /15993.

Koller, Josef S. (2019). *The Future of Ubiquitous, Realtime Intelligence: A GEOINT Singularity*. Arlington, Virginia: Center for Space Policy and Strategy.

Koops, E. J. et al. (2006). "Should ICT Regulation Be Technology-Neutral?" In: *Starting Points for ICT Regulation*. The Hague, Netherlands: TMC Asser Press, pp. 77–108.

Korzeczek, Martin C. and Daniel Braun (2020). "Quantum-Router: Storing and Redirecting Light at The Photon Level." arxiv.org /abs/2003.03363.

Kuo, Lucas and Jason Arterburn (2009). *Lux and Loaded: Exposing North Korea's Strategic Procurement Networks*. Washington, DC. www.c4reports.org/lux-and-loaded.

Kwak, Sean (October 7, 2017). "The Coming Quantum Revolution: Security and Policy Implications, Hudson Institute." www.hudson.org/events/1465-the-coming-quantum-revolution-security-and-policy-implications102017.

Kwiatkowski, Kris (June 20, 2019). "Towards Post-Quantum Cryptography in TLS." *The Cloudflare Blog.* blog.cloudflare.com/towards-post-quantum-cryptography-in-tls/.

Landauer, R. (1961). "Irreversibility and Heat Generation in The Computing Process." *IBM Journal of Research and Development* 5.3, pp. 183–191.

Landauer, Rolf (1982). "Physics and Computation." *International Journal of Theoretical Physics* 21.3, pp. 283–297.

Langenberg, Brandon, Hai Pham, and Rainer Steinwandt (2019). "Reducing The Cost of Implementing AES As a Quantum Circuit." Report 2019/854. eprint.iacr.org/2019/854.

Langley, Adam (November 22, 2011). "Protecting Data for The Long Term with Forward Secrecy." *Google Security Blog.* security.googleblog.com/2011/11/protecting-data-for-long-term-with.html.

Lanzagorta, Marco (2011). *Quantum Radar*. Synthesis Lectures on Quantum Computing. Morgan & Claypool.

— (2013). *Underwater Communications*. Williston, VT: Morgan & Claypool.

— (August 2018). "Envisioning The Future of Quantum Sensing and Communications." Remarks of Marco Lanzagorta at the Conference on Quantum Sensing and Communications held by the National Academies of Sciences, Engineering and Medicine.

Lanzagorta, Marco and Jeffrey Uhlmann (2015). "Space-Based Quantum Sensing for Low-Power Detection of Small Targets." 9461, pp. 946115.

— (2020). "Opportunities and Challenges of Quantum Radar." *IEEE Aerospace and Electronic Systems Magazine* 35.11, pp. 38–56.

Lanzagorta, Marco, Jeffrey Uhlmann, and Salvador E. Venegas-Andraca (2015). "Quantum Sensing in The Maritime Environment." *OCEANS 2015 – MTS/IEEE Washington*, pp. 1–9.

Lerman, Amy E. (June 2019). *Good Enough for Government Work: The Public Reputation Crisis in America (And What We Can Do to Fix It)*. University of Chicago Press.

Levy, Steven (April 1, 1999). "The Open Secret." *Wired*. www.wired .com/1999/04/crypto/.

Lewis, Gilbert N (1926). "The Conservation of Photons." *Nature* 118.2981, pp. 874–875.

Li, Meixiu et al. (2019). "Review of Carbon and Graphene Quantum Dots for Sensing." *ACS Sensors* 4.7, pp. 1732–1748.

Li, T. et al. (2016). "Security Attack Analysis Using Attack Patterns." *2016 IEEE Tenth International Conference on Research Challenges in Information Science (RCIS)*. Grenoble, France, pp. 1–13.

Liao, Sheng-Kai et al. (2018). "Satellite-Relayed Intercontinental Quantum Network." *Physical Review Letters* 120.3, pp. 030501.

Lloyd, Seth (2014). "The Computational Universe." In: *Information and The Nature of Reality: From Physics to Metaphysics*. Cambridge University Press, pp. 118–133.

Loriani, S. et al. (June 2019). "Atomic Source Selection in Space-Borne Gravitational Wave Detection." *New Journal of Physics* 21.6, pp. 063030. iopscience.iop.org/article/10.1088/1367-2630/a b22d0.

Loss, Daniel and David P. DiVincenzo (1998). "Quantum Computation with Quantum Dots." *Physical Review A* 57.1, pp. 120–126.

Ma, Lijun et al. (September 1, 2015). "EIT Quantum Memory with Cs Atomic Vapor for Quantum Communication." *Proceedings of SPIE Optics and Photonics 2015*. www.nist.gov/publicati ons/eit-quantum-memory-cs-atomic-vapor-quantum-communicati on.

Ma, Xiongfeng et al. (2016). "Quantum Random Number Generation." *Npj Quantum Information* 2.1.

Majorana, Ettore and Luciano Maiani (2006). "A Symmetric Theory of Electrons and Positrons." In: *Ettore Majorana Scientific Papers*. Berlin: Springer, pp. 201–233.

Mallapaty, Smriti (2020). "China Bans Cash Rewards for Publishing Papers." *Nature* 579.7798, pp. 18–19.

Manglaviti, Ariana (June 5, 2018). "Exploring Greener Approaches to Nitrogen Fixation." www.bnl.gov/newsroom/news.php?a=212 919.

Manin, Yuri I. (May 1999). "Classical Computing, Quantum Computing, and Shor's Factoring Algorithm." Talk at the Bourbaki Seminar, June 1999, later published in *Astréisque* 266 (2000), exp. no. 862, p. 375–404. arxiv.org/pdf/quant-ph/9903008.pdf.

— (2007). *Mathematics As Metaphor: Selected Essays of Yuri I. Manin*. Providence, RI: American Mathematical Society.

Marcus, Amy Dockser (May 8, 2020). "Covid-19 Raises Questions About The Value of Personalized Medicine." *Wall Street Journal – Online Edition*.

Marks, Paul (October 15, 2007). "Quantum Cryptography to Protect Swiss Election." *New Scientist*. www.newscientist.com/article/dn12786-quantum-cryptography-to-protect-swiss-election/.

Matzke, Doug (1993). "Message From The Chairman." In: *Workshop on Physics and Computation PhysComp '92*. Dallas, TX: IEEE Computer Society Press.

Mazzucato, Mariana (2015). *The Entrepreneurial State: Debunking Public Vs. Private Sector Myths*. New York: PublicAffairs.

McCarthy, J. et al. (1955). "A Proposal for The Dartmouth Summer Research Project on Artificial Intelligence." Last accessed August 23, 2020. www-formal.stanford.edu/jmc/history/dartmouth/dartmouth.html.

McDermott, Roger N. (September 2017). *Russia's Electronic Warfare Capabilities to 2025*. International Centre for Defence. euagenda.eu/upload/publications/untitled-135826-ea.pdf.

McGrew, W. F. et al. (2018). "Atomic Clock Performance Enabling Geodesy Below The Centimetre Level." *Nature* 564.7734, pp. 87–90. doi.org/10.1038/s41586-018-0738-2.

Merkle, Ralph Charles (June 1979). *Secrecy, Authentication and Public Key Systems*. Tech. rep. 1979-1. Information Systems Laboratory, Stanford University. www.merkle.com/papers/Thesis1979.pdf.

Merriam-Webster Incorporated (2020). ""Machine."" www.merriam-webster.com/dictionary/machine.

Metropolis, N. (1987). "The Beginning of The Monte Carlo Method." *Los Alamos Science* 15.

Meyer, David H. et al. (2020). "Assessment of Rydberg Atoms for Wideband Electric Field Sensing." *Journal of Physics B: Atomic, Molecular and Optical Physics* 53.3, pp. 034001.

Meyers, Ronald E. and Keith S. Deacon (2015). "Space-Time Quantum Imaging." *Entropy* 17.3, pp. 1508–1534.

Meyers, Ronald E., Keith S. Deacon, and Yanhua Shih (April 2008). "Ghost-Imaging Experiment by Measuring Reflected Photons." *Physical Review A* 77.4, pp. 041801. link.aps.org/doi/10.1103/PhysRevA.77.041801.

Microsoft Corp. (2013). "Microsoft Security Advisory 2862973: Update for Deprecation of MD5 Hashing Algorithm for Microsoft Root Certificate Program." docs.microsoft.com/en-us/security-up dates/SecurityAdvisories/2014/2862973.

— (2018). "Developing a Topological Qubit." cloudblogs.microsoft.c om/quantum/2018/09/06/developing-a-topological-qubit/.

Miller, Victor S. (1986). "Use of Elliptic Curves in Cryptography." *Advances in Cryptology – CRYPTO '85 Proceedings*, ed. Hugh C. Williams. Berlin, Heidelberg: Springer, pp. 417–426.

Minsky, Marvin (1982). "Cellular Vacuum." *International Journal of Theoretical Physics* 21.6, pp. 537–551.

Mirhosseini, M. et al. (2015). "High-Dimensional Quantum Cryptography with Twisted Light." *New Journal of Physics* 17, pp. 1–12.

MIT (May 29, 2018). "Outside Professional Activities." In: *MIT Policies*. Last accessed March 6, 2021. Chap. 4.5. policies.mit.ed u/policies-procedures/40-faculty-rights-and-responsibilities/45-out side-professional-activities.

MIT Endicott House (2020). "Our History." Last accessed September 28, 2020. mitendicotthouse.org/our-history/.

MIT Institute Archives (2011). "Laboratory for Computer Science (LCS)." Last accessed August 2, 2020. libraries.mit.edu/mithistor y/research/labs/lcs.

Mohseni, M. et al. (2017). "Commercialize Quantum Technologies in Five Years." *Nature* 543.7644, pp. 171–174.

Moller, Violet (May 10, 2019). "How Anti-Immigrant Policies Thwart Scientific Discovery." *Washington Post*.

Möller, Matthias and Cornelis Vuik (2017). "On The Impact of Quantum Computing Technology on Future Developments in High-Performance Scientific Computing." *Ethics and Information Technology* 19.4, pp. 253–269.

Molteni, Megan (September 14, 2017). "With Designer Bacteria, Crops Could One Day Fertilize Themselves." *Wired*. www.wired .com/story/with-designer-bacteria-crops-could-one-day-fertilize-th emselves/.

Monroe, C. et al. (December 1995). "Demonstration of a Fundamental Quantum Logic Gate." *Physical Review Letters* 75.25, pp. 4714–4717. link.aps.org/doi/10.1103/PhysRevLett.75.4714.

Monroe, Christopher, Michael G. Raymer, and Jacob Taylor (2019). "The US National Quantum Initiative: From Act to Action." *Science* 364.6439, pp. 440–442. science.sciencemag.org/content/3 64/6439/440.

Montanaro, Ashley (January 2016). "Quantum Algorithms: an Overview." *Npj Quantum Information* 2.1. dx.doi.org/10.103 8/npjqi.2015.23.

Moore, Gordon E. (1965). "Cramming More Components Onto Integrated Circuits." *Electronics Magazine* 38 (8), pp. 82–85. www.computerhistory.org/collections/catalog/102770822.

— (September 2006). "Progress in Digital Integrated Electronics [Technical Literature, Copyright 1975 IEEE. Reprinted, with Permission. Technical Digest. International Electron Devices Meeting, IEEE, 1975, pp. 11–13.]" *IEEE Solid-State Circuits Society Newsletter* 11.3, pp. 36–37. ieeexplore.ieee.org/document /4804410/.

Morozov, Evgeny (October 13, 2014). "The Planning Machine: Project Cybersyn and The Origins of The Big Data Nation." *New Yorker* 90.31.

Morser, Bruce (2020). "Inertial Navigation." Last accessed October 24, 2020. www.panam.org/the-jet-age/517-inertial-navigation-2.

Mourik, Vincent et al. (May 25, 2012). "Signatures of Majorana Fermions in Hybrid Superconductor-Semiconductor Nanowire Devices." *Science* 336.6084, pp. 1003–1007.

Murph, Paul (December 11, 2019). "Wirecard Critics Targeted in London Spy Operation." *Financial Times*. www.ft.com/content /d94c938e-1a84-11ea-97df-cc63de1d73f4.

Musiani, Francesca et al. (2016). *The Turn to Infrastructure in Internet Governance*. Berlin: Springer.

Nash, Gerald D. (1999). *The Federal Landscape: an Economic History of The Twentieth-Century West*. Tucson: University of Arizona Press. www.h-net.org/review/hrev-a0b4n6-aa.

National Center for Science and Engineering Statistics (December 2019). *Doctorate Recipients From US Universities*. Tech. rep. NSF 21-308. Washington, DC: Directorate for Social, Behavioral and Economic Sciences, National Science Foundation. ncse s.nsf.gov/pubs/nsf21308.

National Coordination Office for Space-Based Positioning, Navigation, and Timing (October 2001). "Frequently Asked Questions About Selective Availability." *GPS.gov.* www.gps.gov/systems/gps/modernization/sa/faq/.

National Institute of Standards and Technology (January 3, 2017). "Post-Quantum Cryptography." Last accessed February 9, 2021. csrc.nist.gov/Projects/post-quantum-cryptography/post-quantum-cryptography-standardization/Call-for-Proposals.

National Security Agency (January 6, 2001). "Groundbreaking Ceremony Held for 1.2 Billion Utah Data Center." *Press Release Pa-118-18.* www.nsa.gov/news-features/press-room/Article/1630552/groundbreaking-ceremony-held-for-12-billion-utah-data-center/.

— (2020). "Quantum Key Distribution (QKD) and Quantum Cryptography (QC)." Last accessed July 25, 2021. www.nsa.gov/what-we-do/cybersecurity/quantum-key-distribution-qkd-and-quantum-cryptography-qc/.

National Security Agency and Central Security Service (2016). "Commercial National Security Algorithm Suite and Quantum Computing FAQ." apps.nsa.gov/iaarchive/library/ia-guidance/ia-solutions-for-classified/algorithm-guidance/cnsa-suite-and-quantum-computing-faq.cfm.

— (2021). "VENONA." Last accessed March 13, 2021. www.nsa.gov/News-Features/Declassified-Documents/Venona/.

Nicholson, T. L. et al. (2015). "Systematic Evaluation of an Atomic Clock at 2×10^{-18} Total Uncertainty." *Nature Communications* 6.1, pp. 6896. doi.org/10.1038/ncomms7896.

NobelPrize.org (October 2019). "The Nobel Prize in Physics 1965." www.nobelprize.org/prizes/physics/1965/summary/.

Nuttall, William J., Richard H. Clarke, and Bartek A. Glowacki (2012). "Stop Squandering Helium." *Nature* 485.7400, pp. 573–575.

O'Mara, Margaret Pugh (2015). *Cities of Knowledge: Cold War Science and The Search for The Next Silicon Valley.* Princeton, NJ: Princeton University Press.

— (2019). *The Code: Silicon Valley and The Remaking of America.* New York: Penguin Press.

Obama, Barack (December 29, 2009). "Executive Order 13526: Classified National Security Information." www.archives.gov/isoo/policy-documents/cnsi-eo.html.

Office of the Secretary of Defense (2020). "Department of Defense Fiscal Year (FY) 2021 Budget Estimates." In: *Defense-Wide Justification Book*. Last accessed February 20, 2021. US Department of Defense. Chap. 3, pp. 1094. comptroller.defense.gov/Portals/45/Documents/defbudget/fy2021/budget_justification/pdfs/03_RDT_and_E/RDTE_Vol3_OSD_RDTE_PB21_Justification_Book.pdf.

Ohm, Paul (2009a). "Broken Promises of Privacy: Responding to The Surprising Failure of Anonymization." *UCLA L. Rev.* 57, pp. 1701–1777.

— (2009b). "The Rise and Fall of Invasive ISP Surveillance." *University of Illinois Law Review*. ssrn.com/abstract_id=1261344.

Olson, Parmy (April 10, 2020). "My Girlfriend Is a Chatbot." *Wall Street Journal*.

Omar, Yasser (May 6, 2015). "Workshop on Quantum Technologies and Industry." *DG Connect*. digital-strategy.ec.europa.eu/en/library/report-workshop-quantum-technologies-and-industry.

Oqubay, Arkebe (2015). "Climbing without Ladders: Industrial Policy and Development." In: *Made in Africa*. Oxford University Press.

Organized Crime and Corruption Reporting Project (March 20, 2017). "The Russian Laundromat Exposed." www.occrp.org/en/laundromat/the-russian-laundromat-exposed/.

Ortega, Almudena Azcárate (January 28, 2021). "Placement of Weapons in Outer Space: The Dichotomy between Word and Deed." *Lawfare*. www.lawfareblog.com/placement-weapons-outer-space-dichotomy-between-word-and-deed.

Padma, T. V. (2020). "India Bets Big on Quantum Technology." *Nature*. www.nature.com/articles/d41586-020-00288-x/.

Palmer, J. (2017). "Technology Quarterly: Here, There and Everywhere." *The Economist* 413.9027.

Pan, Feng and Pan Zhang (2021). "Simulating The Sycamore Quantum Supremacy Circuits." arxiv.org/abs/2103.03074.

Pant, Mihir et al. (March 2019). "Routing Entanglement in The Quantum Internet." *Npj Quantum Information* 5.1, pp. 25. doi.org/10.1038/s41534-019-0139-x.

Patinformatics (2017). "Quantum Information Technology Patent Landscape Reports."

Pawlyk, Oriana (June 10, 2020). "Air Force Will Pit a Drone Against a Fighter Jet in Aerial Combat Test." www.military.com/daily-n

ews/2020/06/10/air-force-will-pit-drone-against-fighter-jet-aerial-combat-test.html.

Pednault, Edwin et al. (October 12, 2019). "On 'Quantum Supremacy'." *IBM Research Blog.* www.ibm.com/blogs/research/2019/10/on-quantum-supremacy/.

Peng, Wang Chun et al. (2019). "Factoring Larger Integers with Fewer Qubits via Quantum Annealing with Optimized Parameters." *Science China Physics, Mechanics and Astronomy* 62.6, pp. 60311. doi.org/10.1007/s11433-018-9307-1.

Peres, Asher (December 1985). "Reversible Logic and Quantum Computers." *Physical Review A* 32 (6), pp. 3266–3276. link.aps.org/doi/10.1103/PhysRevA.32.3266.

Perlroth, Nicole (September 10, 2013). "Government Announces Steps to Restore Confidence on Encryption Standards." *The New York Times.* bits.blogs.nytimes.com/2013/09/10/government-announces-steps-to-restore-confidence-on-encryption-standards/.

— (2021). *This Is How They Tell Me The World Ends: The Cyberweapons Arms Race.* New York: Bloomsbury Publishing.

Perrin, Léo (2019). "Partitions in The S-Box of Streebog and Kuznyechik." Report 2019/092. eprint.iacr.org/2019/092.

Peterson, Scott and Payam Faramarzi (December 15, 2011). "Exclusive: Iran Hijacked US Drone, Says Iranian Engineer." *Christian Science Monitor.* www.csmonitor.com/World/Middle-East/2011/1215/Exclusive-Iran-hijacked-US-drone-says-Iranian-engineer.

Pfaff, W. et al. (2014). "Unconditional Quantum Teleportation between Distant Solid-State Quantum Bits." *Science* 345.6196, pp. 532–535.

Physics Today (February 4, 2019). "Rolf Landauer." *Physics Today.* physicstoday.scitation.org/do/10.1063/PT.6.6.20190204a/full/.

Pirandola, S. et al. (2018). "Advances in Photonic Quantum Sensing." *Nature Photonics* 12.12, pp. 724–733.

Pisana, Simone et al. (March 2007). "Breakdown of The Adiabatic Born – Oppenheimer Approximation in Graphene." *Nature Materials* 6.3, pp. 198–201. doi.org/10.1038/nmat1846.

Plutarch (1921). *Lives. Vol. 10, Agis and Cleomenes, Tiberius and Caius Gracchus, Philopoemen and Flamninius.* Loeb Classical Library. Cambridge, MA: Heinemann.

Popkin, Gabriel (June 16, 2017). "Spooky Action Achieved at Record Distance." *Science* 356.6343, pp. 1110–1111.

Poplavskii, R. P. (1975). "Thermodynamical Models of Information Processing." *Uspekhi Fizicheskikh Nauk, Advances in Physical Sciences* 115.3, pp. 465–501.

Posen, Barry R. (2003). "Command of The Commons: The Military Foundation of US Hegemony." *International Security* 28.1, pp. 5–46.

Prabhakar, Shashi et al. (2020). "Two-Photon Quantum Interference and Entanglement at 2.1 Mm." *Science Advances* 6.13. advances.sciencemag.org/content/6/13/eaay5195.

Preskill, John (2012). "Quantum Computing and The Entanglement Frontier." *WSPC Proceedings*. Rapporteur talk at the 25th Solvay Conference on Physics "The Theory of the Quantum World". Brussels.

— (October 6, 2019). "Why I Called It 'Quantum Supremacy.'" *Wired*. wired.com/story/why-i-coined-the-term-quantum-supremacy.

"ProQuest Dissertations and Theses Global" (n.d.) (). Last accessed February 20, 2021. www.proquest.com/products-services/pqdtglobal.html.

PYMNTS (2018). "The Meal Kits Crowding Problem." *PYMNTS Subscriptions*. www.pymnts.com/subscriptions/2018/meal-kits-crowding/.

Qiu, Longqing et al. (2016). "Development of a Squid-Based Airborne Full Tensor Gradiometer for Geophysical Exploration." *Seg Technical Program Expanded Abstracts*. Society of Exploration Geophysicists.

Quan, Wei, Bikun Chen, and Fei Shu (2017). "Publish or Impoverish: an Investigation of The Monetary Reward System of Science in China (1999–2016)." *Aslib Journal of Information Management* 69 (5), pp. 486–502.

Quantique, ID (2020). "Quantis QRNG Chip." www.idquantique.com/random-number-generation/products/quantis-qrng-chip/.

Rabin, M. O. and D. Scott (April 1959). "Finite Automata and Their Decision Problems." *IBM Journal*, pp. 114–125.

Rabkin, Jeremy A. and John Yoo (2017). *Striking Power: How Cyber, Robots, and Space Weapons Change The Rules for War*. New York: Encounter Books.

RAND (2002). *Space Weapons: Earth Wars*. Santa Monica, CA: RAND.

Reardon, Joel (2016). *Secure Data Deletion*. New York: Springer.

Reece, Andrew G. and Christopher M. Danforth (August 2017). "Instagram Photos Reveal Predictive Markers of Depression." *EPJ Data Science* 6.1, pp. 15. doi.org/10.1140/epjds/s13688-017-0110-z.

Remington Rand (1954). *Sorting Methods for UNIVAC Systems.* www.bitsavers.org/pdf/univac/univac1/UnivacSortingMethods.pdf.

Ren, J. G. et al. (2017). "Ground-To-Satellite Quantum Teleportation." *Nature* 549.7670, pp. 70–73.

Rendell, Paul (2011). "A Universal Turing Machine in Conway's Game of Life." *2011 International Conference on High Performance Computing and Simulation.* IEEE, pp. 764–772.

Repantis, Kate (March 19, 2014). "Why Hasn't Commercial Air Travel Gotten Any Faster Since The 1960s?" *Slice of MIT.* alum.mit.edu/slice/why-hasnt-commercial-air-travel-gotten-any-faster-1960s.

Research, Transparency Market (2017). "Ring Laser Gyroscope Market – Snapshot." www.transparencymarketresearch.com/ring-laser-gyroscope-market.html.

Rich, Ben R. and Leo Janos (1994). *Skunk Works: a Personal Memoir of My Years at Lockheed.* Boston, MA: Little, Brown.

Rid, Thomas (2020). *Active Measures: The Secret History of Disinformation and Political Warfare.* New York: Farrar, Straus and Giroux.

Rideout, Ariel (July 24, 2008). "Making Security Easier." *Official Gmail Blog.*

Rivest, Ronald L. (2011). "The Early Days of RSA: History and Lessons." In: *ACM Turing Award Lectures.* New York: Association for Computing Machinery, pp. 2002. doi.org/10.1145/1283920.1961920.

Rivest, Ronald L., Adi Shamir, and Len Adleman (February 1978). "A Method for Obtaining Digital Signatures and Public-Key Cryptosystems." *Communications of The Association for Computing Machinery* 21.2, pp. 120–126. dl.acm.org/doi/10.1145/359340.359342.

Roberts, Siobhan (April 15, 2020). "John Horton Conway, a 'Magical Genius' in Math, Dies at 82." *New York Times.* www.nytimes.com/2020/04/15/technology/john-horton-conway-dead-coronavirus.html.

Robinson, Andrew (2018). "Did Einstein Really Say That?" *Nature* 557.7703, pp. 30–31.

Robson, David P. (April 1984). "Profile Edwin H. Land." *Chem-Matters*. www.cs.cornell.edu/~ginsparg/physics/Phys446-546/840 412t.pdf.

Rose, Scott et al. (February 2019). *Trustworthy Email*. Tech. rep. SP 800-177 Rev. 1. National Institute of Standards and Technology. csrc.nist.gov/publications/detail/sp/800-177/rev-1/final.

Rowlett, Frank B. (1999). *The Story of Magic: Memoirs of an American Cryptologic Pioneer*. Laguna Hills, CA: Aegean Park Press.

Ruf, M. et al. (2019). "Optically Coherent Nitrogen-Vacancy Centers in Micrometer-Thin Etched Diamond Membranes." *Nano Lett* 19.6, pp. 3987–3992.

Rule, Nicholas O. (January 2017). "Perceptions of Sexual Orientation From Minimal Cues." *Archives of Sexual Behavior* 46.1, pp. 129–139. doi.org/10.1007/s10508-016-0779-2.

Rzetenly, Xaq (September 23, 2017). "Is Beaming Down in Star Trek a Death Sentence?" *Ars Technica*. arstechnica.com/gaming /2017/09/is-beaming-down-in-star-trek-a-death-sentence/.

Sadkhan, S. B. and B. S. Yaseen (2019). "DNA-Based Cryptanalysis: Challenges, and Future Trends." *2019 2nd Scientific Conference of Computer Sciences (SCCS)*, pp. 24–27.

Samuelson, Arielle (June 19, 2019). "What Is an Atomic Clock?" www.nasa.gov/feature/jpl/what-is-an-atomic-clock.

Sandia National Laboratories and National Nuclear Security Administration (2015). *ASCR Workshop on Quantum Computing for Science*. Electronic Book 1194404. www.osti.gov/servlets/pur l/1194404/.

Saxenian, AnnaLee (1996). *Regional Advantage: Culture and Competition in Silicon Valley and Route 128*. Cambridge, MA: Harvard University Press.

Scarani, Valerio, H. Bechmann-Pasquinucci, et al. (2009). "The Security of Practical Quantum Key Distribution." *Reviews of Modern Physics* 81.3, pp. 1301–1350.

Scarani, Valerio and Christian Kurtsiefer (2014). "The Black Paper of Quantum Cryptography: Real Implementation Problems." *Theoretical Computer Science* 560. Theoretical Aspects of Quantum Cryptography – celebrating 30 years of BB84, pp. 27–32. www.sciencedirect.com/science/article/pii/S03043 97514006938.

Schelling, Thomas C. (1980). *The Strategy of Conflict: With a New Preface by The Author*. Cambridge, MA: Harvard University Press.

Schiermeier, Quirin (2019). "Russia Joins Race to Make Quantum Dreams a Reality." *Nature* 577.7788, pp. 14.

Schneier, Bruce (November 2007). "Did NSA Put a Secret Backdoor in New Encryption Standard?" *Wired*. Last accessed May 30, 2020. www.wired.com/2007/11/securitymatters-1115/.

Schofield, Jack (February 11, 2018). "John Perry Barlow Obituary." *The Guardian*. www.theguardian.com/technology/2018/feb/11/john-perry-barlow-obituary.

Schumacher, Benjamin (April 1995). "Quantum Coding." *Physical Review A* 51 (4), pp. 2738–2747. link.aps.org/doi/10.1103/PhysRevA.51.2738.

Sciutto, Jim (May 10, 2019). "A Vulnerable US Really Does Need a Space Force." *Wall Street Journal – Online Edition*.

Scott, James C. (1998). *Seeing Like a State: How Certain Schemes to Improve The Human Condition Have Failed*. The Yale ISPS series. New Haven, CT: Yale University Press. www.gbv.de/dms/sub-hamburg/233487662.pdf.

Shamir, Adi (November 1979). "How to Share a Secret." *Communications of The Association for Computing Machinery* 22.11, pp. 612–613. doi.org/10.1145/359168.359176.

— (1984). "Identity-Based Cryptosystems and Signature Schemes." *Advances in Cryptology: Proceedings of CRYPTO 84*. Vol. 7. Santa Barbara, California, pp. 47–53.

— (2011). "Cryptography: State of The Science." In: *ACM Turing Award Lectures*. New York: Association for Computing Machinery, pp. 2002. doi.org/10.1145/1283920.1961903.

Shankland, Stephen (June 29, 2019). "Startup Packs All 16GB of Wikipedia Onto DNA Strands to Demonstrate New Storage Tech." www.cnet.com/news/startup-packs-all-16gb-wikipedia-onto-dna-strands-demonstrate-new-storage-tech/.

Shannon, Claude Elwood (1948). "A Mathematical Theory of Communication." *The Bell System Technical Journal* 27.3, pp. 379–423.

— (1949). *Communication Theory of Secrecy Systems*. New York: ATT.

Shkel, Andrei M. (2010). "Precision Navigation and Timing Enabled by Microtechnology: Are We There Yet?" *SENSORS, 2010 IEEE.* IEEE. Waikoloa Village, HI, pp. 5–9.

Shor, Peter W. (October 1997). "Polynomial-Time Algorithms for Prime Factorization and Discrete Logarithms on a Quantum Computer." *SIAM Journal on Computing* 26.5, pp. 1484–1509. dx.doi.org/10.1137/S0097539795293172.

Shostack, Adam (August 27, 2009). "The Threats to Our Products." *Microsoft Security Blog.* www.microsoft.com/security/blog/2009/08/27/the-threats-to-our-products/.

— (2014). *Threat Modeling: Designing for Security.* New York: Wiley.

Simonite, Tom (March 23, 2016). "Intel Puts The Brakes on Moore's Law." *MIT Technology Review.*

Singer, P. W. and August Cole (2015). *Ghost Fleet: a Novel of The Next World War.* Boston, MA: Houghton Mifflin Harcourt.

Singh, Simon (August 29, 2000). *The Code Book: The Science of Secrecy From Ancient Egypt to Quantum Cryptography.* London: Anchor.

Sipser, Michael (2012). *Introduction to The Theory of Computatio.* 3rd ed. Independence, KY: Cengage Learning.

Sitz, Greg (February 2005). "Approximate Challenges." *Nature* 433.7025. doi.org/10.1038/433470a.

Smith, Brad (October 26, 2018). "Technology and The US Military." *Microsoft On The Issues.* blogs.microsoft.com/on-the-issues/2018/10/26/technology-and-the-us-military/.

Sola Pool, Ithiel de (1983). *Technologies of Freedom.* Cambridge, MA: Harvard University Press.

Solove, Daniel J. (2007). "'I've Got Nothing to Hide' and Other Misunderstandings of Privacy." *San Diego Law Review* 44.4, pp. 745–772.

Spiegel, Peter (October 25, 2013). "Angela Merkel Eyes Place for Germany in US Intelligence Club." *Financial Times.*

Spinellis, D. (May 2008). "The Antikythera Mechanism: A Computer Science Perspective." *Computer* 41.5, pp. 22–27.

Springer, Paul J. (2020). *Cyber Warfare: A Documentary and Reference Guide.* Santa Barbara, CA: ABC-CLIO. products.abc-clio.com/abc-cliocorporate/product.aspx?pc=A6167C.

Starr, Michelle (January 19, 2014). "Fridge Caught Sending Spam Emails in Botnet Attack." *CNet*. www.cnet.com/news/fridge-cau ght-sending-spam-emails-in-botnet-attack/.

Steele, Beth Anne (November 26, 2019). "Oregon FBI Tech Tuesday: Securing Smart TVs." *FBI Portland*. www.fbi.gov/contact-us/field-offices/portland/news/press-releases/tech-tuesdaysmart-t vs.

Stern, Jessica (1999). *The Ultimate Terrorist*. Cambridge, MA: Harvard University Press.

Stevens, Hallam (January 30, 2018). "Hans Peter Luhn and The Birth of The Hashing Algorithm." *IEEE Spectrum*. spectrum.iee e.org/tech-history/silicon-revolution/hans-peter-luhn-and-the-birt h-of-the-hashing-algorithm.

Storbeck, Olaf and Guy Chazan (June 28, 2020). "Germany to Overhaul Accounting Regulation after Wirecard Collapse." *Financial Times*.

Strunsky, Steve (August 8, 2013). "N.J. Man Fined $32K for Illegal GPS Device That Disrupted Newark Airport System." *NJ Advance Media*. www.nj.com/news/2013/08/man_fined_32000_for _blocking_newark_airport_tracking_system.html.

Susskind, Leonard (2008). *The Black Hole War: My Battle with Stephen Hawking to Make The World Safe for Quantum Mechanics*. Boston, MA: Little, Brown and Company.

Svoboda, Karel and Ryohei Yasuda (2006). "Principles of Two-Photon Excitation Microscopy and Its Applications to Neuroscience." *Neuron* 50.6, pp. 823–839.

Swire, Peter (July 15, 2015). "The Golden Age of Surveillance." *Slate*. slate.com/technology/2015/07/encryption-back-doors-arent -necessary-were-already-in-a-golden-age-of-surveillance.html.

Symul, T., S. M. Assad, and P. K. Lam (2011). "Real Time Demonstration of High Bitrate Quantum Random Number Generation with Coherent Laser Light." *Applied Physics Letters* 98.23, pp. 231103.

Takemoto, Kazuya et al. (September 2015). "Quantum Key Distribution Over 120 km Using Ultrahigh Purity Single-Photon Source and Superconducting Single-Photon Detectors." *Scientific Reports* 5.1, pp. 14383. doi.org/10.1038/srep14383.

Tambe, Milind (2012). *Security and Game Theory: Algorithms, Deployed Systems, Lessons Learned*. Cambridge: Cambridge University Press.

Tapley, B. D. et al. (2004). "The Gravity Recovery and Climate Experiment: Mission Overview and Early Results." *Geophysical Research Letters* 31.9.

Taylor, Michael A. and Warwick P. Bowen (2016). "Quantum Metrology and Its Application in Biology." *Physics Reports* 615, pp. 1–59.

Temperton, James (January 26, 2017). "Got a Spare $15 Million? Why Not Buy Your Very Own D-Wave Quantum Computer." *Wired UK.*

Thorton, Will (October 16, 2018). "Selective Availability: A Bad Memory for GPS Developers and Users." *Spirent Blog.* www.spirent.com/blogs/selective-availability-a-bad-memory-for-gps-developers-and-users.

Tierney, Tim M. et al. (2019). "Optically Pumped Magnetometers: From Quantum Origins to Multi-Channel Magnetoencephalography." *NeuroImage* 199, pp. 598–608. www.sciencedirect.com/science/article/pii/S1053811919304550.

Tirosh, Ofer (January 8, 2020). "Top Translation Industry Trends for 2020." *Tomedes Translator's Blog.* www.tomedes.com/translator-hub/translation-industry-trends-2020.

Toffoli, Tommaso (1977). *Journal of Computer and System Sciences* 15, pp. 213–231.

— (1982). "Physics and Computation." *International Journal of Theoretical Physics* 21.3, pp. 165–175.

Tretkoff, Ernie (December 2007). "This Month in Physics History: December 1938: Discovery of Nuclear Fission." *APS News* 16.11. www.aps.org/publications/apsnews/200712/physicshistory.cfm.

Tsividis, Yannis (December 1, 2017). "Not Your Father's Analog Computer." *IEEE Spectrum.* spectrum.ieee.org/computing/hardware/not-your-fathers-analog-computer.

Turing, Alan M. (1936). "On Computable Numbers, with an Application to The Entscheidungsproblem." *Proceedings of The London Mathematical Society* 2.42, pp. 230–265. www.cs.helsinki.fi/u/gionis/cc05/OnComputableNumbers.pdf.

Uhlmann, Jeffrey, Marco Lanzagorta, and Salvador E. Venegas-Andraca (2015). "Quantum Communications in The Maritime Environment." *OCEANS 2015 – MTS/IEEE Washington,* pp. 1–10.

Union of Concerned Scientists (May 1, 2021). "UCS Satellite Database." Last accessed July 25, 2021. www.ucsusa.org/resources/satellite-database.

United Nations (1986). *Principles Relating to Remote Sensing of The Earth From Outer Space.* Resolution adopted by the General Assembly / United Nations, 41/65. New York: United Nations.

US Agency for International Development, Bureau for Africa (July 2019). "Government Complicity in Organized Crime." pdf.usaid.gov/pdf_docs/PA00TSH2.pdf.

US Air Force Scientific Advisory Board (2016). *Utility of Quantum Systems for The Air Force Study Abstract.* Tech. rep. US Air Force Scientific Advisory Board. web.archive.org/web/20170427005155/www.scientificadvisoryboard.af.mil/Portals/73/documents/AFD-151214-041.pdf?ver=2016-08-19-101445-230.

US Census Bureau (2021). "US and World Population Clock." Last accessed January 1, 2021. www.census.gov/popclock/.

US Congress (2018). *National Quantum Initiative Act.* [US Government Publishing Office]. purl.fdlp.gov/GPO/gpo126751.

US Congress, House Permanent Select Committee on Intelligence (2016). "Executive Summary of Review of The Unauthorized Disclosures of Former National Security Agency Contractor Edward Snowden." purl.fdlp.gov/GPO/gpo75954.

US Federal Communications Commission (April 2020). "Jammer Enforcement." www.fcc.gov/general/jammer-enforcement.

Vandersypen, Lieven M. K. et al. (2001). "Experimental Realization of Shor's Quantum Factoring Algorithm Using Nuclear Magnetic Resonance." *Nature* 414.6866, pp. 883–887.

Venegas-Andraca, Salvador E., M. Lanzagorta, and J. Uhlmann (2015). "Maritime Applications of Quantum Computation." *OCEANS 2015 – MTS/IEEE Washington,* pp. 1–8.

Vidas, Timothy, Daniel Votipka, and Nicolas Christin (2011). "All Your Droid Are Belong to Us: A Survey of Current Android Attacks." *Proceedings of The 5th USENIX Conference on Offensive Technologies.* WOOT'11. San Francisco, CA: USENIX Association, pp. 10.

von Neumann, John (1945). *First Draft of a Report on The ED-VAC.* Tech. rep. United States Army Ordnance Department and the University of Pennsylvania.

— (1951). "Various Techniques Used in Connection with Random Digits." *Journal of Research, Applied Math Series* 3. Summary written by George E. Forsythe, pp. 36–38. mcnp.lanl.gov/pdf_fil es/nbs_vonneumann.pdf.

von Neumann, John and Arthur W. Burks (1966). *Theory of Self-Reproducing Automata.* Champaign, IL: University of Illinois Press.

Wagner, Michelle (2006). "The Inside Scoop on Mathematics at The NSA." *Math Horizons* 13.4, pp. 20–23.

Walden, David (2011). "Early Years of Basic Computer and Software Engineering." In: *A Culture of Innovation: Insider Accounts of Computing and Life at BBN.* East Sandwich, MA: Waterside Publishing.

Wallace, D. and J. Costello (July 2017). "Eye in The Sky: Understanding The Mental Health of Unmanned Aerial Vehicle Operators." *Journal of The Military and Veterans' Health* 25.3. jmvh.org/article/eye-in-the-sky-understanding-the-mental-health-of-unmanned-aerial-vehicle-operators/.

Wang, Hai-Tian et al. (August 2019). "Science with The TianQin Observatory: Preliminary Results on Massive Black Hole Binaries." *Physical Review D* 100.4, pp. 043003. link.aps.org/doi/10.1 103/PhysRevD.100.043003.

Wang, Y. and M. Kosinski (2018). "Deep Neural Networks Are More Accurate Than Humans at Detecting Sexual Orientation From Facial Images." *Journal of Personality and Social Psychology* 114.2, pp. 246–257.

Wang, Yunfei et al. (2019). "Efficient Quantum Memory for Single-Photon Polarization Qubits." *Nature Photonics* 13.5, pp. 346–351. doi.org/10.1038/s41566-019-0368-8.

Watson, James D. and Francis H. C. Crick (1953). "Molecular Structure of Nucleic Acids: a Structure for Deoxyribose Nucleic Acid." *Nature* 171.4356, pp. 737–738.

Wehner, S., D. Elkouss, and R. Hanson (2018). "Quantum Internet: A Vision for The Road Ahead." *Science* 362.6412.

Weinbaum, Cortney et al. (2017). *SIGINT for Anyone: The Growing Availability of Signals Intelligence in The Public Domain.* RAND Perspective; 273. Santa Monica, CA: RAND. www.rand .org/pubs/perspectives/PE273.html.

Weiner, S. et al. (2020). "A Flight Capable Atomic Gravity Gradiometer With a Single Laser." *2020 IEEE International Sym-*

posium on Inertial Sensors and Systems (INERTIAL), pp. 1–3.

Wertheimer, Michael (2015). "Encryption and The NSA Role in International Standards." *Notices of The AMS*. Note: At the time of publication, Michael Wertheimer was the Director of Research at the US National Security Agency. Last accessed May 30, 2020. www.ams.org/notices/201502/rnoti-p165.pdf.

Weyers, Stefan (2020). "Unit of Time Working Group 4.41." Last accessed October 9, 2020. www.ptb.de/cms/en/ptb/fachabteilung en/abt4/fb-44/ag-441/realisation-of-the-si-second/history-of-the-unit-of-time.html.

Wheeler, John Archibald (1982). "The Computer and The Universe." *International Journal of Theoretical Physics* 21.6, pp. 557–572.

— (1983). "On Recognizing 'Law Without Law,' Oersted Medal Response at The Joint APS–AAPT Meeting, New York, 25 January 1983." *American Journal of Physics* 51.5, pp. 398–404. aapt.scitation.org/doi/pdf/10.1119/1.13224.

Whitfield, Stephen E. and Gene Roddenberry (1968). *The Making of Star Trek*. New York: Ballantine Books.

Whitten, Alma and J. D. Tygar (1999). "Why Johnny Can't Encrypt: A Usability Evaluation of PGP 5.0." *Proceedings of The 8th USENIX Security Symposium*. Washington, DC.

Wiesner, Stephen (January 1983). "Conjugate Coding." *SIGACT News* 15.1. Original manuscript written circa 1970, pp. 78–88. doi.org/10.1145/1008908.1008920.

Winner, Langdon (2018). "Do Artifacts Have Politics?" *Daedalus* 109:1, pp. 121–136.

Winterbotham, F. W. (1974). *The Ultra Secret*. New York: Harper and Row.

Wolfram, Stephen (2002). *A New Kind of Science*. English. Champaign, IL: Wolfram Media. www.wolframscience.com.

Woo, Jesse, Peter Swire, and Deven R. Desai (2019). "The Important, Justifiable, and Constrained Role of Nationality in Foreign Intelligence Surveillance." *Hoover Institution Aegis Series Paper* 1901.

Wood, Laura (March 14, 2019). "Global $15.6Bn Signals Intelligence (SIGINT) Market by Type, Application and Region – Forecast to 2023 – ResearchAndMarkets.com." *BusinessWire*.

Wright, Robert (April 1988). "Did The Universe Just Happen?" *The Atlantic*. www.theatlantic.com/past/docs/issues/88apr/wrigh t.htm.

Wu, Jun et al. (2016). "The Study of Several Key Parameters in The Design of Airborne Superconducting Full Tensor Magnetic Gradient Measurement System." *2016 SEG International Exposition and Annual Meeting*. Dallas, Texas: Society of Exploration Geophysicists, pp. 1588–1591.

Xu, Nanyang et al. (March 2012). "Quantum Factorization of 143 on a Dipolar-Coupling Nuclear Magnetic Resonance System." *Physical Review Letters* 108.13, pp. 130501. link.aps.org/doi/10 .1103/PhysRevLett.108.130501.

Yan, Wei-Bin and Heng Fan (April 2014). "Single-Photon Quantum Router with Multiple Output Ports." *Scientific Reports* 4.1, pp. 4820. doi.org/10.1038/srep04820.

Yardley, Herbert O. (1931). *The American Black Chamber*. London: The Bobbs-Merrill Company.

Yin, Juan et al. (June 16, 2017). "Satellite-Based Entanglement Distribution Over 1200 Kilometers." *Science* 356.6343, pp. 1140–1144.

Yoo, J. (2020). "Rules for The Heavens: The Coming Revolution in Space and The Laws of War." *University of Illinois Law Review* 2020.1, pp. 123–194.

Yuan, Z. S. et al. (2008). "Experimental Demonstration of a BDCZ Quantum Repeater Node." *Nature* 454.7208, pp. 1098–101.

Zach, Dorfman (December 21, 2020). "China Used Stolen Data to Expose CIA Operatives in Africa and Europe." foreignpolicy.co m/2020/12/21/china-stolen-us-data-exposed-cia-operatives-spy-n etworks/.

Zelnio, Ryan (January 6, 2006). "The Effects of Export Control on The Space Industry." *The Space Review*.

Zetter, Kim (January 12, 2018). "Google to Stop Censoring Search Results in China After Hack Attack." *Wired*. www.wired.com/20 10/01/google-censorship-china/.

Zhang, Hao et al. (2018). "Quantized Majorana Conductance." *Nature* 556.7699, pp. 74–79. doi.org/10.1038/nature26142.

Zhong, Han-Sen et al. (2020). "Quantum Computational Advantage Using Photons." *Science* 370.6523, pp. 1460–1463. science.s ciencemag.org/content/370/6523/1460.

Zissis, Carin (February 2007). "China's Anti-Satellite Test." www.cf r.org/backgrounder/chinas-anti-satellite-test.

Zuboff, Shoshana (2019). *The Age of Surveillance Capitalism: The Fight for a Human Future at The New Frontier of Power*. New York: PublicAffairs.

Zuckoff, Mitchell (2005). *Ponzi's Scheme: The True Story of a Financial Legend*. New York: Random House.

Zukav, Gary (1979). *The Dancing Wu Li Masters*. William Morrow.

Zweben, Stuart and Betsy Bizot (2019). *2019 Taulbee Survey*. cra.o rg/resources/taulbee-survey/.

Zyga, Lisa (November 28, 2014). "New Largest Number Factored on a Quantum Device Is 56,153." phys.org/news/2014-11-largest-factored-quantum-device.html.

Index

Colophon

This book was designed and typeset by Simson Garfinkel using X∄LATEX version 2.6-0.999992 on a 2018 Mac mini (3 GHz 6-Core Intel Core i5 with 32 GB 2667 MHz DDR4 RAM). The input consists of 20 source files, and 87 image files (42 jpegs, 30 PDFs, and 15 PNGs). A few of the PDF illustrations were created using OmniGraffle by Simson Garfinkel; other illustrations were designed using TikZ and PGF by Chris Hoofnagle and rendered by X∄LATEX. A total of 71 LATEXpackages were used, some of which conflicted, requiring a separate pass for the tree illustration in the preface, which was then imported as a PDF. Total compilation time was 731.58 s user, 3.37 s system utilizing 99% cpu—that is, fully utilizing a single Intel core and ignoring the others. (This is more evidence of the need to implement parallelism in the various LATEX engines.) The main text is typeset in Latin Modern Roman, the listings are in Latin Modern Mono, and much of the math is in Latin Modern Math, all free fonts.

CPSIA information can be obtained
at www.ICGtesting.com
Printed in the USA
LVHW082141221221
707016LV00008B/304